Advances in Spatial Science

Editorial Board
David F. Batten
Manfred M. Fischer
Geoffrey J. D. Hewings
Peter Nijkamp
Folke Snickars (Coordinating Editor)

Springer
*Berlin
Heidelberg
New York
Barcelona
Hong Kong
London
Milan
Paris
Singapore
Tokyo*

Titles in the Series

C. S. Bertuglia, M. M. Fischer and G. Preto (Eds.)
Technological Change,
Economic Development and Space
XVI, 354 pages. 1995. ISBN 3-540-59288-1
(out of print)

H. Coccossis and P. Nijkamp (Eds.)
Overcoming Isolation
VIII, 272 pages. 1995. ISBN 3-540-59423-X

L. Anselin and R. J. G. M. Florax (Eds.)
New Directions in Spatial Econometrics
XIX, 420 pages. 1995. ISBN 3-540-60020-5
(out of print)

H. Eskelinen and F. Snickars (Eds.)
Competitive European Peripheries
VIII, 271 pages. 1995. ISBN 3-540-60211-9

J. C. J. M. van den Bergh, P. Nijkamp and P. Rietveld (Eds.)
Recent Advances in
Spatial Equilibrium Modelling
VIII, 392 pages. 1996. ISBN 3-540-60708-0

P. Nijkamp, G. Pepping and D. Banister
Telematics and Transport Behaviour
XII, 227 pages. 1996. ISBN 3-540-60919-9

D. F. Batten and C. Karlsson (Eds.)
Infrastructure and the Complexity
of Economic Development
VIII, 298 pages. 1996. ISBN 3-540-61333-1

T. Puu
Mathematical Location and Land Use Theory
IX, 294 pages. 1997. ISBN 3-540-61819-8

Y. Leung
Intelligent Spatial Decision Support Systems
XV, 470 pages. 1997. ISBN 3-540-62518-6

C. S. Bertuglia, S. Lombardo and P. Nijkamp (Eds.)
Innovative Behaviour in Space and Time
X, 437 pages. 1997. ISBN 3-540-62542-9

A. Nagurney and S. Siokos
Financial Networks
XVI, 492 pages. 1997. ISBN 3-540-63116-X

M. M. Fischer and A. Getis (Eds.)
Recent Developments in Spatial Analysis
X, 434 pages. 1997. ISBN 3-540-63180-1

R. H. M. Emmerink
Information and Pricing
in Road Transportation
XVI, 294 pages. 1998. ISBN 3-540-64088-6

F. Rietveld and F. Bruinsma
Is Transport Infrastructure Effective?
XIV, 384 pages. 1998. ISBN 3-540-64542-X

P. McCann
The Economics of Industrial Location
XII, 228 pages. 1998. ISBN 3-540-64586-1

L. Lundqvist, L.-G. Mattsson and T. J. Kim (Eds.)
Network Infrastructure
and the Urban Environment
IX, 414 pages. 1998. ISBN 3-540-64585-3

R. Capello, P. Nijkamp and G. Pepping
Sustainable Cities and Energy Policies
XI, 282 pages. 1999. ISBN 3-540-64805-4

M. M. Fischer and P. Nijkamp (Eds.)
Spatial Dynamics of European Integration
XII, 367 pages. 1999. ISBN 3-540-65817-3

M. M. Fischer, L. Suarez-Villa and M. Steiner (Eds.)
Innovation, Networks and Localities
XI, 336 pages. 1999. ISBN 3-540-65853-X

J. Stillwell, S. Geertman and S. Openshaw (Eds.)
Geographical Information and Planning
X, 454 pages. 1999. ISBN 3-540-65902-1

G. J. D. Hewings, M. Sonis, M. Madden and Y. Kimura (Eds.)
Understanding and Interpreting Economic Structure
X, 365 pages. 1999. ISBN 3-540-66045-3

A. Reggiani (Ed.)
Spatial Economic Science
XII, 457 pages. 2000. ISBN 3-540-67493-4

D. G. Janelle and D. C. Hodge (Eds.)
Information, Place, and Cyberspace
XII, 381 pages. 2000. ISBN 3-540-67492-6

P. W. J. Batey and P. Friedrich
Regional Competition
VIII, 290 pages. 2000. ISBN 3-540-67548-5

B. Johansson, Ch. Karlsson and R. R. Stough (Eds.)
Theories of Endogenous Regional Growth
IX, 428 pages. 2001. ISBN 3-540-67988-X

Graham Clarke · Moss Madden
Editors

Regional Science in Business

With 79 Figures
and 61 Tables

Springer

Dr. Graham Clarke
University of Leeds
School of Geography
Leeds LS2 9JT
United Kingdom

Prof. Dr. Moss Madden

ISBN 3-540-41780-X Springer-Verlag Berlin Heidelberg New York

Library of Congress Cataloging-in-Publication Data applied for
Die Deutsche Bibliothek - CIP-Einheitsaufnahme
Regional science on business: with 61 tables / Graham Clarke; Moss Madden ed.
- Berlin; Heidelberg; New York; Barcelona; Hong Kong; London; Milan; Paris;
Singapore; Tokyo: Springer, 2001
  (Advances in spatial science)
  ISBN 3-540-41780-X

This work is subject to copyright. All rights are reserved, whether the whole or part of the material is concerned, specifically the rights of translation, reprinting, reuse of illustrations, recitation, broadcasting, reproduction on microfilm or in any other way, and storage in data banks. Duplication of this publication or parts thereof is permitted only under the provisions of the German Copyright Law of September 9, 1965, in its current version, and permission for use must always be obtained from Springer-Verlag. Violations are liable for prosecution under the German Copyright Law.

Springer-Verlag Berlin Heidelberg New York
a member of BertelsmannSpringer Science+Business Media GmbH
http://www.springer.de

© Springer-Verlag Berlin · Heidelberg 2001
Printed in Germany

The use of general descriptive names, registered names, trademarks, etc. in this publication does not imply, even in the absence of a specific statement, that such names are exempt from the relevant protective laws and regulations and therefore free for general use.

Hardcover-Design: Erich Kirchner, Heidelberg

SPIN 10782963   42/2202 - 5 4 3 2 1 0 - Printed on acid-free paper

Moss Madden

During the editing of this book Moss died suddenly and tragically from septicaemia. Many tributes have already been made to his massive contribution to regional science and everyone in our community will miss him enormously. I was delighted when he agreed to become a co-editor of this book. He believed passionately in the importance of applied research and I am sure he was pleased with the assemblage of papers in the book. It has always been a pleasure to work with Moss. I became secretary of the British and Irish section of the RSA when Moss was Chairman. We enjoyed a good working relationship and many a pint of real ale at one of Moss's many favourite watering holes around the UK. I remember my trips to Moss's hometown of Falmouth (to organise the British & Irish Conference in 1997) with particular fondness. I would like to think that our friendship went far beyond work. I once told him of my ambition to travel on all the major airlines of the World. As a collector of many things Moss thought this a splendid idea and decided to set up in competition. Most of our e-mails from then on were not about RSA matters but who had booked another airline. Given his senior years he was miles ahead of me and relished in flying on some obscure South American Airline (especially if it was likely to go bust and thus prevent me from ever using it!).

I hope Moss would be pleased with the final collection and that it will be one of many lasting tributes to him.

Graham Clarke
Leeds, Xmas 2000/2001

Acknowledgements

The editors would like to thank our commissioning editor Marianne Bopp for her advice and guidance throughout the production of this book. Thanks also to David Appleyard for drawing many of the figures, Adam Davenport for his skilful manipulation of the text and graphics and an anonymous referee for his/her invaluable comments. Inevitably, the death of Moss put delivery behind schedule and hence thanks also to all the contributors for their patience.

Contents

1	Introduction GRAHAM CLARKE AND MOSS MADDEN	1
2	Creating and Expanding Trade Partnerships Within the Chicago Metropolitan Area: Applications Using a Miyazawa Accounting System GEOFF HEWINGS, YASUHIDE OKUYAMA AND MICHAEL SONIS	11
3	Socio-economic Impact Assessment: Meeting Client Requirements PETER BATEY AND MOSS MADDEN	37
4	Multiregional Computational General Equilibrium and Spatial Interaction Trade Modelling: An Empirical Example MARTIN SCHNEIDER AND MANFRED FISCHER	61
5	The Regional Cambridge Multisectoral Dynamic Model of the UK Economy TERRY BARKER, BERNIE FINGLETON, K. HOMENIDOU AND R. LEWNEY	79
6	Discrete Choice Modelling: Basic Principles and Application to Parking Policy Assessment HARMEN OPPEWAL AND HARRY TIMMERMANS	97
7	Applied Population Projection for Regional and Local Planning JOHN STLLWELL AND PHIL REES	115
8	Applied Spatial Interaction Modelling GRAHAM CLARKE AND MARTIN CLARKE	137
9	The Objectives and Design of a New Land-use Modelling Package: DELTA DAVID SIMMONDS	159

10	Transport and Urban Development GUNTER HAAG AND KATHRIN GRUETZMANN	189
11	Boundary-Swapping Optimisation and Regional Design BILL MACMILLAN	211
12	Optimal Distribution Strategies MARK BIRKIN AND RICHARD CULF	223
13	An Applied Microsimulation Model: Exploring Alternative Domestic Water Consumption Scenarios PAUL WILLIAMSON	243
14	Fuzzy Geodemographic Targeting LINDA SEE AND STAN OPENSHAW	269
15	Analysing Access to Hospital Facilities with GIS MARTIN CHARLTON, STEWART FOTHERINGHAM AND CHRIS BRUNDSON	283
16	GIS and Large-scale Linear Programming: Evolution of a Spatial Decision Support System for Land Use Management RICHARD CHURCH	305
17	Crime Pattern Analysis, Spatial Targeting and GIS: The Development of New Approaches for Use in Evaluating Community Safety Initiatives ALEX HIRSCHFIELD, DAVID YARWOOD AND KATE BOWERS	323
	Subject Index	343
	Figures	349
	Tables	353
	Author Index	355
	Contributors	361

1 Introduction

Graham Clarke and Moss Madden

1.1 Background

In the mid 1990s there were a number of papers in regional science that questioned the relevance and purpose of the entire sub-discipline. Bailly and Coffey (1994) for example, talked of 'regional science in crisis'. They argued that there were two fundamental problems. First, regional science was too theoretical in the sense that many of its products were models that could neither be calibrated (too complex) or operationalised (too abstract) in the real world. They suggested that regional science had not sufficiently demonstrated that it can address real-world problems and subsequently lacked a focus on relevant policy issues. Second, they argued that regional science had become too narrow in focus and had moved away too far from real people and their daily concerns or struggles in life. This was not the first time we had witnessed these sorts of arguments, both from outside the discipline and from within. Sayer (1976) was perhaps the first to argue for a shift from a model-based focus in regional science to one based on political economy. Breheny (1984) criticised the 'deep ignorance among regional scientists of the nature of practical policy making and implementation' (see also Rodwin (1987) for similar views in the mid 1980s). Such self-reflection is a feature of many disciplines as they reach maturity. There have been many similar reflections in geography (Johnston 1996, Barnes 1996) and economics (see the collection in the January edition of the Economic Journal 1991).

The article by Bailly and Coffey stimulated an interesting debate on 'regional science in crisis' in a special edition of 'Papers in Regional Science' in 1994. Some agreed with the general thrust of their arguments but others were more defensive. Plane (1994) and Vickerman (1994) both felt more optimistic and urged regional scientists not to get too preoccupied with these matters. Plane noted the healthy attendance at regional science conferences, and suggested that this was a useful barometer on vitality (see also Isserman 1995).

So what has happened to regional science since 1994 in terms of the two issues raised by Bailly and Coffey? The move towards a more people-focused discipline has been slow in coming. For example, a glance at the programme of papers at the European meeting in Dublin and the North American meeting in Montreal, both in

1999, reveals very little new in this area. However, there has been a renewed debate on the potential links between regional science and social theory/political economy. The opportunities for regional science to engage more fully with social theory are discussed eloquently by Rees (1999). He argues that 'regional science has much to gain from further engagement with social theory and postmodernism', although there is little discussion on how that might come about or what the focus should be. We shall have to reserve judgement on the benefits of this move if and when it takes place. Many in geography might agree with Rees but there are conflicting views. Given the focus on theory and methods within regional science it is perhaps difficult to ever envisage the sort of fundamental shift from spatial analysis to social theory which has taken place in a discipline such as geography (Sheppard 1995). Indeed, the focus on spatial analysis and modelling is the very reason many geographers have become interested in regional science (Plane 1994). Some rightly argue that postmodernism is the antithesis of applied work: "In terms of real world problems, postmodern thought would condemn us to inaction while we reflect on the nature of the issue" (Pacione 1999 p6). More alarmingly perhaps, postmodernism can be identified as rejecting theory and representation, surely the cornerstones of any social science (Rosenau 1992), and some may even ask how a social science like regional science could possibly gain anything from an anti-scientific approach such as postmodernism.

But what of relevance? Since Bailly and Coffey's attack it is undoubtedly true that there have been many more real-world applications in regional science. This has coincided with a general move back to the question of applied research in both geography and economics. Michael Pacione in a recent volume (Pacione 1999) has put together an impressive range of applied work in modern geography. The aim of this book is to attempt a similar exercise in regional science. It is the first attempt to put together a series of papers based on funded research from business or Government; that is, research specifically tendered by these organisations rather than resulting from a general application to a Government research council. Table 1.1 summarises the range of organisations that have sponsored work reported in this book. We hope therefore it is a major contribution to the 'regional science in crisis' debate.

1.2 Background and Motivation

It is very difficult, if not impossible, to write a comprehensive review of applied spatial analysis, especially those projects sponsored by a commercial organisation. Very often the results of such research are confidential as clients would see any commercial value to the project vanish if competitors were able to get hold of the results. Indeed, the authors of a number of chapters in this book have anonymised client names in order to safeguard confidentiality. In some cases it may also be the case that academics have not been keen to publish the fact that they are doing consultancy work. In many institutions consultancy has been seen as a dirty word, as

Table 1.1. Major sponsors of applied regional science described in this book

Public sector (or quasi public /private sector agencies):

Chicago United
London Chamber of Commerce and Industry
Austrian Government
Brabant Chamber of Commerce
Yorkshire and Humberside Planning Conference group
UK Regional Office of the Department of the Environment
West Yorkshire Planning Authorities
Greater Manchester Planning Transport Executive
UK 'Environment Resources Management for the Regions'
Ministry for Umvelt and Verkehr Baden-Wurrttemberg
A national Electoral Agency
US Forest Service
Various UK Health Authorities
Merseyside Police Authority

Private sector:

British Aerospace
ASDA
Sainsbury
ESSO
Toyota
Ford
Halifax
Yorkshire Water PLC

[a major UK bank forms the basis of chapter 14 whilst chapter 5 is based on many private sector and public sector agencies]

such analysis (in the eyes of many other academics) removes the 'objective' status of academia and, if it helps the organisation to make extra profits, can be seen as 'selling the soul to the devil'. Academics who see colleagues substantially increasing their salary through undertaking such activities may also question the validity of such an exercise (in terms perhaps of perceived increases in workloads needed to cover for colleagues working on such activities). There is also the perception that most consultancy work is likely to be generally of a poor standard, since such research needs to be completed in days or weeks rather than months or years. For example, West and Jensen (1995) observed that commercial regional modelling lacks transparency and is associated with practices "that are often more in keeping with the used car industry than with academia". Cole (1997) cites the same paper and goes on to conclude:

"These tendencies have contributed to the dichotomy between the 'conventional' models that typically are those used by policy actors in planning offices, community action groups and the like, and the more elaborate and impenetrable models of professional academics" (p37-38)

One of the central aims of the rest of the book is to show that this dichotomy need not exist. Many of the authors are well known regional scientists using state of the art spatial models. Indeed, very often the models themselves are, or have been, improved through empirical testing and evaluation. If academics are to make lasting partnerships with commercial or government agencies then they must show that their models actually work. Hence, calibration, model extensions and evaluations are common themes explored in many of the following chapters.

However, another important change has been a move towards promoting consultancy work in universities rather than frowning upon it. We shall explore the reasons for this below. It is useful to place this discussion in the wider context of why more recent times have been kinder to applied spatial analysis and academics that wish to test their models in the real world. So first, it is interesting to reflect on why the mid to late 1990s has been a period of greater activity in applied analysis. Clarke and Wilson (1987) provide a useful checklist of enabling factors which relate to the rise of applied spatial interaction modelling in the mid 1980s, perhaps the first major methodology in regional science to be commercialised effectively (see Birkin *et al.*, 1996, and the discussion in chapters 8 and 12). It is useful to revisit this list and extend the debate into regional science and interesting to try and articulate which are the most important factors.

The first and perhaps most important factor would seem to be that there is a new generation of policy makers who are computer-aware (if not always literate) and more open to the use of models to help improve their business. This is certainly the case in the private sector but it is increasingly apparent also in the public sector. Perhaps the older generation of public sector planners (whether they be in local government, health, education etc), who may have been brought up in the era of large-scale modelling that was so heavily criticised by academics and planners alike (Lee, 1973 and Sayer, 1976 for example), have now retired and a new breed more open to the benefits of spatial modelling has emerged. What do such planners (public or private sector) hope to get from regional science? Well, clearly location problems have never gone away. Although models were heavily criticised in the 1970s and 1980s planners still have to grapple with classic what-if questions: what will happen if I build a new airport, new shopping centre, open a new bank branch, reorganise the location of fire stations etc? Although many social scientists left modelling for the worlds of social theory, marxism and postmodernism, regional scientists, on the whole, did not abandon this research agenda. In addition, new policy issues have emerged such as urban sustainability, transport problems in cities, regional devolution, international trade and migration. Regional scientists are now beginning to respond to these issues perhaps more effectively than many other academic disciplines. Van Geenhuizen and Nijkamp (1996) provide an excellent review of this new policy agenda and the response of the European regional science community in particular.

At the same time, the field of regional science has matured and there is a greater appreciation of the strengths and limitations of our modelling methods. More progress has probably been made with applied sub-system models than with comprehensive ones. The former are in many senses easier to operationalise and calibrate. Clarke and Wilson (1987) suggest such models have potential for greater control and relatively fewer degrees of freedom. These new sub-system models now provide an attractive suite of models for applied spatial analysis. However, we should note that as computers get more powerful we are likely to see a resurgence of interest in large-scale models (see Boyce 1988 for the first arguments that this was taking place in the late 1980s). Indeed, a number of chapters in this book are testimony to the fact that large-scale models can be increasingly operationalised.

The advances made in computer hardware is undoubtedly another crucial factor in understanding the increase in applications. Modelling work in the 1960s and 1970s was often undertaken using mainframe computers and suffered through slow processing times, an inability to handle large data sets and very poor graphical output. Today, all modelling work can be done on PCs or workstations that can sit happily on the desks of key strategic managers. If the output of the models can be converted into useful and easy to interpret performance indicators so much the better (Bertuglia *et al.*, 1994).

Another important push factor has been the changing University environment that now encourages and rewards academics that connect to business and commerce. All university departments realise that the publicly funded expansion days of higher education have long gone and that resources from traditional Government sources will be much harder to obtain in the future. This includes resources associated with the traditional block grants to cover teaching and research and those that come additionally from the various research councils. Many universities have set ambitious targets for increasing revenues from non-traditional Government sources (the University of Leeds for example labels this 3rd arm activity).

Another major enabling factor has been access to better data. This includes public and private sector sources. There is often a common misconception that private sector companies would have better data (in terms of quantity and quality) than public sector sources. However, this has often not been the case, although the situation is now changing rapidly. There is now a much clearer understanding that better data can lead to better information. This has obviously been a boon to the modelling community who can use these data to calibrate and test their models more fully.

The collection of papers in this book would not have materialised however without the will to do applied research from the individuals or groups of individuals concerned. Perhaps then we have begun to take the sorts of comments from Bailly and Coffey seriously. Many of us do want to see our models in operation and we take great pride when others find them useful and informative. Not only does it add credibility to our research credentials (especially with the outside world) it can provide more research staff and papers and it can, in some circumstances, put something extra in our back pockets. Such research also enriches our teaching. Students see the worth of the methodologies they are being taught and lectures can be enriched through good empirical applications. As the higher education sector becomes increasingly more competitive, it is clear that parents and potential students are looking far more closely at the vocational element of the courses on offer. For many geography and economic departments to survive they must emphasise the applied nature of the discipline. If they do not then many will undoubtedly struggle to attract students (look at the growth of economic departments which emphasise business studies – or degree courses generally with the word business or marketing attached).

In the remaining chapters of this book we present a series of papers on applied regional science. Even though regional science is largely about models and spatial analysis there is a wide range of methodologies in the kit-bag. Isard *et al.,* (1998) provide an excellent summary of the richness and diversity of methods and techniques available to the regional scientist. We have tried to incorporate many of these regional science modelling techniques within this collection. Thus we hope these books will compliment each other well. Isard *et al.,* provide much more detail on the methodologies and their foundations whilst we provide examples of their use in the real-world.

A final word on applied research. Although we believe such work enriches the discipline, we do not, of course, devalue theoretical research. Many academics see the primary role of academia as being one of searching for new theoretical insights. We would never want to lose that 'blue skies' element. What we are advocating here is a balance between theoretical explorations and applied policy work. There must be sufficient room for both!

1.3 Contents

Without the desire to be too prescriptive, we have asked each author to address a number of key issues. Generally, there is first a description of the methodology and the particular requirements of each of the clients. Data issues are important and it is interesting to see where data for applied work originates. The results of the work then generally follow as well as a discussion on how the application has modified the model or the theoretical assumptions of the model (if appropriate).

Chapter 2 by Geoff Hewings *et al.,* looks at an economic input-output model with a Miyazawa-styled social accounting system for Chicago United. Their client was concerned to examine how economic development in Chicago would impact on very different regions of the City and to ascertain what formed existing impediments to further trade expansion. In chapter 3 Peter Batey and the late Moss Madden develop extended input-output models for the impact assessment of extending both Liverpool and Heathrow airports in the UK. Their sponsors were British Aerospace and the London Chamber of Commerce and Industry. Chapter 4 by Martin Schneider and Manfred Fischer uses a multi-regional computable general equilibrium model (CGE model) with a spatial interaction model of bilateral trade flows. They were sponsored by a major EU grant and by the Austrian Government to look at the trade implications of the growth of economies in Central and Eastern Europe on the Austrian economy. Chapter 5 combines regional input-output models and CGE models. Terry Barker *et al.*, describe the Cambridge Multisectoral Dynamic Model of the UK economy. Confidentiality prevents them naming clients but they have a long and successful business serving many public and private sector clients interested in forecasting trends in the economy and their regional impacts. In chapter 6 Harmen Oppewal and Harry Timmermans introduce applied discrete choice models. Their client is the Brabant Chamber of Commerce who were interested in finding sustainable solutions to retail growth by commissioning a study of the impact of various car parking pricing and provision policies on spatial shopping behaviour. The forecasting theme is continued in chapter 7 where John Stillwell and Phil Rees explore the methodologies for population and household small-area forecasts. Their work has been sponsored by a number of public sector agencies. Spatial interaction models form the basis of chapter 8 and much of the applied work of the School of Geography and GMAP at the University of Leeds. Graham Clarke and Martin Clarke explore the development of these models for applications with a number of blue-chip retail clients. Chapters 9 and 10 illustrate the progress made with more ambitious dynamic urban models. In chapter 9 David Simmonds describes the DELTA land use modelling package which consists of discrete choice models, extended input-output models and utility maximising consumer behaviour models. Their clients are the Greater Manchester Passenger Transport Executive (interested in the impacts of a wide range of possible impacts on alternative land use and transport strategies in the City of Manchester) and a partnership group of MVA Ltd, DSC Ltd and the Environment Resources Management for the Regions (interested to examine the impacts of land use developments in the Trans-Pennine corridor of Northern England). In chapter 10 Gunter Haag and Katherin Gruetzmann describe a similar land use/transportation model to evaluate many alternative strategies for new land use and road networks in Stuttgart (sponsored by the local regional Ministry).

The focus in the next two chapters switches to optimisation. In chapter 11 Bill Macmillan describes his work for a national Electoral Agency on optimal zone design associated with redistricting. This theme is taken up by Mark Birkin and Richard Culf in chapter 12. They also describe models for optimising distribution networks (for the motor industry in particular). The theme of chapter 13 is microsimulation. Paul Williamson describes the advantages of the micro approach for estimating water demand (and subsequent household pricing strategies) for Yorkshire Water. Chapter 14 introduces applied geodemographics. Linda See and Stan Openshaw describe how fuzzy geodemographics has helped a major UK bank to improve its marketing activities. The final major theme is geographical information systems (GIS). In chapter 15 Martin Charlton *et al.*, describe the use of GIS for network analysis in their study of accessibility to hospitals in central England. In chapter 16 Rick Church couples GIS to large-scale linear programmes to evaluate various planning policies for the US Forest Service. Finally, Alex Hirschfield *et al.*, examine the use of GIS for crime pattern analysis in Merseyside, work sponsored by the Merseyside Police Authority.

We hope this first collection will stimulate other attempts to show the applied worth of the discipline. We have drawn on only a handful of well-known regional scientists in this collection – there must be many more examples out there! As we mentioned earlier, it is an interesting debate as to whether good science involves the creation of new theoretical knowledge or whether it enables practical real world problems to be solved. Hopefully many of the chapters in this book show that these need not be conflicting aims.

References

Bailly A.S. and Coffey W.J. (1994) Regional science in crisis: a plea for a more open and relevant approach, *Papers in Regional Science, 73*, 3-14

Barnes T. (1996*) Logics of dislocation: models, metaphors and meanings of economic space,* Guildford Press, New York.

Bertuglia C.S, Clarke G.P. and Wilson A.G. (1994) *Modelling the city*, Routledge, London.

Birkin M., Clarke G.P., Clarke M. and Wilson A.G. (1996) *Intelligent GIS*, Geoinformation, Cambridge.

Boyce D. (1988) Renaissance of large-scale models, *Papers of the Regional Science Association, 65*, 1-10

Breheny M.J. (1984) The relevance of regional science to regional policy, Paper presented to the annual meeting of the Australian and New Zealand section of the Regional Science Association, Melbourne.

Clarke M. and Wilson A.G. (1987) Towards an applicable human geography, *Environment and Planning A, 19*, 1525-1541

Cole S. (1997) Closure in Cole's reformulated Leontief model, *Papers in Regional Science, 76(1)*, 29-42

Isard W., Azis I.J., Drennan M.P., Miller R.E., Saltzman S. and Thorbecke E. (1998) *Methods of interregional and regional analysis*, Ashgate, New York.

Isserman A. (1995) The history, status and future of regional science: an american perspective, *International Regional Science Review, 17(3)*, 249-296

Johnston R.J. (1996) *Geography and geographers*, 5th edition, Edward Arnold, London.

Lee D.B. (1973) Requiem for large-scale models, *Journal of the American Institute of Planners, 3*, 163-178

Pacione M. (ed, 1999) *Applied geography: principles and practice,* Routledge, London.

Plane D.A. (1994) Comment: on discipline and disciplines in regional science, *Papers in Regional Science, 73*, 19-23

Rees J. (1999) Regional science: from crisis to opportunity, *Papers in Regional Science, 78*, 101-110

Rodwin L. (1987) On the education of urban and regional specialists: a comparative perspective, *Papers of the Regional Science Association, 62*, 1-11

Rosenau P.M. (1992) *Post-Modernism and the Social Sciences: insights, inroads and intrusions* Princeton University Press, Princeton

Sayer R.A. (1976) *A critique of urban modelling: from regional science to urban and regional political economy*, Pergamon Press, Oxford.

Sheppard E. (1995) Dissenting from spatial analysis, *Urban Geography, 16*, 283-303

Van Geenhuizen M. and Nijkamp P. (1996) Progress in regional science: a European perspective, *International Regional Science Review, 19(3)*, 223-246

Vickerman R. (1994) Comment: regional science in crisis? A European view, *Papers in Regional Science, 73*, 33-36

West G. and Jensen R. (1995) Validation issues in regional modelling, Paper presented to the 42nd North American Regional Science Meeting, Cincinnati, November.

2 Creating and Expanding Trade Partnerships Within the Chicago Metropolitan Area: Applications Using a Miyazawa Accounting System

Geoffrey J.D. Hewings, Yasuhide Okuyama and Michael Sonis

2.1 Introduction

Consider the following scenario; a business news report highlights the discovery of a new country in which wage and salary income amounts to slightly less than $10 billion. This country has escaped notice by entrepreneurs since very little was known about it – including its location. In addition, the news report continues that a second country with purchasing power exceeding $8 billion has been found close by. How could this be possible in an age of globalization, internet, and rapid diffusion of information? The cause for these countries being overlooked is due, in part, to lack of information about their location but the greater problem is due to casual dismissal of the size of their markets – based on little or no information.

Where are these countries? It turns out that these are not countries but parts of the Chicago metropolitan area; the presentation of their characteristics in this form was made to drive home an important point, namely that there are large, untapped markets within North America that have been ignored for decades. Many of these markets lie within large, metropolitan regions like Chicago. That they have been ignored speaks to a major problem in intra-metropolitan development, namely the lack of information about how the various geographic parts of a metropolitan area are linked together.

In this chapter, the ideas and approaches adopted in understanding how regions within a country or countries themselves are linked with each other are used to explore the complex interdependencies that exist *within* the Chicago metropolitan area (hereafter, Chicago).[1] The methodology that has been developed will form the basis for analysis that can explore the benefits to all parts of the area from economic initiatives generated in one region and begin to provide the basis for the notion that the gains from free trade that are promoted at the international level can also be realized within metropolitan areas. The analysis will be based in part on two models – a comprehensive econometric-input-output model for Chicago and a Miyazawa-styled social accounting system for a four-fold division of this area (see Hewings *et*

---
[1] The Chicago metropolitan region comprises the six counties, Cook, Will, Dupage, McHenry, Lake and Kane.

al., 1999, for a collection of papers illustrating applications of the Miyazawa framework). Given time constraints and lack of reliable data, a full social accounting system could not be developed; however, for the clients' needs, the addition of income flows proved to be the single most valuable source of additional information beyond that provided in standard input-output models.

The research was provided to *Chicago United*, an organization that was formed in the 1960s in the wake of race riots in the city; the goal was to bring together business leaders of different ethnic backgrounds from major companies to focus on enhancing economic development in parts of the region in which a large percentage of the residents were African-American and, in later years, Hispanic. These regions had been overlooked from a pro-active development standpoint, had suffered disproportionately from the dramatic downturn in manufacturing activity in the 1970s and 1980s and were felt to offer little in the way of attractive rates of return to businesses contemplating locating a plant or retail outlet in one of these regions. The negative business perceptions were exacerbated by rapidly deteriorating physical infrastructure, a poor public school system and, most importantly, a lack of hard data, information or analysis to support a countervailing viewpoint.

Hence, for Chicago United, there were three major objectives:

1. Analysis of the nature and strength of existing economic interdependence between four regions in Chicago, with particularly attention directed to predominantly minority areas on the south and west parts of the city of Chicago;
2. Assessment of the potential gains to all regions from economic development within either of the two targeted regions;
3. Identification of existing impediments to further trade expansion

The focus of this chapter will be on the first two objectives; the third objective is still being explored and only some summary remarks can be provided at this stage. The chapter is organized as follows; a brief summary of the development of the areas and the methodology employed in the analysis will occupy section 2.2. In the following three sections, attention will be directed, in turn, to flows of goods and services, employment and income/journey-to-work, between a four-region division of Chicago. In the sixth section, some preliminary scenario development will be conducted to examine the economic impacts of proposed initiatives that focus on the Pullman district within the South region. The chapter concludes with some summary evaluations.

2.2 Methodology

Chicago was divided into four areas, as shown in figure 2.1; the geographic disparity in the size of the regions created some problems and concerns; however, the major objective was to demonstrate an appropriate methodology. The relative size of the regions may be gauged by employment numbers for place of work for 1995; these are shown in table 2.1.

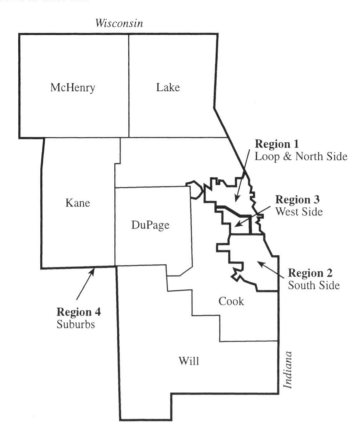

Figure 2.1. Four-fold division of Chicago

Table 2.1. Employment by region based on place of work, 1995

Region	Employment Totals	Percentage of Region Total
1 Loop/North Side	769,829	24.20%
2 South Side	184,131	5.79%
3 West Side	123,874	3.89%
4 Suburbs	2,102,881	66.11%
Total Region	3,180,715	100.00%

Source: Illinois Department of Employment Security

Further refinement can be employed to divide the region into smaller areas that are closer in size (in terms of population). These areas were treated as though they were separate, yet interdependent economies – rather like the way one might treat trade between states within the US.

Data were drawn in large part from the Chicago Region Econometric Input-Output Model developed by the Regional Economics Applications Laboratory (REAL)[2] and data generated from journey-to-work information initially collected by the Chicago Area Transportation Study (CATS) and further modified by a research team at the University of Illinois at Chicago lead by Professor David Boyce.

The Regional Economics Applications Laboratory (REAL) has developed a comprehensive impact and forecasting model for Chicago; this model provides detail for 53 different sectors of the economy and enables analysis of "what if…" questions, such as those associated with the expansion or contraction of economic activity. For this present project, part of the model has been divided into four regions – the Loop, South Side, West Side and the Suburbs and analysis has focused on generating estimates of the interaction between those four regions. The model development details are provided next.

2.2.1 Miyazawa's Framework

Miyazawa's (1976) concept of the interrelational income multiplier was designed to analyze the structure of income distribution by endogenizing consumption demands in the standard Leontief model. In an interregional context, this inclusion of the income formation process has clear advantages for linking the location of production (or wage earning) and the location of consumption. In some sense, Miyazawa's system may be considered the most parsimonious in terms of the way it extends the familiar input-output formulation. Miyazawa considered the following system:

$$\begin{pmatrix} X \\ Y \end{pmatrix} = \begin{pmatrix} A & C \\ V & 0 \end{pmatrix} \begin{pmatrix} X \\ Y \end{pmatrix} + \begin{pmatrix} f \\ g \end{pmatrix} \qquad (2.1)$$

where X is a vector of output, Y is a vector of total income for some r-fold division of income groups, A is a block matrix of direct input coefficients, V is a matrix of value-added ratios for r-fold income groups, C is a corresponding matrix of consumption coefficients, f is a vector of final demands except households consumption, and g is a vector of exogenous income for r-fold income groups. Solving this system yields:

$$\begin{pmatrix} X \\ Y \end{pmatrix} = \begin{pmatrix} B(I + CKVB) & BCK \\ KVB & K \end{pmatrix} \begin{pmatrix} f \\ g \end{pmatrix} \qquad (2.2)$$

---

[2] A description of the model may be found in Israilevich et al., 1996, 1997.

where:

$B = (I - A)^{-1}$ is the Leontief inverse matrix,
BC is a matrix of production induced by endogenous consumption,
VB is a matrix of endogenous income earned from production,
$L = VBC$ is a matrix of expenditures from endogenous income, and
$K = (I - L)^{-1}$ is a matrix of the Miyazawa interrelational income multipliers.

Sonis and Hewings (1993) extended this framework using the following perspective:

$$\left(\begin{array}{cc} B(I + CKVB) & BCK \\ KVB & K \end{array} \right) = \left(\begin{array}{cc} I & 0 \\ V & I \end{array} \right) \left(\begin{array}{cc} \Delta & 0 \\ 0 & I \end{array} \right) \left(\begin{array}{cc} I & C \\ 0 & I \end{array} \right) = \left(\begin{array}{cc} \Delta & \Delta C \\ V\Delta & I + V\Delta C \end{array} \right) \quad (2.3)$$

where:

$$\Delta = (I - A - CV)^{-1} = B(I + CKVB) \quad (2.4)$$

is an enlarged Leontief inverse. From (2.2) and (2.3), the following presentation of the Miyazawa interrelational multiplier matrix may be revealed:

$$K = I + V\Delta C \quad (2.5)$$

and, moreover, the Miyazawa fundamental equations of income formation will be:

$$\begin{cases} V\Delta = KVB \\ \Delta C = BCK \end{cases} \quad (2.6)$$

In his empirical analysis using the framework introduced above and applied to a three-region[3] case study for Japan with the 1960 interregional input-output table, Miyazawa (1976) examined the interrelational income multipliers for these regions in order to reveal the interrelational connections among regions. This analysis was recently updated by using the 1985 interregional table (see Okuyama et al., 1999). Table 2.2 displays the comparisons of the interrelational income multipliers, $K = I + V\Delta C$, between 1960 and 1985[4].

The tendencies among multipliers are similar between 1960 and 1985, although the absolute values in 1985 are larger than in 1960 with one exception (Middle to Northeast). As Miyazawa (1976) pointed out, the Middle, including Tokyo

---

[3] The Japanese economy is divided into three regions: Northeast, Middle, and West. The Northeast includes Hokkaido and Tohoku, the Middle consists of Kanto, Chubu, and Kinki, and in the West are Chugoku, Shikoku, Kyushu, and Okinawa (Miyazawa (1976) did not include Okinawa in the West).

and Osaka regions, has slightly smaller values than the other regions among the column sum values – induced effects originated from each region – and this is noticeable in both years. On the other hand, the Middle received significantly larger amount of induced effects – row sum values – from the regions than was the case for the other two regions, and the difference between the Middle and other regions has become larger during 1960 to 1985. This may indicate that the concentration of income formation in the Middle region has increased during this period.

Table 2.2. Comparison of interrelational income multipliers between 1960 and 1985

	Region of Income Origin			
	Northeast	Middle	West	Total
Region of Income Receipt				
Northeast	1.70	0.06	0.05	1.80
	(1.55)	(0.07)	(0.04)	(1.66)
Middle	0.61	1.88	0.56	3.05
	(0.32)	(1.57)	(0.29)	(2.18)
West	0.08	0.09	1.70	1.88
	(0.06)	(0.07)	(1.60)	(1.73)
Total	2.39	2.03	2.31	6.73
	(1.93)	(1.71)	(1.93)	(5.57)

Note: Values in parentheses are estimates for 1960 developed by Miyazawa (1976)

Similarly, the Chicago analysis was focused on the role of income flows across regions, but the regions were part of a large metropolitan economy; in addition, the income flow analysis was complemented by attention to other, more traditional flows-goods and services and people (journey-to-work).

2.2.2 The Model

The Chicago Metropolitan area (including six counties: Cook, Du Page, Kane, Lake, McHenry, and Will) was divided into four sub-regions [(Region 1 – Loop and North Side; Region 2 – South Side; Region 3 – West Side; and Region 4 – Suburbs); Regions 1, 2, and 3 consist of the City of Chicago] to investigate the spatial interaction of

---

[4] Using the 1960 interregional input-output table and the income statistics available at that time, Miyazawa (1976) divided the consumption demand of the household sector into endogenous and exogenous (expenditure from transfer income) items, and also divided the household income into endogenous and exogenous (retirement allowance and transfer income) items. In the 1985 study, however, these exogenous items are not subtracted from household consumption and income. This may cause slight differences in the results.

these sub-regions. In this section, a four-region input-output model is constructed based on the REAL's Chicago Region Econometric Input-Output Model, and is extended to Miyazawa's framework for analysis of income formation among the sub-regions.

Basic Framework. The data for the Chicago multiregional table at hand are the 1992 Chicago Input-Output table from the Chicago Region Econometric Input-output Model (CREIM), with 53 industrial sectors (see Israilevich *et al.*, (1996, 1997), and the employment data for the four regions, based on the data by the Illinois Department of Employment Security (1996). Since the input-output data, including input coefficients, final demand, imports, and exports, are only for the Chicago Metropolitan region (total of four regions), and the only data to indicate the structural differences among the four regions is the employment data, the location quotient technique was employed to disaggregate the one-region model to a four-sub-region model. A typical location quotient may be calculated as follows:

$$LQ_i = \left[e_i / e \right] / \left[E_i / E \right] \tag{2.7}$$

where e_i is the employment of sector i in the region, e is the total employment in the region, E_i is the employment of sector i in benchmark (usually state or nation), and E is the total employment of benchmark. In this case, the benchmark was set as the total employment of the Chicago region as a whole (equal to total of four regions), because structural differences from the whole system, the Chicago Metropolitan region, may determine the relationship among the four sub-regions. Then, the export share by sector for the four regions are derived as follows:

$$ex_i = \left(1 - \frac{1}{LQ_i} \right) \tag{2.8}$$

where ex_i is the export share in sector i.

Estimation of Regional Trade Coefficients. The export shares indicate the export portion of employment, and also of output assuming the tendency of output follows employment's; hence, these can be easily seen as row trade coefficients in the row coefficient model in the interregional input-output literature (Polenske, 1970). Because of its nature in location quotient and export share, the sum of local share and export share must be equal to unity over a region. However, at this point, a region has only one value (export share) to other regions; therefore, this must be divided into three receiving regions. In order to do so, the total employment of the regions and the distance factor among the regions are utilized based on the assumption of a gravity model-type distribution of export share. Consequently, the row trade coefficients (by sector) of each region are derived. By definition, the row trade coefficients add up to unity over the purchasing regions as follows:

$$\sum_q r_i^{pq} = 1 \tag{2.9}$$

Based on Polenske (1970), the row coefficient model is given by:

$$\sum_p r_i^{pq} x_i^p = \sum_j a_{ij}^q x_j^q + y_i^q \tag{2.10}$$

In matrix form, equation (2.10) can be transformed as:

$$R' \Delta X = \hat{A} \Delta X + \Delta Y \tag{2.11}$$

or

$$\Delta X = \left(R' - \hat{A}\right)^{-1} \Delta Y \tag{2.12}$$

where R is a square matrix filled with diagonal matrices of r_i^{pq}. This model, however, frequently generates negative estimation of elements in $(R' - \hat{A})^{-1}$ (Bon 1975 and 1984, and Toyomane 1988), and for the Chicago four-region case numerous negative values, especially in interregional parts of the table, were derived.

With this set of trade coefficients, the next step may be to transform these row trade coefficients to column trade coefficient imposing the following constraints:

$$\sum_p c_i^{pq} = 1 \tag{2.13}$$

where c_i^{pq} is the column trade coefficient of sector i. This transformation was carried out by calculating the following value:

$$c_i^{pq} = \frac{r_i^{pq} . emp_i^p}{\sum_p r_i^{pq} . emp_i^p} \tag{2.14}$$

where emp_i^p is the employment of sector i in region p.

Derivation of Four-Region Model. Now, we have a column coefficient model given as follows:

$$x_i^p = \sum_q \sum_j c_i^{pq} a_{ij}^q x_j^q + \sum_q c_i^{pq} y_i^p \tag{2.15}$$

In matrix form, the column coefficient model can be transformed as:

$$\Delta X = C(\hat{A}\Delta X + \Delta Y) \tag{2.16}$$

or

$$\Delta X = (I - C\hat{A})^{-1} C\Delta Y \tag{2.17}$$

However, the problem arose because $\hat{A}$ is the block diagonal matrix with square matrices of input coefficient of each region and the only matrix of input coefficient that can be used is for the Chicago region's (total of four regions). With this data constraint and assuming that the economic structure of these four regions may not be very different from the system as a whole, $\hat{A}$ is constructed employing the input coefficient matrix of Chicago region as their regional matrices. The structure, therefore, is given as:

$$\hat{A} = \begin{bmatrix} A & 0 & 0 & 0 \\ 0 & A & 0 & 0 \\ 0 & 0 & A & 0 \\ 0 & 0 & 0 & A \end{bmatrix} \tag{2.18}$$

where A is the input coefficient matrix from CREIM. Using this $\hat{A}$, the multiregional input-output system is derived by calculating $(I - C\hat{A})^{-1}$.

Extended Model using Miyazawa's Framework. In order to extend the four-region input-output model to Miyazawa's framework, wages and salaries, and consumption coefficients in each region need to be determined. The data for these coefficients are derived utilizing 1990 CATS Household Travel Survey by the Chicago Area Transportation Study (1994). From this data, two sets of trip data for the Chicago Metropolitan region were extracted: home-to-work, and home-to-shop. Value-added coefficients are derived using the home-to-work trip data of eight income groups adjusted by county income data from REIS CD-ROM. Consumption coefficients are calculated utilizing the CATS home-to-shop trip data and the consumption vector from CREIM. These coefficients are estimated as interregionally specified so that value-added coefficients form a 4 by 36 (four regions, nine sectors each) matrix and consumption coefficients become a matrix with 36 by 4 dimension. Forming a partitioned matrix, with $\hat{A}$ and these value-added and consumption matrices, the model now becomes the Miyazawa's extended input-output formulation.

The underlying assumption in the model is that a change in any activity will generate further impacts (ripple effects) elsewhere in the economy; the size of these ripple effects and the specific sectors that will be affected will depend on the nature of the sector which experiences the initial change. Furthermore, since economic activity is not spread uniformly throughout the regions, a change in one sector in one region can create very different spatial ripple effects than a change in another sector in the same region or the same sector in another region. In other words the geography of impacts do vary.

2.3 Trade Flows of Goods and Services Between Regions within Chicago

2.3.1 Introduction

The growth of trade within NAFTA has commanded a great deal of attention in the U.S. Midwest economies.[5] The whole state of Illinois' exports to Mexico amounted to $1.4 billion in the mid 1990s; the annual growth in this trade captures front-page headlines in the business sections of newspapers. Yet, as Hewings *et al.*, (1998) have shown, interregional trade within the Midwest approaches in volume the combined US-Canada-Mexico trade figures. Since international trade data are issued monthly, they command more attention in the business press; in contrast, data on interregional trade and certainly on intra-metropolitan trade are rarely gathered, and, not surprisingly, there is little appreciation for the magnitude of these trade relationships. As a result, attention is first directed to flows of goods and services between sectors within the region; the estimation of these flows involved the development of a matrix of interactions for all 53 sectors. The final dimensions of this matrix were thus 212 x 212. Attempting to show this matrix in a convenient fashion presents a daunting problem; instead, some summary aggregate analysis is provided and then attention is focused on two sectors to illustrate the potential of the methodology. This matrix was then used to generate the spatial distributions of impacts highlighted in the case study in section 2.6.

2.3.2 Trade Flows: Aggregate Analysis

Tables 2.3 through 2.5 show the aggregate trade flows between the four regions expressed in $1996; these are the flows for all sectors of the economy. Tables 2.4 and 2.5 portray the flows in terms of dependencies. The entries on the diagonal reveal the degree to which the sales and purchases among sectors occurs within the region; note that regions 4 and 1 are much more self-contained than regions 2 and 3. Region 2 is more dependent on trade with the rest of the economy (especially region 4). Region 4 turns out to be the major source for trade outside each region (see table 2.4).

---

[5] The states of Wisconsin, Illinois, Indiana, Ohio and Michigan

Table 2.3. Aggregate trade flows between the regions [$1996 billions]

	Region 1	Region 2	Region 3	Region 4	Total
Region 1	19,343	401	293	2,886	22,924
Region 2	137	3,770	67	400	4,375
Region 3	74	110	3,072	246	3,501
Region 4	1,516	786	395	60,265	62,962
Total	21,071	5,067	3,827	63,798	93,762

Table 2.4. Origin of in-flows as percentage of total

	Region 1	Region 2	Region 3	Region 4
Region 1	91.8%	7.9%	7.7%	4.5%
Region 2	0.7%	74.4%	1.8%	0.6%
Region 3	0.4%	2.2%	80.3%	0.4%
Region 4	7.2%	15.5%	10.3%	94.5%
Total	100.0%	100.0%	100.0%	100.0%

Table 2.5. Destination of outflows as a percentage of total

	Region 1	Region 2	Region 3	Region 4	Total
Region 1	84.4%	1.7%	1.3%	12.6%	100.0%
Region 2	3.1%	86.2%	1.5%	9.2%	100.0%
Region 3	2.1%	3.1%	87.7%	7.0%	100.0%
Region 4	2.4%	1.2%	0.6%	95.7%	100.0%

2.3.3 The Impact of Trade Flows

As noted earlier, with 53 sectors, the volume of information about flows between areas is enormous. To help provide a sense of the capability of the methodology, only two illustrations will be presented, one focusing on the aggregated interactions and one on the impacts of the construction sector. In both cases, the analysis was designed to reveal the impact of an injection of $1 million into each region in the first case and into one specific sector in the latter case.

Table 2.6. Impact multipliers for aggregate trade flows [The aggregate (one-sector) interregional leontief inverse matrix]

	Region 1	Region 2	Region 3	Region 4
Region 1	1.400	0.041	0.045	0.027
Region 2	0.004	1.295	0.012	0.004
Region 3	0.002	0.014	1.375	0.002
Region 4	0.047	0.093	0.071	1.416

Table 2.7. Construction impacts by region (Assumes a $1 million injection in new construction activity in the region at the top of the column)

Impact on	Region 1	Region 2	Region 3	Region 4
Region 1	$1.47 m	$50,000	$50,000	$30,000
Region 2	$10,000	$1.41 m	$10,000	<$5,000
Region 3	<$10,000	<$10,000	$1.41m	<$5,000
Region 4	$110,000	$130,000	$120,000	$1.56 m

Table 2.6 provides the details for the aggregate trade flows, while table 2.7 reveals the spatial distribution of impacts from an expansion of $1 million in construction activity in each region in turn. The asymmetries are readily apparent; while a significant portion of the impact remains in each of the regions of origin, interregional spillovers are not so evenly divided. The significant difference in size of region 4 provides a large measure of the explanation but, in addition, the region does capture feedback effects that are larger than one might assume given the initial distributions shown in table 2.1.

2.4. Employment Impact Analysis

2.4.1 Examples for Five Sectors

Using data, by specific sector, of a similar kind to that shown in table 2.3, it is possible to trace the impacts of changes in economic activity in one region on the region itself and the rest of the metropolitan economy. Recall that table 2.3 is an aggregate over 53 sectors; however, there is considerable variation in the trade dependencies by sector and the impacts of changes in five sectors, banking, fabricated metals, health services, retail and trucking, will be used to illustrate these variations.

Table 2.8. Employment impacts associated with an injection of $2 million in each sector in each region in turn

Sector	Injection in Region 1	Region 2	Region 3	Region 4
Banking	95.95%	0.42%	0.27%	3.36%
Fabricated Metal	90.56%	1.49%	0.65%	7.31%
Health Services	95.30%	0.46%	0.24%	3.99%
Retail	96.76%	0.41%	0.17%	2.65%
Trucking	91.91%	0.38%	0.13%	7.58%

Sector	Region 1	Injection in Region 2	Region 3	Region 4
Banking	21.76%	73.66%	0.29%	4.30%
Fabricated Metal	11.38%	77.96%	1.51%	9.15%
Health Services	3.78%	91.25%	0.30%	4.67%
Retail	11.31%	84.97%	0.25%	3.46%
Trucking	9.33%	81.31%	0.21%	9.15%

Sector	Region 1	Region 2	Injection in Region 3	Region 4
Banking	23.22%	0.46%	72.35%	3.97%
Fabricated Metal	12.62%	1.90%	77.65%	7.83%
Health Services	4.65%	0.45%	90.88%	4.02%
Retail	11.64%	0.38%	85.02%	2.96%
Trucking	12.42%	0.43%	78.45%	8.70%

	Region 1	Region 2	Region 3	Injection in Region 4
Banking	9.01%	0.17%	0.11%	90.70%
Fabricated Metal	4.31%	0.85%	0.39%	94.46%
Health Services	1.76%	0.24%	0.11%	97.89%
Retail	4.04%	0.19%	0.08%	95.69%
Trucking	3.90%	0.21%	0.09%	95.80%

Table 2.8 shows the impact of a shock in each region in turn for five different sectors; the shock was assumed to be $2 million. The entries in the table record the percentage number of jobs created in each region. Several features may be highlighted:
- For each of the sectors undergoing shocks, the distribution of impacts (across regions) varies significantly, depending on the locus of the initial location.
- There is similar variation across sectors
- Regions 1 (Loop) and 4 (Suburbs) retain more of the impacts that occur when expansion takes place initially within these two regions; this concentration is far less marked in the case of the other two regions (West and South sides).

Contrast Region 1 and 2; almost 96% of the banking sector expansion in region 1 will remain in that region while only 74% would remain locally if the expansion were to take place in region 2. For the case of region 1, the majority of the impact outside this region would be found in region 4 (and similarly for the case in which the expansion took place in region 4).

There is a greater accumulation of impacts within regions 2 and 3 when the sector being shocked is health services, retail or trucking; however, the intra-regional impacts, on a percentage basis, are still smaller than those recorded in regions 1 and 4 for the same sectors. How can this be interpreted? First, the metropolitan region is the spatial scale at which firms optimize production, impact purchases and sales when these firms are engaged in manufacturing activities – hence, there is likely to be greater spillover to other regions. However, while regions 2 and 3 do exhibit these tendencies, this is not the case for the other two regions. Secondly, a great deal of retail, health and trucking activity is more locally organized, with greater attention being paid to local demands and suppliers (with a region).

Table 2.9 provides a summary of what may be considered a net benefit or balance of impact (analogous to trade). Take the banking sector again; the entries should be read as follows. The diagonal entries record the percentage of the impacts that remain in the region of origin. Reading down column 1, the entries should be interpreted as follows:

Positive net balance of impacts with region 2 of 21.34% (this is the impact of an expansion in banking in region 2 on region 1 minus the impact of region 1 on 2); since the matrix is antisymmetric, the entry in column 2, row 1 will be –21.34%, indicating that region 2 has a negative net balance of impacts with region 1. For the most part, regions 1 and 4 enjoy a positive net impact balance with regions 2 and 3 and a varied net balance with each other. Region 2 enjoys a small positive net balance with region 3.

Table 2.9. Net impacts by sector and region

Banking

Originating Region	Net Benefit to Region			
	1	2	3	4
1	90.56%	-21.34%	-22.95%	-5.66%
2	21.34%	73.66%	-0.18%	4.30%
3	22.95%	0.18%	72.35%	4.19%
4	5.66%	-4.30%	-4.19%	90.70%

Fabricated Metal

Originating Region	Net Benefit to Region			
	1	2	3	4
1	90.56%	-9.89%	-12.35%	3.00%
2	9.89%	73.66%	-0.39%	4.84%
3	12.35%	0.39%	77.65%	7.44%
4	-3.00%	-4.84%	-7.44%	0.00%

Health Services

Originating Region	Net Benefit to Region			
	1	2	3	4
1	95.30%	-3.32%	-4.41%	2.23%
2	3.32%	91.25%	-0.15%	4.43%
3	4.41%	0.15%	90.88%	3.91%
4	-2.23%	-4.43%	-3.91%	97.89%

Retail

Originating Region	Net Benefit to Region			
	1	2	3	4
1	96.76%	-10.90%	-11.47%	-1.39%
2	10.90%	84.97%	-0.13%	3.27%
3	11.47%	0.13%	85.02%	2.88%
4	1.39%	-3.27%	-2.88%	95.69%

Trucking

Originating Region	Net Benefit to Region			
	1	2	3	4
1	91.91%	-8.95%	-12.29%	3.68%
2	8.95%	81.31%	-0.22%	8.94%
3	12.29%	0.22%	78.45%	8.62%
4	-3.68%	-8.94%	-8.62%	95.80%

Note: Diagonal entries are the % of impact that remains in the region of initial impact

2.4.2 Interpretation

Part of the reason that a smaller percentage of the ripple effects remains in either region 2 or 3 has a great deal to do with the nature of the industrial structures in those two regions. For the most part, they are much more specialized than the region as a whole and there are fewer opportunities for local residents to consume goods and services within the region. Thus, the hoped-for trickle down effect does not apply in Chicago – regions 2 and 3 receive relatively little benefit from expansions of sector activity in regions 1 and 4.

The major implications of these findings are the need for development initiatives to explore the creation of locally driven complexes of activity that could offer the opportunity to keep a larger percentage of the impacts within the region. Single, disconnected activities will not generate as much impact as would be the case if some set of interconnected sectors could be lured into the region.

2.5. Income Flows and Journey-to-Work

2.5.1 Journey-to-Work

Using the CATS data, it has been possible to aggregate the journey-to-work information to conform to the four regions chosen; the summary data are shown in tables 2.10 through 2.12.

Table 2.10. Journey-to-work flows

From/To	Region 1	Region 2	Region 3	Region 4	total
Region 1	251,416	15,145	39,933	89,623	396,117
Region 2	79,466	137,940	39,411	51,264	308,081
Region 3	43,034	9,626	37,725	28,877	119,262
Region 4	123,093	50,859	36,300	1,378,078	1,588,329
Total	497,008	213,570	153,369	1,547,842	2,411,790

Table 2.11. Destination flows from each region as a percentage of the total outflow

From/To	Region 1	Region 2	Region 3	Region 4	Total
Region 1	63.47%	3.82%	10.08%	22.63%	100%
Region 2	25.79%	44.77%	12.79%	16.64%	100%
Region 3	36.08%	8.07%	31.63%	24.21%	100%
Region 4	7.75%	3.20%	2.29%	86.76%	100%

Table 2.12. Origin flows from each region as a percentage of the total inflow

From/To	Region 1	Region 2	Region 3	Region 4
Region 1	50.59%	7.09%	26.04%	5.79%
Region 2	15.99%	64.59%	25.70%	3.31%
Region 3	8.66%	4.51%	24.60%	1.87%
Region 4	24.77%	23.81%	23.67%	89.03%
Total	100.00%	100.00%	100.00%	100.00%

Table 2.13. Trade balance in work trips

From/To	Region 1	Region 2	Region 3	Region 4
Region 1	0	-64,321	-3,101	-33,470
Region 2	64,321	0	29,785	405
Region 3	3,101	-29,785	0	-7,423
Region 4	33,470	-405	7,423	0

The tables reveal some striking differences in work-trip patterns, ignoring for the moment, the differences in the volumes of flows. The entries along the main diagonal in table 2.11 show the percentage of residents in a region who work there; note that all regions save one have more than 50% of the residents working in the same region. The exception, region 3 (West), has only 31% of its residents employed within the region. This difference is brought out even more in table 2.12 wherein the origin flows are portrayed. Region 3 is much more 'open' to in/out flows than any of the other regions, drawing an almost equal percentage of its employees from the other three regions and itself. Note also that 23% of region 2's employees originate in region 4.

In table 2.13, an attempt has been made to provide the idea of trade balances by summarizing the entries in table 2.10 in net terms. Positive entries down each column indicate that the region at the top of the column has a surplus of work trips (inflows are greater than outflows); thus, region 1 is the only region in that category. In contrast, region 2 has all negative entries while region 3 has more in-flows than outflows with all regions except region 1. Surprisingly, region 4 has only one, small positive balance, and that is with region 2.

The data in these tables reveal that the employment and trade flows provide a misleading indication of the nature of the interdependencies within the Chicago metropolitan region; there is far more *cross-hauling* than the trade data would reveal. By *cross-hauling*, one is referring to the simultaneous in-flow and out-flow of people on a daily basis. The differences in the roles played by regions 2 and 3 are striking. While different, they indicate a strong degree of involvement in the economic activities of the other two regions – providing important sources of labour in firms located outside their region of residence.

2.5.2 Gross Income Flows

Table 2.14 takes the work trip data one stage further; the survey information provides details on the number of employees per household; from these data, it is possible to generate income flows in total between the regions. The table should be read as follows; each entry records the income flow from the region at the left to the region at the top of the column. Tables 2.15 and 2.16 convert the data in table 2.14 into percentages to and from each region. Note, from table 2.14, that a much smaller percentage of the income generated by employment in regions 2 and 3 remains in the region of origin (around 35%) in contrast to region 1 (64%) and region 4 (84%). Between 17 and 19% of the income earned in regions 1 through 3 ends up in region 4.

Table 2.14. Gross income flows between the regions [$ billion]

From/To	Region 1	Region 2	Region 3	Region 4	Total
Region 1	13,413	1,128	2,106	4,146	20,792
Region 2	4,466	5,147	2,062	2,655	14,331
Region 3	1,692	313	1,528	726	4,258
Region 4	9,032	3,112	2,603	82,681	97,428
Total	28,603	9,700	8,299	90,208	136,810

Table 2.15. Destination of income originating in work performed in each region

From/To	Region 1	Region 2	Region 3	Region 4	Total
Region 1	64.51%	5.42%	10.13%	19.94%	100%
Region 2	31.17%	35.92%	14.39%	18.53%	100%
Region 3	39.72%	7.35%	35.87%	17.05%	100%
Region 4	9.27%	3.19%	2.67%	84.86%	100%

Table 2.16. Origin of income brought into each region

From/To	Region 1	Region 2	Region 3	Region 4
Region 1	46.89%	11.63%	25.37%	4.60%
Region 2	15.62%	53.07%	24.85%	2.94%
Region 3	5.91%	3.23%	18.41%	0.81%
Region 4	31.58%	32.08%	31.37%	91.66%
Total	100.00%	100.00%	100.00%	100.00%

The dependencies in terms of the origin of income are more varied; almost one-third of region 1's income is generated in region 4 while region 1 provides only 4% of region 4's income. For region 3, the distribution of the origin of income is even less than the distribution of employment (see table 2.12), with most of the differences accounted for by region 4 (i.e., incomes generated in region 4 by region 3 residents are higher than the incomes earned by residents in region 3 from jobs located in region 3).

From these data, it is possible to estimate the size of the potential market in each region; income flows created purchasing power of $9.7 billion in the south and $8.2 billion in the west. If income generated outside the region had been ignored, the corresponding values would have been reduced by 46% (region 2) and 75% (region 3) respectively, seriously underestimating the size of the regional spending power.

2.5.3 Selected Sector Impacts

Three sectors, construction, durable manufacturing and trade, have been chosen to illustrate the income impacts at the sectoral level across the regions. These are shown in tables 2.17 through 2.19. Each table shows the percentage allocation of income generated by a $1 million increase in the output of the sector in question in each region in turn.

Table 2.17. Construction income impacts

From/To	Region 1	Region 2	Region 3	Region 4
Region 1	43%	12%	24%	5%
Region 2	15%	48%	23%	3%
Region 3	5%	3%	17%	1%
Region 4	37%	37%	36%	91%
Total	100%	100%	100%	100%

Table 2.18. Durable manufacturing income impacts

From/To	Region 1	Region 2	Region 3	Region 4
Region 1	43%	12%	24%	5%
Region 2	15%	47%	23%	3%
Region 3	6%	3%	16%	1%
Region 4	36%	38%	37%	91%
Total	100%	100%	100%	100%

Table 2.19. Trade income impacts

From/To	Region 1	Region 2	Region 3	Region 4
Region 1	46%	13%	25%	5%
Region 2	15%	51%	24%	3%
Region 3	6%	3%	18%	1%
Region 4	37%	33%	33%	91%
Total	100%	100%	100%	100%

For the most part, there are few significant variations across sectors; they tend to follow the pattern of results described in tables 2.12 and 2.15

Table 2.20 provides a complementary picture that reveals some important differences; the income flows have been converted to a *per household* basis (thus, there may be more than one wage-earner in any household). The table must be read as a to/from table (in contrast to the other tables; thus each entry shows the household average income flow from person(s) working in the region at the top of the column who reside in the region at the left). Even though there are relatively few work-trips from region 1 to 2, the per household value of those flows is very high ($74,466). Region 3 is the only region in which those resident and working there earn more than others who commute to other regions to work.

Table 2.20. Income flows on household basis

TO\FROM	Region 1	Region 2	Region 3	Region 4	Average
Region 1	$53,350	$74,466	$52,730	$46,259	$52,490
Region 2	$56,206	$37,316	$52,329	$51,798	$46,519
Region 3	$39,310	$32,502	$40,496	$25,150	$35,707
Region 4	$73,379	$61,180	$71,708	$59,997	$61,340
Average	$57,551	$45,416	$54,109	$58,280	$56,725

Finally, some summary interpretation of the strengths of interactions between the regions can be gained by inspection of table 2.21. Essentially, what is shown here is the way in which income spent by residents of one region creates additional income in the other regions – analysis that is similar to the ripple effect for trade but, in this case, the source of the stimulus is the set of households resident in each region.

One of the most interesting findings here is that the income multiplier is higher for region 2 than any other region – in other words, income spent by residents of region 2 creates a greater amount of additional income in the Chicago metropolitan area than a similar expenditure by residents of other regions. Further, the major beneficiaries, outside the region of origin, are those residing in region 4. The diagonal entries show how much additional income is generated within the region of origin – note the significant differences between region 4 – where an additional $0.774 is generated for each $1 of income that is spent to region 3 where the intraregional ripple effect is only $0.063).

Table 2.21. Interrelational income multipliers for the four regions

Region of income receipt	Region of income origin			
	Region 1 Loop/North	Region 2 South	Region 3 West	Region 4 Suburbs
Region 1	**1.230**	0.119	0.157	0.067
Region 2	0.106	**1.279**	0.134	0.048
Region 3	0.034	0.027	**1.063**	0.012
Region 4	0.445	0.561	0.496	**1.774**
Total	1.815	1.986	1.851	1.900

The patterns of interdependency suggest that careful attention needs to be paid to as many dimensions of exchange in a metropolitan area as possible; of course, an important pattern, shopping trips, has not been presented. This dimension would provide yet another indicator of the degree to which these regions are even more integrated; the data presented in table 2.21 suggest that shopping trips may be a very strong circulator of impacts throughout the metropolitan region.

In the next section, the analytical framework established for this project will be illustrated with an example based on an economic programme for the Pullman district.

2.6 Pullman Impact Analysis

While a number of proposals have been advanced to assist in the economic development and revitalization of the South-side, attention will be directed to a package of development initiatives that appear to be close to realization.

The activities that have been discussed are shown in table 2.22; note that the projects have been phased over time to simulate the actual, annual expenditures. Furthermore, where appropriate, a distinction has been made between the construction activities and the operations (i.e., the direct impacts once a building, firm, or activity is able to produce a good or service).

While the major impacts (see table 2.23) occur in region 2 (the location of the project), when attention is directed to the indirect effects (i.e. the ripple or multiplier effects), the projects' impacts are more widely diffused throughout the metropolitan region. In particular, region 4 benefits, garnering from 8 to 13% of the indirect impacts. Very little of the impacts diffuse to region 3 while less than 5% accrue in region 1.

Table 2.22. Projects under consideration for the pullman district within region 2

PROJECT	[$'000] 1999	2000	2001	2002	2003	total
Industrial						
Family TV	$2,000	$25,000	$25,000			$52,000
Chemical	$3,000	$3,000				$6,000
Steel	$500	$1,000				$1,500
Office Building	$500	$1,500				$2,000
Total	$6,000	$30,500				$36,500
Commercial						
Drug Store	$500	$5,500				$6,000
Comm. Strip						$0
Theaters						$0
Food Store						$0
Clock Tower	$200	$800	$2,000			$3,000
Total	$700	$6,300	$2,000			$9,000
Housing						
Senior Housing	$2,500	$5,000				$7,500
Market Rate Hs	$500	$500	$500	$500		$2,000
Affordable Hs	$200	$200	$200	$400		$1,000
Total	$3,200	$5,700	$700	$900		$10,500
Public Facilities						
Library		$3,000				$3,000
Cultural Center		$3,000	$5,000	$12,000		$20,000
Total		$6,000	$5,000	$12,000		$23,000
Total Project	$9,900	$48,500	$7,700	$12,900		$79,000
Operations						
Wages		$8,000	$12,000	$14,000		$34,000
Employment	[Jobs]	400	200	100		700

In summary, the projects would have the desired goal of creating significant economic activity in region 2; at the same time, the indirect benefits certainly move to other parts of the Chicago metropolitan economy. A different combination of activities might produce a different spatial pattern of impacts; however, it is likely that the most significant source of variation will be in the residential location of employees in these new enterprises. The more that local (region 2) residents are employed, the greater will be the impact on region 2.

Table 2.23. Impact analysis of the proposed projects: total output
[*Volume of Goods and Services*]

TOTAL IMPACT (Current $ million)	1999	2000	2001	2002	2003	Total
Total Output						
region 1	0.86	4.23	2.70	1.58	0.00	9.37
region 2	27.49	135.66	86.42	50.81	0.03	300.40
region 3	0.19	0.92	0.59	0.34	0.00	2.04
region 4	2.55	12.60	8.02	4.72	0.00	27.89
total	31.09	153.40	97.73	57.45	0.03	339.70
Indirect only	1999	2000	2001	2002	2003	total
region 1	0.86	4.23	2.70	1.58	0.00	9.37
region 2	15.48	76.40	48.64	28.54	0.03	169.08
region 3	0.19	0.92	0.59	0.34	0.00	2.04
region 4	2.55	12.60	8.02	4.72	0.00	27.89
total	19.08	94.14	59.94	35.19	0.03	208.38
indirect only (%)	1999	2000	2001	2002	2003	total
region 1	4.49%	4.49%	4.50%	4.50%	2.76%	4.50%
region 2	81.15%	81.15%	81.14%	81.11%	88.43%	81.14%
region 3	0.98%	0.98%	0.98%	0.98%	0.60%	0.98%
region 4	13.38%	13.38%	13.39%	13.41%	8.21%	13.38%
total	100.00%	100.00%	100.00%	100.00%	100.00%	100.00%

2.7 Summary Evaluation

The analysis has revealed that interactions among regions vary depending on whether the focus is on production, employment or income. While the circulation of economic activity generates impacts outside the region of origin of any economic stimulus, the greatest source of variation would appear to be in the residence-work trips. The West-side seems to have the most open economy – it is more dependent on the rest of area in that a very high percentage of the workforce commutes into the region during the day. On the other hand, the suburbs are more self-contained - but this finding is also a reflection of the much larger geographic region that was defined for region 4.

Notwithstanding these limitations, the most striking finding is in the size of the market (in terms of potential spending power) in both the South and West side. Further, income spent by residents of these two regions has a far higher probability of generating additional income to residents of other regions. In large part, this probably stems from home or work to shop trips that involve destinations outside region 2 or 3.

Given this finding, it would appear important to focus next on shopping trips. In addition, the analysis has not been able to account for differences in consumption tastes, savings rates and differences in households across the regions. Further, there is enough evidence to suggest that regions 2 and 3 have much stronger interactions with the close-in suburbs of region 4 than with the region in toto. Hence, it would appear appropriate, as a next step, to disaggregate region 4 into several smaller regions. However, the client, Chicago United, was most interested in ascertaining the size of the untapped markets on the south and west side of Chicago in the near term and was less interested in a more academic (and prolonged analysis) that might have initially divided the metropolitan economy into a larger number of regions. In retrospect, it was a wise decision, given the enormous problems involved in matching data collected by many different agencies. The academic-business interaction landscape is littered with many unfinished projects that promised a great deal and often failed to deliver anything of value. On the other hand, while it might have been tempting to adopt a *back-of-the-envelope* calculation in place of the formal analysis conducted here, the stakes involved were high; the client's advisory panel consistently advocated the type of research program that was undertaken with the expectation that the results could be presented not only in the form of this present chapter but also in a form more appropriate for a non-academic audience.[6]

Returning to the initial theme – trade analysis – there would appear to be significant opportunities to enhance the economies of the South and West side while, at the same time, enhancing the region-wide economy. There is a strong implication from international and interregional trade theory that as per capita incomes rise, there is a tendency for consumers to demand higher quality products and more product differentiation – often leading to enhanced trade among regions. A stronger, more vibrant South-side can contribute to enhancing demand for goods and services provided by firms and businesses in other parts of the metropolitan area; further, the growth and development of the South side can enhance Chicago's competitive position in the Midwest economy and thus further stimulate trade between states and with other countries in the world.

Economic development in inner-city metropolitan America has always been hampered by the absence of strategic information about the nature of the markets in these regions, and their internal and external interdependencies. The creation of an analytical framework of this kind offers an important input in the attraction of new investment. Of particular importance is the notion of income circulation – involving consideration of the spatial origin of this income and the way in which its spending generates impacts throughout a metropolitan area. The Miyazawa framework offers an important analytical tool to begin this exploration.

---

[6] One member of the advisory panel commented that while he enjoyed reading the full report, he did encourage that the material be "translated into English for wider dissemination." Several thousand copies of such a translation have been widely distributed to the business community and public sector decision-makers.

Acknowledgements.
This project was made possible with generous support from the John D. and Catherine T. MacArthur Foundation. Professor David Boyce (University of Illinois at Chicago) and his research assistants provided significant contributions in preparing the journey to work data for use in the analysis. The Project Steering Committee appointed by Chicago United and the enthusiastic support of Carolyn Nordstrom, Charles Brown and their staff provided invaluable guidance in helping shape the analysis.

References

Bon R. (1975) Some conditions of macroeconomic stability in multiregional models. DOT Report No.10. Washington D.C.: University Research Program, U.S. Department of Transportation.

Bon R. (1984) Comparative stability analysis of multiregional input-output models: column, row, and Leontief-Strout gravity coefficient models, *Quarterly Journal of Economics. 99* (4): 791-825.

Chicago Area Transportation Study (1994) CATS 1990 Household Travel Survey, Technical Documentation for the Household, Person and Trip Files.

Hewings G.J.D., Sonis M., Guo J., Israilevich P. R. and Schindler G. R. (1998) The hollowing-out process in the Chicago economy, 1975-2011. *Geographical Analysis 30*, 217-233.

Illinois Department of Employment Security (1996) Where workers work in the Chicago Metro area, Supplemental Report: 1993-1995p.

Israilevich P.R., Hewings G.J.D., Schindler G.R. and Mahidhara R. (1996). The choice of input-output table embedded in regional econometric input-output models. *Papers in Regional Science 75*,103-119.

Israilevich P.R., Hewings G.J.D., Sonis M. and Schindler G.R. (1997) Forecasting structural change with a regional econometric input-output model, *Journal of Regional Science 37*, 565-90.

Miyazawa K. (1976) *Input-output analysis and the structure of income distribution*. New York, NY; Springer-Verlag.

Okuyama Y., Sonis M. and Hewings G.J.D.(1999) Economic impacts of an unscheduled, disruptive event: A Miyazawa multiplier analysis, in G.J.D. Hewings, M. Sonis, M. Madden and Y. Kimura (eds.) *Understanding and Interpreting Economic Structure*. Heidelberg, Springer-Verlag.

Polenske K.R. (1970) An empirical test of interregional input-output models estimation of 1963 Japanese production, *American Economic Review. 60*, 76-82.

Sonis M. and Hewings G.J.D. (1993) Hierarchies of regional sub-structures and their multipliers within input-output systems: Miyazawa revisited. Hitotsubashi Journal of Economics. 34(1): 33-44.

Toyomane N. (1988) *Multiregional input-output models in long-run simulation*. Dordrecht: Kluwer Academic.

3 Socio-economic Impact Assessment: Meeting Client Requirements

Peter W.J. Batey and Moss Madden

3.1 Introduction

Socio-economic impact assessment is a field of analysis, simulation and forecasting dating back in the USA to the mid 1960s, where early examples such as the Susquehanna River Basin study (Hamilton *et al.*, 1969) relied upon a sequential simulation structure to model the effects of the implementation of different policy packages upon the economy and population structure of their study area. Other studies, such as that of the Boulder economy (Miernyk *et el.,* 1967), took as their starting point a Leontief input-output model, and used this as the driver for an investigation of potential shifts in demographic variables as a result of economic changes.

To some extent, socio-economic impact assessment has over the years since its inception become the poor relation of environmental impact assessment (Glasson *et al.,* 1994), but a number of more recent primers on the latter topic, such as Glasson *et al.*, and Morris and Therivel (1995) make explicit reference to the importance of the socio-economic aspect of environmental impact assessment. Indeed, Morris and Therivel devote substantial space to the use of input-output methods amongst other economic assessment tools available to impact assessors. Earlier work by Leistritz *et al.*, (1986) also provides a useful typology of socio-economic assessment models, again with some references to input-output amongst other economic modelling techniques.

In this chapter we follow the input-output approach, and report on the application of extended input-output models to the impact assessment of two large-scale infrastructure developments associated with airport expansion. These applications have been carried out on a commercial basis, in order to assist the clients with the preparation of a case, in one instance to gain the permission to carry out the proposed development and in the second instance to be used to assist a third party, the potential developer, to gain such a permission. The models elaborated were both based on the Leontief input-output structure, but were extended to include a number of demographic and labour market variables to enable the clients to assess the impact upon, for example, unemployment rates in the areas under study and in neighbouring areas.

3.2 Client Requirements

3.2.1 British Aerospace

The client in this case was proposing an extremely large-scale infrastructure development on the outskirts of the City of Liverpool, Merseyside, North West England. The proposal involved the expansion of the existing airport, essentially a small-scale operation with an annual passenger throughput of around 0.65 million passengers per annum.

The client intended to submit a planning application to the local authority for an expansion of the airport over a period of time to up to 45 mppa (millions of passengers per annum). This expansion involved a number of very large-scale physical changes to the existing airport. New terminal buildings were planned; approach roads and access roads from nearby motorways and trunk roads were proposed; a new railway line was suggested; and the client proposed the development of a second runway, and the complete realignment of both runways from the present position. This last item included the reclamation of a large tract of land in the (tidal) estuary of the River Mersey, and the extension of the runways out on to this reclaimed land.

Clearly, a proposal of this size, enlarging a relatively modest local, sub-regional airport into a hub-spoke operation of international scale, could not be expected to be handled by a local authority, on its own, as a straightforward planning application. The proposal was large enough and contentious enough to be guaranteed to be called in by the Secretary of State for the Environment in London, and for there to be a public enquiry. For these reasons the client wanted to build up a store of facts and projections favourable to the airport development.

At the same time, it was known that the nearby (35 miles away) airport at Manchester was planning a second runway. The Secretary of State was expected to look at both airports together when the applications were called in, and probably to make a decision to permit expansion at one or the other but not both airports.

Of crucial importance to the client's case was the likely impact on local employment of the development. Liverpool and its hinterland of Merseyside are amongst the most economically depressed areas of the UK, and rate fairly high on EU-wide criteria for depressed areas. If the client could show that the airport development and expansion would bring dramatic employment possibilities to the local labour market, it would be likely that any adverse aspects of the development would be judged by the local authority and more importantly the UK government as being overridden by the positive aspects. If the debate could, furthermore, be focused as exclusively as possible upon the employment issue, the Liverpool expansion could be argued to be of much greater socio-economic importance than that at Manchester, which has fewer social and employment problems than Liverpool, and whose airport is located physically further from those problems that do exist.

A further important issue for the client was the desirability of demonstrating to the Exchequer in central government the beneficial fiscal impacts that the airport development would have in terms of reductions in unemployment benefit, increases in income tax receipts etc.

There were also a number of extremely serious environmental considerations. These included the reclamation of some of the Mersey Estuary, which was bound to disturb a site of Special Scientific Interest which a number of bird populations hosted at various stages in their annual migration, feeding and mating routines. Flight paths were also a serious problem, as the airport at Liverpool is close to the city; the desire to route aircraft away from heavily populated areas was the reason behind the proposed realignment of the runway(s).

The client therefore offered a contract to a consortium of consultants. The contract included an environmental analysis of the airport expansion proposals, and a full socio-economic assessment of the entire development and operation of the expansion. The environmental analysis was undertaken by a private sector consultancy, which was linked to the University of Liverpool, from which the two groups undertaking the socio-economic assessment were drawn. This assessment was divided into two more or less separate exercises. The first of these was an analysis of the labour market of Liverpool and the surrounding Merseyside County, and the second was a socio-economic impact assessment of the airport development, using mathematical modelling techniques.

3.2.2 London Chamber of Commerce and Industry

The London Chamber of Commerce and Industry (LCCI) is one of the principal organisations promoting the interests of the business community in the Greater London area, within the South East of England. Part of its remit is to lend support to policies and projects that are designed to strengthen London's position *vis a vis* other world cities. Improvements to physical infrastructure, including airports, are seen as an important way of maintaining competitiveness.

In 1991, the LCCI began a four-year study of the airport sector in London. This study was intended to draw attention to the economic benefits of airport operations to the Greater London economy. A threefold approach was followed, consisting of a literature survey based on similar studies carried out elsewhere, to elicit impact multipliers that might be applicable to the London case; a business survey within Greater London to establish the current and prospective importance of airports for business operations; and a series of forecasts at local, regional and national levels intended to inform impact studies for each of London's main airports.

In the case of Heathrow airport, there was a need to consider the impact of an additional terminal, Terminal 5 (T5), which would increase the passenger handling capacity of the airport from 50 to 80 million passengers per annum by 2016. This would be achieved using the current runways, with a small (5 per cent) increase in the number of passenger air transport movements. Larger aircraft would use the new terminal, thus enabling a significantly larger number of passengers to use the airport. The proposal involved the provision of new passenger terminal facilities, comprising a core building and up to three satellites, additional airside facilities, including taxiways and aircraft stands, ancillary activities, a maintenance hangar and a noise pen. A dual carriageway spur road would be built linking the terminal with the M25 motorway and the existing rail link connecting the airport with central London would be extended to provide public transport access to the terminal.

The proposal to build T5 was the subject of an outline planning application submitted by the airport's owners, the British Airports Authority (BAA), in 1993. Accompanying the planning application was an Environmental Statement (BAA, 1993a) describing the results of a series of impact studies carried out by consultants; among these studies was a detailed assessment of the employment impact of the airport by 2016, with and without T5. Using 1988 as a base year, the employment study focused on impacts within the local labour market, an area covering those parts of Greater London, Surrey, Berkshire and Buckinghamshire lying within approximately fifteen miles of the airport (BAA, 1993b).

LCCI's airport study was intended to produce separate reports on each of London's airports, starting with Heathrow (LCCI, 1994). The expectation was that there would be a public inquiry on the T5 proposal, beginning in mid-1995, at which interested parties would have the opportunity of giving evidence for or against the proposal. LCCI wanted to support the proposal and aimed to use the airport study for this purpose. By commissioning its own impact study, LCCI would, it hoped, be able to show that it had reached the same conclusions as the developer, BAA, but by an independent route.

The brief from LCCI was very specific in relation to method and required outputs. The commissioned study would be expected to examine the economic impacts of Heathrow in terms of employment, output, taxation and unemployment, on the London area and beyond. It was also felt to be important to take account of commuting flows and of impacts on a sector-by-sector basis. The study would be able to use more up-to-date information than BAA's Environmental Statement, particularly in relation to Heathrow airport itself where a detailed employment survey had been carried out in 1992. This year would serve as the base against which employment impacts could be measured.

In parallel with this impact modelling study, the LCCI commissioned other consultants to carry out economic forecasts of the local, regional and national economies over the period of airport expansion covered by the T5 proposal. These forecasts were intended to inform the detailed impact studies, with and without the T5 development. The forecasts would be particularly important in ensuring that realistic assumptions were made about long-term increases in labour productivity in the main sectors affected by the airport development.

3.3 Model Choice

In both the cases described in the previous section it was decided that the best mathematical modelling tool for the socio-economic impact assessment elements of the clients' requirements was the extended input-output model. Extensions to input-output models have a long history, going back to the earliest work on the concept by Leontief (1941). The first development that can be described as an extension was the closure of the model with respect to households. This is very well documented throughout the literature, and involves quite simply the expansion of the simple Type I Leontief model:

$$\mathbf{x} = (\mathbf{I} - \mathbf{A})^{-1}.f \qquad (3.1)$$

where $\mathbf{x}$ = gross outputs
  $\mathbf{A}$ = technical coefficients
  f = final demand

to the Type II formulation where $\mathbf{x}$, $\mathbf{A}$ and f are extended to include households as well as industries, in the simplest case by the addition of one row to the $\mathbf{A}$ matrix to represent income to labour, and the incorporation of the household consumption portion of final demand into the $\mathbf{A}$ matrix as a new column.

This has the effect of increasing the multiplier in the $(\mathbf{I} - \mathbf{A})^{-1}$ matrix as induced effects are added to the Type I direct and indirect effects. The problem here (see, e.g., Batey and Madden, 1983, Madden and Batey, 1980) is that households are treated in exactly the same way as industries, and expansions in the economy assume workers to appear from nowhere (as, perhaps, in-migrants), while economic decline causes workers to disappear (as, perhaps, out-migrants). These problems were initially addressed by Miernyk *et al.*, (1967), at least in so far as expanding economies were concerned, and later in a series of papers by the authors of this chapter (e.g., Batey and Madden, 1981,1999; Batey, Madden and Weeks, 1987; Madden and Batey, 1983, 1986; Madden, Batey and Thomson, 1996) and others (e.g., Oosterhaven and Dewhurst, 1990; Phibbs, 1989).

The nature of the problem can be addressed simply by recognising the fact that workers taking up new jobs often come from an unemployed labour pool associated with the economy in question, and workers losing their jobs tend by and large to enter the unemployed labour pool, at least in the first instance, and in labour forces with relatively low mobility such as that in the UK to stay in that pool perhaps for a considerable period of time. In the first case the new consumption displayed by workers taking up new jobs is attenuated by the amount they were consuming whilst unemployed, and in the second case their decline in consumption is partly offset by the consumption they continue to finance from welfare benefits when they enter unemployment.

The authors of this chapter have identified this aspect of the employment/consumption nexus as the redistribution effect, and have developed a number of formulations designed to model this effect along with the usual direct, indirect and induced components of the multiplier matrix. A simple version of such a model is given by:

$$\begin{bmatrix} x \\ e \\ u \end{bmatrix} = \begin{bmatrix} I-A & -\alpha & -\beta \\ -l & 0 & 0 \\ 0 & 1 & 1 \end{bmatrix} \cdot \begin{bmatrix} f \\ 0 \\ P \end{bmatrix} \qquad (3.2)$$

where α is the consumption vector of employed workers,
β is the consumption vector of unemployed workers,
l is the vector of labour demand coefficients,
e is the number of employed workers,
u is the number of unemployed workers and
P is the total number of workers.

The equations of this system are simple:

$$(I - A)x - \alpha e - \beta u = f \qquad (3.3)$$
$$-lx + e = 0 \qquad (3.4)$$
$$e + u = P \qquad (3.5)$$

and are discussed in some detail in Batey and Madden (1983).

An alternative formulation, which avoids the problem of equations (3.2) - (3.5) of mixing economic and demographic data in the same model, is given by

$$\begin{bmatrix} x_I \\ x_H \\ s.u \end{bmatrix} = \begin{bmatrix} I-A & -h_c^e & -h_c^u \\ -h^a & 1 & 0 \\ s.l & 1 & 1 \end{bmatrix} \cdot \begin{bmatrix} f_I \\ f_H \\ s.P \end{bmatrix} \qquad (3.6)$$

where x_I is industrial gross output
x_H is total income to employed workers
s is the welfare benefit payable to one employed worker
h_c^e is the consumption propensity vector of employed workers
and h_c^u is the consumption propensity vector of unemployed workers.

The equations here are:

$$(I - A)x_I - h_c^e \cdot x_H - h_c^u \cdot s.u = f_I \qquad (3.7)$$
$$-h^a \cdot x_I + x_H = f_H \qquad (3.8)$$
$$l \cdot x_I + u = P \qquad (3.9)$$

In both these simple extended input-output models, categorised as Type IV, where Miernyk *et al.*, (1967) used a Type III formulation, we find that the multiplier values fall between those of the Type I and Type II formulation. This is to be expected: in Type I the income/consumption nexus is ignored completely, and in Type II the existence of consumption from welfare benefits is ignored, thereby returning multipliers with artificially large values.

The Type IV model, on the other hand, represents the real world in a much closer way than do the earlier formulations, with its explicit representation of direct, indirect, induced and redistributive effects. Accordingly, it has become an exemplar of the sort of input-output model used to analyse and predict demographic-economic systems in many different countries, sometimes on its own and sometimes as an element within or derived from a Social Accounts Matrix (e.g., Oosterhaven, 1981; Phibbs, 1989; Hewings and Madden, 1995).

The particular requirements of the client in the two cases discussed here prescribed developments of the basic Type IV model to take account of the circumstances in which the impact assessments were to be carried out.

3.4 Model Development

3.4.1 Model Specification

Although superficially similar, the requirements of the clients indicated the development of models that were ultimately rather different from each other. The first issue concerned the definition of the study area in each model. In the Liverpool study, the client was particularly concerned, as mentioned in Section 3.2 above, to demonstrate the employment impact of the proposed airport development. It was obvious that this impact would not be limited to the area of Liverpool alone. This would be true of any large industrial expansion in almost any area of the UK, but in the case of Liverpool airport it was compounded by the physical location of the airport on the extreme south-east border of the city. A location of this nature means the likelihood of newly employed workers at the airport itself coming from outside the local authority area is much increased. Of course, the location of workers in industries 'down the line' is less likely to be affected by the exact location of the airport, but these workers are, because of the extremely open nature of the Liverpool economy, unlikely in any event to be restricted to the Liverpool local authority area.

Accordingly, we derived a geographical area we called Greater Merseyside. As well as Liverpool, this included the other four districts of Merseyside County (Knowsley, St.Helens, Wirral and Sefton) and five contiguous districts (Chester, Ellesmere Port and Neston, Halton, Vale Royal and Warrington). These districts were identified as being those most likely to be affected in economic and demographic terms by the expansion of the airport. Many of the data needed to derive the model empirically were available at local authority district level, and could easily be agglomerated to form a database for the Greater Merseyside economy. Although it was possible to identify from this grouping two lower level agglomerations (Inner and Outer Merseyside) we did not develop a model for each of these groups, retaining only one model for the whole of Greater Merseyside.

For Heathrow we adopted a different approach, although the location of the airport was also of paramount importance in this case. Heathrow airport is on the extreme periphery of what used to be Greater London County and is now the collection of boroughs that make up London. The client in this case, the London Chamber of Commerce and Industry, clearly had a prime interest in the effect of the development of Terminal Five upon Greater London itself. However, it was also important in making a case at the public enquiry that the overall effect could be assessed upon the South-East Region as a whole, within which Greater London is located. Further, the Chamber of Commerce had a potential interest in assessing the impacts of other airports in the South-East Region, such as Gatwick, Stansted and Luton, and required a model which was capable of being used for these other purposes.

For these reasons, we decided to develop two models, one for Greater London and one for the South-East Region. Unlike the Liverpool case most data were available already aggregated to the spatial levels we had identified meaning that there was less difficulty in obtaining data for this model.

The next issue involved the industrial sectoring schemes that were adopted for the models. Our concerns were threefold: to represent the Greater Merseyside, Greater London and South East economies using a sectoring scheme which would a) be tailored to suit the particular regional economies that we were modelling; b) would explicitly represent in a fairly disaggregated form sectors of the economy in which the impacts of the proposed airport developments would be most immediately felt and c) would be amenable to construction by simple aggregation from the sectoring scheme used in the UK input-output tables.

To achieve these objectives we developed for both research projects a 34-sector aggregation of the Standard Industrial Classification shown in Appendix 3.A. The main characteristics of the sectoring scheme are the aggregation of Agriculture, Forestry and Fishing, and Energy and Water Supply, into single sectors, with the great majority of manufacturing industries retaining their identities as individual sectors. Service sectors, on the other hand, were treated in greater aggregation with the exception of those sectors, particularly transport industries, which we judged to be closely related to the operation of an airport. In practice, this sectoring scheme meant that for the Greater Merseyside table, which was based on the 1984 national table, we aggregated from 101 to 34, and in the case of the Greater London/South East region tables we aggregated the 1990 UK table from 123 sectors down to 34.

The third important issue involved establishing appropriate household disaggregations. In both subjects we adopted what was then a novel disaggregation for household consumption. We divided the workforce into three sub-groups: the employed; the short-term unemployed and the long-term unemployed. In the Greater Merseyside model we separated out income received by workers from employment, welfare payments made to short-term unemployed and welfare payments made to long-term unemployed and inactive. We assumed that consumption propensities of short-term unemployed workers were the same as those of the employed, but that the long-term unemployed/inactive consumed with different propensities from the other groups. In the London/South East region model we made a further disaggregation: employed and short-term unemployed workers were divided into manual and non-manual categories, enabling us to discriminate between the rather different consumption and income patterns of such workers expressed particularly in expenditure variations on housing, motoring and leisure. This manual/non-manual difference persisted across the employed and short-term unemployed categories, but did not manifest itself in the inactive category.

3.4.2 Model Structure

The requirements of the clients, coupled with the evidence and availability of the data, led us to develop two similarly structured but internally somewhat distinct models. The Greater Merseyside model took the form shown in equation 3.10:

$$\begin{bmatrix} I-A & -h_c^i & -h_c^1 & -h_c^2 & 0 & 0 & 0 & 0 \\ -h^a & 1 & 0 & 0 & 0 & 0 & 0 & 0 \\ 0 & 0 & 1 & 0 & 0 & -s^s & 0 & 0 \\ 0 & 0 & 0 & 1 & 0 & 0 & -s^l & 0 \\ -\rho l & 0 & 0 & 0 & 1 & 0 & 0 & 0 \\ 0 & 0 & 0 & 0 & (1-\alpha) & 1 & 0 & 0 \\ 0 & 0 & 0 & 0 & 1 & 1 & 1 & 0 \\ -(1-\rho)l & 0 & 0 & 0 & 0 & 0 & 0 & 1 \end{bmatrix} \begin{bmatrix} \Delta x_1 \\ \Delta x_H \\ \Delta x_u^s \\ \Delta x_u^l \\ \Delta e^r \\ \Delta u^s \\ \Delta u^l \\ \Delta e^m \end{bmatrix} = \begin{bmatrix} \Delta f_1 \\ \Delta f_H \\ 0 \\ 0 \\ 0 \\ 0 \\ \Delta P \\ -\Delta e^c \end{bmatrix} \quad (3.10)$$

where
h_c^1 is the consumption profile of employed and short-term unemployed workers,
h_c^2 is the consumption profile of long-term unemployed workers,
s^s is the annual unemployment benefit payable to a short-term unemployed worker,
s^l is the annual unemployed benefit payable to a long-term unemployed worker,
ρ is the proportion of jobs taken up by indigenous workers,
α is the ratio of long-term unemployed to short-term unemployed taking up new jobs,
x_u^s is total short-term unemployed benefits,
x_u^l is total long-term unemployed benefits,
e^r is jobs taken up by locally resident formerly unemployed workers,
u^s is the number of short-term unemployed,
u^l is the number of long-term unemployed,
e^m is jobs taken up by in-migrant workers
e^c is net number of in-commuting workers

This structure enabled us to use as inputs to the system changes in industrial final demand, changes in exogenous worker income, changes in population and assumptions about in-commuting. These inputs produced a range of outputs that were geared to the client's requirements. In particular, they enabled us to predict changes in industrial gross outputs, jobs taken up by previously unemployed local residents, jobs taken up by in-migrants, and changes in the numbers of long-term and short-term unemployed.

The Heathrow study adopted a very similar approach, but took the route of developing two identically structured but really different models. One was set up for the Greater London area, and the other for the South East Region as a whole. We also, from close examination of the Family Expenditure Survey, 1990 (CSO, 1991), identified particularly important spending differentials between non-manual and manual households. These centred on housing, motoring and leisure, and persisted between employed and unemployed households, although they were not present in inactive households. Accordingly, we decided to disaggregate the consumption vectors into non-manual and manual categories, which obviously had impacts upon the structure of the rest of the model.

Both the Greater London and South East Region models, then, took a form somewhat different from the Greater Merseyside model. Equation 3.11 shows the matrix form of these two models to highlight the difference between them and the Merseyside formulation:

$$\begin{bmatrix} I-A & -h_c^{ln} & -h_c^{lm} & -h_c^{ln} & -h_c^{lm} & -h_c^2 \\ -h^{an} & 1 & 0 & 0 & 0 & 0 \\ -h^{am} & 0 & 1 & 0 & 0 & 0 \\ s^{sn}.\rho.l^n & 0 & 0 & 1 & 0 & 0 \\ s^{sm}.\rho.l^m & 0 & 0 & 0 & 1 & 0 \\ s^l.\rho.l^l & 0 & 0 & 0 & 0 & 1 \end{bmatrix} \begin{bmatrix} \Delta x_1 \\ \Delta x_H^n \\ \Delta x_H^m \\ s^{sn}\Delta u^{sn} \\ s^{sm}\Delta u^{sm} \\ s^l\Delta u^l \end{bmatrix} = \begin{bmatrix} \Delta f_1 \\ \Delta f_H^n \\ \Delta f_H^m \\ s^{sn}\Delta P^n \\ s^{sm}\Delta P^m \\ s^l\Delta P^l \end{bmatrix} \quad (3.11)$$

Note that in this model the unemployment benefits are located directly on the row of the matrix relating to the labour market of the relevant type of worker, and there is no opportunity for long-term unemployed/inactive workers to take up employment. All movements into employment are assumed to come from the short-term manual and non-manual unemployed. Long term unemployed and inactive workers' consumption is built into the model, however, via the h_c^2 parameter.

This model, then, generates a fixed number of jobs for non-manual and manual short-term unemployed and long-term unemployed for each unit of increase in industrial gross output. The relationship between the long-term and short-term unemployed, recognised explicitly in the Merseyside model by the sixth and seventh equations of the model, is implicit in this model, lying embedded within the coefficients l^n, l^m and l^l.

3.4.3 Data Issues

The Greater Merseyside model and the Heathrow models clearly have many similar characteristics, not least their feasibility in terms of data availability. None of the three models could be developed on the basis of an existing input-output model for the local area, nor was there time or money to build a survey-based model. Accordingly, a non-survey input-output model was developed for each of the applications, relying mainly on published data. The stages in this are fairly straightforward: 1) aggregate the sectors of the most recent UK table, obtained from the Central Statistical Office, as it then was, as described in Section 3.4.1 above; 2) update the aggregated national table to the most recent year possible, relying on Commodity Flow Accounts (CFA) purchased from the Department of Trade and Industry for the Merseyside model and providing gross output and intermediate demands. Intermediate purchases were derived from the CFA and the Blue Book of National Accounts. These data enabled the RAS technique (Miller and Blair, 1985) of bi-proportional matrix adjustment to be applied to generate a national table for the most recent year possible.

This national table was then regionalised using the semi-logarithmic location quotient approach developed by Round (1978), which required gross output data for the national and regional levels. The former were derived from the CFA, and the latter from special purchases from the Business Statistics Office and the Annual Census of Production. In the case of service industries, where local gross output figures could not be obtained, national figures were factored down by applying employment ratios, using data from NOMIS, the National Online Manpower Information System.

Consumption vectors in each model were derived from the Family Expenditure Survey (FES) (CSO, 1988, 1991). An average household consumption, and income, were attributed to a notional individual worker in each category. For Merseyside we related the expenditure headings of the FES to those of the Input-Output Tables following Henderson (1984). For the Greater London/South East region models, we were able to make use of a consumption table that accompanied the CSO-provided 'Make and Use' matrices to assign aggregate expenditures by type of household to Input-Output commodity sectors.

Income from employment was calculated in the Merseyside model from NOMIS and Business Statistics Office data and regional gross weekly earnings figures available from the Department of Employment. Multiplying the sectoral wage rate by a sectoral ratio of employment to gross output generates sectoral gross output-income from employment flows. The Greater London/S.E. Region models used wage rates taken from the New Earnings Survey 1990 (DoEmp, 1990). Unemployment benefits were calculated from Department of Social Security figures, from the FES and from Regional Trends. In the absence of any data to advise on values for I and Ψ (the proportion of employment drawn from previously unemployed people), they were hypothesised and sensitivity testing carried out to check the models' robustness in response to shifts in their values.

3.4.4 Modelling Impacts

Two basic approaches are available in using the model to measure impacts. The first, mentioned already in relation to equation 3.10, may be referred to as a simulation approach. It requires airport expenditure to be expressed in financial units as a vector of final demand. Construction and operational expenditure will generally be identified separately, and be specified by industrial sector and by relevant time period. By using the final demand vector to post-multiply the inverse of the model's matrix of coefficients, a vector of activity levels may be calculated. This will express the change that has occurred in each of a number of activity variables as a result of the change in airport expenditure. The activity variables provide, or can easily be converted to, estimates of changes in employment, unemployment, industrial output and National Exchequer payments/receipts.

The second approach makes use of the impact multipliers that can be derived from the model inverse. Several different multipliers can be calculated all of which summarise, on a sector-by-sector basis, the direct, indirect and (where appropriate) induced effects on a given activity variable of a unit direct change in that variable (Batey and Weeks, 1989).

Employment multipliers are particularly useful in this respect. Assuming that estimates are available of direct employment in airport operation or construction by sector, employment multipliers may then be applied to these estimates in order to calculate the total, region-wide employment impacts. These employment impacts can be converted into other activity variables, such as industrial output and National Exchequer payments/receipts.

The choice of which approach to follow is strongly dependent on the nature and quality of survey data and exogenous forecasts that are available. In practice, data are provided by individuals from organisations unfamiliar with sectoring schemes and with input-output analysis. The analyst is faced with the problem of allocating data items to sectors, of determining which items are exogenous and which leak out of the economy being modelled. Frequently data will be incomplete or in the wrong form to be included in the model.

A good example of the sector allocation problem is provided by the Heathrow study. Here there was a need to compile employment data not only on the current operation of the airport, but also on its future operation with and without T5. In all cases the data needed to be compatible with the model's sectoring scheme.

The main source of data for the current operation of the airport was the 1992 Heathrow Employment Survey (BAA, 1993b). This document contained the results of two separate surveys, one of employers based at the airport, and the second of employees working at the airport. Employment figures were presented in a number of different ways in this survey: by employment category, occupational grouping and airport-related functions. Cross-tabulations were available showing the correspondence between these classification schemes.

The 1992 Employment Survey provided an adequate basis for allocating base year employment by industrial sector, area of worker residence (Greater London, rest of South East Region, or beyond South East Region), and skill category (manual or non-manual). A much more difficult task involved the generation of equivalent data for the future operation of the airport. The T5 Environmental Statement (BAA, 1993a) provided employment estimates of operational employment levels at Heathrow both with and without T5. However, these figures were coarsely disaggregated by airport-related functions only (passenger-related, air traffic movement-related, cargo-related, traffic-related support and non-traffic-related support). It was necessary, therefore, to develop an allocation procedure to convert these estimates into employment figures that conformed to the industrial sectoring scheme used in the model.

To do this, cross-tabulations from the employment survey were translated into coefficient form and used to create an allocation matrix to distribute airport-related functions to employment categories compatible with both the survey and the model, divided into non-manual and manual classes. As for residence, the 1992 Employment Survey made the point that "patterns of residence among Heathrow's employees showed only marginal changes when compared with results from the 1986 Survey" (BAA, 1993b, p.48) and there was no evidence to suggest that this would change significantly before 2016. Data from the 1992 Survey were therefore used to create a coefficients matrix to distribute workers into Greater London and non-Greater London residential groups, by industrial sector.

A second example illustrates the problem of assembling financial data in the case where a simulation approach is adopted. In the Liverpool airport study, consultants were engaged to collect financial data on the operation of three other UK airports with passenger throughputs, in 1990, of 3, 10 and 21 millions of passengers per annum. These comparator airports were intended to represent the types of airport to which Liverpool airport would be similar at various stages in its development, ranging from a regional airport to a major international gateway.

Expenditure and employment data were gathered from the operational activities at the airports. Three categories of expenditure were judged to be exogenous : (1) payments by airlines to the airport company; (2) receipts of on-airport companies receiving revenue directly from airlines and passengers; and (3) wages and salaries paid by airlines and government agencies to their employees based at the airport.

Some items under category (2) were judged to be prone to leakage out of the regional economy. In the absence of any information about the extent of this, it was assumed that half would pass through the local economy and that half would be lost through leakage. Expenditure data were not available for many items under categories (1) and (2). In these circumstances employment figures provided a means of estimating expenditures. The l coefficients, measuring the ratio between labour demand and gross output by sector, were used to derive expenditure from employment data for those items of operation-related expenditure for which financial data were not available. In the case of wages and salaries paid to employees of airlines and government agencies, only the numbers of employees were available. Annual wages and salaries were calculated using the sectoral wage rates used in estimating the income from employment coefficients used in the model.

These two examples show clearly that whichever approach is adopted in measuring impacts, there will be problems in assembling data on airport activities. While the employment multiplier approach probably has the edge over the simulation approach in terms of data availability, the use of expenditure data to create a vector of final demand does bring the benefit of a stricter definition of endogenous and exogenous activity.

3.5 Impact Measurement

3.5.1 Measuring Employment Impacts

In both the Liverpool and the Heathrow studies, the primary purpose was to produce estimates of the employment impacts of the expansion proposals. It is common in studies of this kind to make separate estimates of construction and operational employment, given that the profile of employment will differ over time and by sector (Glasson *et al.*, 1994). It is also likely that the studies will be used to answer a number of 'what if ...?' questions related to whether the proposal is assumed to go ahead, or not, and to what extent individual factors are influencing the results.

The Heathrow study provides a good example of the way in which employment estimates can be presented to yield the maximum amount of information about the role of individual factors. Table 3.1 shows the estimates of employment impacts in summary form. The table distinguishes between Greater London and the South East Region, the two geographical areas that were of particular interest to the client and which were each modelled separately. The input to the impact measurement exercise was an estimate of direct on-airport operational employment, shown in the first column. For the base year this was derived from the 1992 Heathrow Employment Survey (BAA, 1993b) and for the design year (2016) it was obtained from exogenous forecasts made by the developer and presented in the Environmental Statement that accompanied the planning application (BAA, 1993a).

Two forms of input-output model were used to calculate the direct and induced impacts. In the case of the indirect impacts, employment multipliers were derived from a simple Type 1 Leontief model. These multipliers were used to calculate the direct and indirect effects of a direct change in employment. It was a straightforward matter, therefore, to obtain the indirect effect by subtracting the direct effect from this figure.

A similar approach was used to calculate the induced impacts. Here employment multipliers from the extended input-output model (equation 3.11) were used to obtain the direct, indirect and induced impacts. By subtracting the results of the Type I model, measuring direct and indirect impacts, the induced impacts could be derived. By dividing total impacts into these three types of impact, it is possible to show clearly the relative importance of inter-industry linkages (the indirect impact), reflecting jobs generated as a result of a supply chain, and of additional household income and expenditure (the induced impact).

The same table also shows how the employment impacts are affected by the construction of T5. Estimates are presented for the base year and for 2016, with and without the terminal. Differences are expressed between each pair of estimates so that the effect of varying one factor at a time can be measured. In this particular case, the results show that without T5, total employment associated with the airport is likely to fall by between 14,000 and 17,000 jobs, depending on which regional definition is adopted. The main reason for this decline is that labour productivity at the airport is assumed to increase substantially over the 24-year period under consideration. If the terminal is built, the additional employment generated (between 16,000 and 18,000 jobs) is sufficient to compensate for this reduction and so there is a small net increase in total employment associated with the airport.

Table 3.1. Summary of employment impacts of Heathrow Airport: 1992 Operation, 2016 Operation Without Terminal 5, 2016 Operation with Terminal 5

A. Greater London - On-airport Direct Employment				
	Direct	Indirect	Induced	Total
1992	51072	28384	10549	90005
2016 Without T5	43280	23838	8568	75686
Difference from 1992	*-7792*	*-4546*	*-1981*	*-14319*
2016 With T5	51710	29676	10585	91971
Difference from 1992	*+638*	*+1292*	*+36*	*+1966*
Difference from 2016 W/o T5	*+8430*	*+5838*	*+2017*	*+16285*

B. South East Region - On-airport Direct Employment				
	Direct	Indirect	Induced	Total
1992	51271	29499	19690	100460
2016 Without T5	43280	24171	26012	83464
Difference from 1992	-7791	-5328	-3678	*-16966*
2016 With T5	51710	30095	19770	83464
Difference from 1992	*+439*	*+596*	*+80*	*+1115*
Difference from 2016 W/o T5	*+8430*	*+5924*	*+3758*	*+18111*

One of the main reasons for using an input-output model is to estimate the impact on employment at a sector level. To perform this calculation, it is helpful first to compute a direct and indirect labour requirements matrix (DILRM), a disaggregated form of employment multiplier (Miernyk *et al.*, 1970). The DILRM is based on a similar concept to a Leontief inverse in that each column in the matrix represents the direct and indirect (and, where appropriate, induced) effects on individual sectors of a unit change in a given sector's final demand. The DILRM may be calculated as **L.B** where **L** is a diagonal matrix of labour demand : gross output ratios, by sector, and **B** is a Leontief inverse. In the case of an extended input-output model, only the first n entries in **L** will be non-zero, while **B** will represent the entire extended model's inverse.

In this form, the DILRM may be used to calculate the total employment impact of a change in final demand. It can be converted to a disaggregated multiplier capable of being applied to direct employment estimates by post-multiplying by $\mathbf{L}^{-1}$, a diagonal matrix of reciprocal labour demand: gross output ratios by sector. The modified DILRM is therefore given by $\mathbf{L.B.L}^{-1}$. When post-multiplied by a column vector of direct employment by sector, it will yield a column vector of total employment impacts, also by sector.

Table 3.2 provides an example of the results obtained using this disaggregated employment multiplier. It shows how direct employment in 10 of the 34 industrial sectors generates indirect and induced employment across the whole range of sectors. (The separation between indirect and induced employment was achieved by applying the same technique as was used in constructing Table 3.1). The same approach can be used to calculate employment impacts by skill category. In this case, **L** would be replaced by $\mathbf{L}_{nm}$ and $\mathbf{L}_{m}$, in calculation of non-manual and manual employment impacts, respectively. $\mathbf{L}^{-1}$ would remain unchanged.

3.5.2 Calculation of Fiscal Impacts

Large development projects of the kind described in this chapter inevitably involve the assembly of packages of capital investment drawn from a range of different sources. The developer may be dependent to some extent on the provision of public funds especially if there are infrastructure improvements required beyond the boundaries of the development site. This would apply, for example, if the proposal included plans to build a link road to an existing motorway or to divert a railway line to allow access from the development site.

To support the argument for public funding, it may be necessary to demonstrate the positive fiscal impacts of the development project. This will involve a calculation of the additional tax revenue likely to flow as a result of the project, as well as an estimate of the reduction in social security payments caused by more unemployed workers taking up employment.

Table 3.2. Employment impacts of Heathrow in Greater London 2016 operation with terminal 5 – On-airport Direct Employment

Sector	Industry	Direct	Indirect	Induced	Total
1	Agriculture etc.	0	1196	261	1457
2	Energy + Water	0	1583	407	1990
3	Metalliferous Ores	0	0	0	0
4	Metal Manufacture	0	1061	346	1407
5	Mineral Extraction	0	599	328	927
6	Mineral Manufacture	0	1245	501	1746
7	Chemicals	0	922	318	1240
8	Man-made Fibres	0	0	0	0
9	Metal Goods	0	828	289	1117
10	Mechanical Engin'g	0	629	301	929
11	Office Machinery etc	0	662	353	1015
12	Electrical etc Engin'ng	0	479	229	708
Sector	Industry	Direct	Indirect	Induced	Total
13	Motor Vehicles	0	1714	448	2162
14	Other Transport	0	336	153	489
15	Instrument Engin'ng	0	374	256	631
16	Food, Drink etc	0	973	276	1249
17	Textiles	0	193	207	400
18	Leather	0	545	202	748
19	Footwear + Clothing	0	242	194	436
20	Timber etc	0	747	310	1058
21	Paper etc	0	1284	334	1618
22	Rubber + Plastics	0	905	323	1228
23	Other Manufacturing	0	266	201	467
24	Construction	66	2145	590	2801
25	Distribution + Repairs	1394	436	218	2048
26	Hotels + Catering	3016	287	147	3450
27	Railways etc	115	1348	381	1844
28	Sea Transport	0	3493	565	4058
29	Air Transport	40139	1894	343	42376
30	Transport Support	1510	947	280	2738
31	Postal Services	42	336	596	974
32	Telecommunications	80	680	248	1009
33	Banking, Finance etc	1298	925	815	3038
34	Other Services	4050	401	165	4615
Total		51710	29676	10585	91971

Such a calculation was performed in estimating the fiscal impacts likely to result from expansion at Liverpool airport. Fiscal impacts were calculated to reflect the state of development expected to be reached by 2005, when 16.55 million passengers were forecast to use the airport. At this size the operation of the airport was expected to generate a total of 19,420 jobs within the Greater Merseyside economy.

Fiscal impacts were considered under four main headings : additional income tax payments and employee's National Insurance Contributions; additional employers' National Insurance Contributions; reductions in unemployment benefit and social security payments; and reductions in Community Charge rebates.

To calculate the additional yield of income tax, an estimate was made first of the average annual income among employed workers (£10,070 at 1988 prices). An average tax rate on earnings (32%) was derived using the most recent data then available. This enabled an estimate to be obtained of the income tax yield per employed worker. Employers' National Insurance Contributions were assumed to be made at the (then current) rate of 10.45%.

The reduction in payments of unemployment and social security benefits was calculated by reference to three components – unemployment benefit, supplementary benefit and housing benefit – using the first and third components for the short-term unemployed and the second and third for the long-term unemployed and economically inactive. An 80% rebate was assumed for Community Charge (the system of local taxation then in operation); unemployed workers who obtained employment would forego this rebate and would instead pay the full amount of £350.

Of the 19,420 operational jobs generated by the year 2005, 13,189 were assumed to be filled by previously unemployed people. This figure was derived by combining two main elements: jobs in airlines and government agencies; and jobs in other industrial sectors. In the case of the former, it was assumed that 50% of jobs would be taken up by unemployed people from Greater Merseyside, reflecting the fact that, in contributing to the airport's growth, airlines and government agencies relocating at the airport are likely to bring a significant number of staff with them. For other sectors, accounting for about half of the jobs that are generated, the equivalent figure was assumed to be 85%, reflecting a much higher take-up of workers from the local unemployment pool. The combined take-up rate worked out to be 67.9%.

Of the 13,189 previously unemployed persons, 80% were assumed to have been short-term unemployed, and the remaining 20% long-term unemployed or economically inactive. Table 3.3 shows the results of the four calculations and gives the positive fiscal impact from employment generated by airport operations in 2005.

Calculations of this kind can never be precise and must, of necessity, reflect assumptions about tax and benefit regimes that are current at the time of the study. They do, nevertheless, provide a useful negotiating tool when a developer is seeking a contribution from public funds.

Socio-economic Impact Assessment: Meeting Client Requirements   55

Table 3.3. Fiscal impacts of Liverpool Airport in Greater Merseyside resulting from Airport operations in 2005

Category of Impact	Impact (in £m)
Increase in Income Tax Revenue and Employees' NIC	62.58
Increase in Employers' NIC	20.44
Reduction in Unemployment Benefit and Social Security Payments	43.97
Reduction in Community Charge Rebates	3.69
Total Fiscal Impact	130.68

3.5.3 Sensitivity Analysis

In the discussion of employment impacts, emphasis was placed on the need to construct an analytical framework to aid the interpretation of model outputs. This framework would, it was hoped, enable the influence of individual factors to be determined. Tables 3.1 and 3.2 showed how this was done in the context of the Heathrow study.

A similar form of sensitivity analysis can be applied to the mathematical model itself, in relation to assumptions about key parameter values. Indeed, analysis of this kind is essential if the client is to be satisfied that the models' outputs are reliable and not unduly influenced by arbitrary assumptions without a firm empirical foundation.

Two examples, selected from the Heathrow study, illustrate how this may be accomplished. In this study it was regarded as important to apply sensitivity tests to variations in two factors about which assumptions had to be made, consumer expenditure and the proportion of job vacancies filled by previously unemployed local people.

The base model for the South East Region assumed no leakage of consumer expenditure, while the Greater London base model assumed that leakage would amount to a loss of one-fifth of such expenditure. The South East model was tested by reducing consumer expenditure by 25%, a factor of 0.75. The Greater London model was also varied to examine a decrease of 25%, or a factor of 0.75, in household consumption, compared to the base assumption. In a second variation for Greater London, the effects of assuming no leakage, equivalent to a factor of 1.25 compared to the selected base model, were also calculated. The results of these manipulations are given in Table 3.4.

These results suggested that the models were robust to changes in assumptions about leakage of household consumption, with the Greater London model faring slightly better than that for the South East Region.

In Section 3, the parameter Ψ, expressed the proportion of employment, or change in employment, which is drawn from previously unemployed local people. Ψ can take on values between zero, indicating that all the employment in question is filled from outside the region, and one, which indicates that all such employment is filled from within the region.

The T5 Environmental Statement examined labour demand and labour supply projections for the Heathrow local labour market. These projections were used as a guide in the selection of values for Ψ.

In the year 1990, for which labour demand and supply were known, there was a small excess of supply. The application of three different projection techniques (extrapolation, constant share and shift-share) generated estimated deficiencies in labour supply by the year 2016 which ranged from 3.3% up to 17.1%, with a mean value of 9.2%. Values for Ψ corresponding to the mean and maximum labour supply deficits were tested, yielding 0.908 and 0.829, respectively. In order to allow for the possibility of some recruitment beyond the modelled areas, the variation generated by a value of Ψ of 0.700 was also examined. The outcome of these calculations is presented in Table 3.5.

These sensitivity tests show that the models also performed in a robust fashion when Ψ was varied. Again, the Greater London model showed greater stability than that for the South East Region. The Heathrow study sensitivity analysis results are consistent with similar tests carried out for the Greater Merseyside model as part of the Liverpool airport study (Batey *et al.*, 1993).

Table 3.4. Outcome of sensitivity testing : Heathrow study consumer expenditure

Parameter Value	Greater London	South East Region
Base Index	1.000	1.000
Base **x** 0.75	0.968	0.945
Base **x** 1.25	1.035	-

Table 3.5. Outcome of sensitivity testing : Heathrow study local recruitment

Parameter Value	Greater London	South East Region
Base Index	1.000	1.000
$\Psi = 0.908$	1.004	1.009
$\Psi = 0.829$	1.008	1.016
$\Psi = 0.700$	1.014	1.028

3.6 Conclusions

In this chapter we have identified a number of interesting issues surrounding the development and implementation of socio-economic impact assessment models for clients. Client requirements clearly have a very powerful impact upon such model development and implementation; the other main constraint is data availability. We show how, in two examples of model development for socio-economic impact assessment of airport expansion, client requirements, taken together with existing expertise in extended input-output modelling, determined both the spatial scale of the models developed and also their internal structure. The Greater Merseyside model was built to cover an area which effectively covered the likely journey-to-work radius of the airport, and would therefore meet the client's requirement to show the total impact of the airport, and for Heathrow two models were developed to take account of the client's needs to demonstrate impacts separately upon London and the South East Region, and to use the model in future to assess the impacts of other airports in the Region.

Client requirements also directed us to carrying out a rather unusual exercise in models of this type, in which the fiscal impacts of the proposed airport expansion in Liverpool were estimated. This arose from the client's perceived need to convince the UK Exchequer of the overall benefit of the development, in order to swing Central Government opinion behind the proposal, and to maximise the probability of successfully levering public investment in associated infrastructure.

Data availability, as always in model construction, had a powerful determining effect upon model structure. We were obliged to balance the clients' needs for particular outputs with the capacity of the data to support model structures that met those needs. We were also informed by different data sets of different characteristics of, for example, personal consumption, which persuaded us to disaggregate consumption in the Heathrow model differently from that in the Merseyside model. Had we been aware of the consumption differentials when constructing the Merseyside model we might well have included a non-manual/manual split in consumption in that model as well. Another important aspect of data availability was centred around the lack of guidance from the data on what values to allocate to the I and Ψ parameters. Since it was important to keep these parameters in the models, and there was little guidance on their correct values, we were obliged to carry out sensitivity testing to establish the robustness of the models to variations in these parameters. results from the sensitivity tests on Ψ in the Heathrow model are reported in Section 3.5.3, and show that the model is robust in response to quite large variations in the parameter.

Data availability, both in published and privately-provided form, is also crucial in determining the type of implementation carried out. For example, in the Heathrow model data were available in employment form rather than financial form, which determined the use of this model in employment multiplier form rather than the more usual output multiplier form used in the Merseyside model, which was then translated into employment using labour demand coefficients.

In general terms, we conclude that the influence of client requirements on model development has both a constraining and a coercive effect. Constraints are obviously reflected in model structure, while the coercive aspect can lead to interesting theoretical constructions, such as the modelling of the relationships between short-term and long-term unemployed in both models, and to unusual applications such as the fiscal impact work done in Merseyside. Data availability, as always, has a powerful inhibitory effect upon both model construction and development, but of course leads to the development of interesting devices for circumventing such inhibitions.

References

Batey P.W.J. and Madden M. (1981) Demographic-economic forecasting within an activity-commodity framework: some theoretical considerations and empirical results, *Environment and Planning A 13 (9)*, 1067-1083.

Batey P.W.J. and Madden M. (1983) The modelling of demographic-economic change within the context of regional decline: analytical procedures and empirical results, *Socio-Economic Planning Sciences 17* (5-6), 315-328.

Batey P.W.J. and Madden M. (1999) Interregional employment multipliers in an extended input-output modelling framework, in Hewings G. J. D., Madden M., Sonis M. and Kimura Y. (eds), *Festschrift for ken'ichi miyazawa* (Springer-Verlag, Berlin).

Batey P.W.J., Madden M. and Scholefield G. P. (1993) Socio-economic impact assessment of large-scale projects using input-output analysis: a case study of an airport, *Regional Studies, 27(3)*, 179-191.

Batey P.W.J., Madden M. and Weeks M. (1987) Household income and expenditure in extended input-output models: a comparative theoretical and empirical analysis, *Journal of Regional Science,* 27 (3), 341-356.

Batey P.W.J. and Weeks M.J. (1989) The effects of household disaggregation in extended input-output models, in Miller R.E., Polenske K.R. and Rose A.Z. (eds) *Frontiers of Input-Output Analysis,* Oxford University Press, New York, 119-133.

British Airports Authority (1993a) *Terminal 5 Heathrow Environmental Statement*, Volume 1, BAA, London.

British Airports Authority (1993b) *Heathrow Employment Survey: March-May 1992* BAA, London.

CSO (1988) *Family Expenditure Survey 1987,* HMSO, London.

CSO (1991) *Family Spending: a Report on the 1990 Family Expenditure Survey* HMSO, London

Department of Employment (1990) *New Earnings Survey,* HMSO, London.

Glasson J. Therival R. and Chadwick A. (1994) *Introduction to environmental assessment,* UCL Press, London.

Hamilton H.R., Goldstone H.E., Milliman J.W., Pugh A.L., Roberts E.B. and Zellner A. (1969) *Systems simulation for regional analysis: an application to river basin planning,* MIT Press, Cambridge, MA.

Henderson D.S. (1984) *Scottish input-output tables for 1979: volume 3: further results and analyses,* Industry Department for Scotland, Edinburgh.

Hewings G.J.D and Madden M. (1995) *Social and Demographic Accounting,* Cambridge University Press, Cambridge.

Leistritz F.L., Chase R.A. and Murdock S.H. (1986) Socioeconomic impact models: a review of analytical methods and policy implications in Batey P. W. J. and Madden M. (eds) *Integrated analysis of regional systems*, Pion, London.

Leontief W. (1941) *The structure of the American economy: 1919-1929*, Oxford University Press, New York.

London Chamber of Commerce and Industry (1994) *Heathrow Economic Impact Study*, LCCI, London.

Madden M. and Batey P.W.J. (1983) Linked population and economic models: some methodological issues in forecasting, analysis and policy optimisation, *Journal of Regional Science 23*, 141-164.

Madden M. and Batey P.W.J. (1986) A demographic-economic model of a Metropolis in Woods R. I. and Rees P. (eds), *Population Structure and Models*, Allen and Unwin, London, 173-192.

Madden M. and Batey P. W. J. (1980) Achieving consistency in demographic-economic forecasting, *Papers of the Regional Science Association 44*, 91-106.

Madden M., Batey P.W.J. and Thomson J. (1996) Decomposing input-output tables to establish the effects of temporal changes in demographic and economic characteristics *Economic Systems Research 8, (3)*, 209-224.

Miernyk W.H., Bonner E.R., Chapman J.H.J.R. and Shellhammer K. (1967) *Impact of the Space Program on a local economy*, West Virginia University Library, Morgantown, WV.

Miernyk W.H., Shellhammer K., Brown D.M., Coccari R.L., Gallagher C.J. and Wineman W.H. (1970) *Simulating regional economic development*, Health, Lexington, Mass..

Miller R.E. and Blair P.D. (1985) *Input-output analysis: foundations and extensions*, Prentice-Hall, Englewood Cliffs, NJ.

Morris P. and Therivel R. (1995) *Methods of environmental impact assessment*, UCL Press, London.

Oosterhaven J. (1981) *Interregional input-output analysis and Dutch regional policy problems*, Gower, Aldershot.

Oosterhaven J. and Dewhurst J.H.Ll. (1990) A prototype demo-economic model with an application to Queensland, *International Regional Science Review 13*, 51-64.

Phibbs P.J. (1989) Demographic-economic impact forecasting in Non-Metropolitan regions: an Australian example in Congdon P. and Batey P.W.J. (eds) *Advances in Regional Demography*, Belhaven, London, 150-166.

Round J.I. (1978) An inter-regional input-output approach to the evaluation of non-survey methods, *Journal of Regional Science 18*, 607-661.

4. Multiregional Computational General Equilibrium and Spatial Interaction Trade Modelling: An Empirical Example

Martin Schneider and Manfred M. Fischer

4.1 Introduction and Client Needs

The aim of this chapter is to examine the new patterns and volumes of trade flows between Austria and the growing economies of Central and Eastern Europe. The modelling approach is sequential in nature and relies on two types of classic regional science models:

- The Fischer-Johansson model of bilateral trade flows that serves to estimate the direction, volume and composition of Austrian trade from and to Czech Republic, Slovak Republic, Hungary and Slovenia, referring to a hypothetical successful integration in the long run.
- Using these predictions, the second model aims to estimate regional welfare and sectoral production effects within a multiregional computable general equilibrium (CGE) framework for the Austrian economy.

The attractiveness of the EU as a whole has been dramatically illustrated in recent years by the requests for assistance following the rebirth of democracy in Central and Eastern Europe. Further enlargements or deeper forms of association with neighbouring countries are, thus, likely to happen in the near future. The integration of these economies will provide newly supplied goods and export market opportunities at a scale and speed that is quite unprecedented in modern history. In the EU, especially in Austria, there is some concern among politicians about the spatial implications of this process within Western Europe. Successful economic reforms and further integration of the Central and Eastern European countries (such as the Czech Republic, Slovak Republic, Hungary and Slovenia) are expected to generate spatial repercussions that might contradict the cohesion objective of EU regional policy. It is argued that the economies of the Central and Eastern European countries are more similar to those of the less developed nations than to those of the highly developed regions in the EU. Thus, well developed regions would gain from exploiting comparative advantages, while less developed regions lose market shares to the new Central and Eastern European competitors (Bröcker and Jäger-Roschko 1996).

Though this argument sounds plausible, uncertainty exists on the question of what changes will be brought about by integration. This chapter offers answers to a subset of these questions by focusing on trade effects and regional welfare and sectoral production in Austria. Trade implications are analysed with respect to the Visegrad countries Czech Republic, Slovak Republic, Hungary and Slovenia (called CEEC-4 in the following). The study tries to give some quantitative estimates of this impact at the NUTS-II classification level in Austria, based on a two-stage modelling approach. The first stage examines likely developments on the trade front. There are at least two kinds of questions that need to de addressed. First, how much, if any at all, of Austria's trade volume with the CEEC-4 is likely to increase? Second, what will be the product composition of the bilateral trade flows? The focus is on the very long-run outcome in which the above countries will have become thoroughly integrated into the EU. The questions are answered using a simple, but evidently robust model of bilateral trade flows. The predictions are utilized as input for a multiregional computable general equilibrium model for the Austrian economy to assess the trade effects on regional welfare and sectoral production. Section 4.2 and 4.3 serve to outline the details of the modelling approach, while sections 4.4 to 4.6 present the data and results. A final section draws some conclusions.

4.2 The Fischer-Johansson Model of Bilateral Trade Flows

The Fischer-Johansson model of bilateral trade flows lies in the tradition of spatial interaction modelling (see Hamilton and Winters 1992; Fischer and Rammer 1993, 1995; Fischer and Johansson 1996) and describes trade flows x_{ij} from country i to country j. It is assumed that x_{ij} primarily depends on three types of determinants:

- first, economic forces in the country of origin, say *i*, that represent the total potential supply of the exporting country,
- second, economic forces in the country of destination , say *j*, that represent the size of potential demand, and
- third, link-specific variables that either aid or resist the volume of trade flows between these two countries.

The major factor determining the potential supply of exports of country *i* is its gross domestic product. The potential demand for imports of country *j* is governed by the same economic forces as those determining the potential export supply. A higher gross domestic product suggests higher demand for imports.

The group of link-specific variables includes those that reflect influences that either aid or resist trade between *i* and *j*. They can be divided into trade resistance variables and trade preference variables. An important barrier to international trade is transportation and associated transaction costs. It can be assumed that the volume of trade between two countries will decrease with increasing transportation and transaction costs, which are measured in terms of route distance. In addition, two further trade resistance variables are taken into account: aggregate price levels, approximated in terms of export unit values, and the dissimilarity of demand structures between i and j. One can expect that the volume of trade flows is negatively correlated with the aggregate price level of exports from country i to j (King 1993; Marsh and Tokarick 1996). The similarity of the demand structures of the trading partners is measured by the similarity of incomes. The more similar the demand structures are (i.e. the smaller the difference of incomes between i and j), the higher the trade flows. Two trade preference variables are considered in this study, namely the adjacency of the trading countries and the membership of a trade preference area. Trade between neighbouring countries should be, other things being equal, higher than trade between non-neighbours. This may be due to several reasons: the existence of a common border may contribute to greater familiarity with laws, institutions, habits and language of the partner country (Balassa and Bauwens 1988). As the purpose of a trade preference area is to stimulate trade between its member countries, one can expect that trade between two countries should be higher, if both belong to the same trade preference area. Membership to the European internal market (i.e. the European Union and the European Free Trade Association) is represented by a trade preference dummy variable.

Table 4.1 summarises the variables included in the bilateral trade model along with the corresponding hypothesis. Formally, the model can be described as:

$$x_{ij} = c_i \ y_i^{\alpha_1} \ y_j^{\alpha_2} \ f_{ij} \ \varepsilon_{ij} \qquad i,j = 1,...,N \qquad (4.1)$$

where x_{ij} denotes the aggregate volume of trade flows from country i to country j ($i \neq j$), measured in terms of US dollar. c_i is a constant disaggregated by country of origin. y_i denotes the nominal GDP of the country of origin i (measured in US dollar), reflecting the potential supply of the exporting country. y_j is the corresponding variable for the country of destination, representing potential demand of the importing country j. α_1 and α_2 denote the parameters to be estimated. ε_{ij} is the error term. f_{ij} stands for the link-specific variables, which are specified as

$$f_{ij} = \Delta y_{ij}^{\alpha_3} \ p_{ij}^{\alpha_4} \ exp(\alpha_5 d_{ij} + \alpha_6 d_{ij}^B + \alpha_7 d_{ij}^E) \qquad i,j = 1,...,n \qquad (4.2)$$

where Δy_{ij} denotes the difference of GDP (reflecting the similarity of demand structures) between countries i and j, p_{ij} the f.o.b price of commodities produced in i and delivered to j (measured by export unit values; US dollar per ton), d_{ij} transportation and transaction costs (measured via route distance). d_{ij}^B reflects the adjacency between i and j (1 if i and j have a common border, 0 else), d_{ij}^E is a trade preference dummy (1 if i and j belong to the European Union or to the EFTA, 0 else). α_3 to α_4 are the parameters to be estimated.

Table 4.1. The gravity trade model of bilateral trade flows

Variable	Proxy Variable	Hypotheses
A) Country Specific Variables		x_{ij} correlates ...
Potential supply of the exporting country i	Country i's income (y_i)	**positively** with the average GDP in i
Potential demand of the importing country j	Country j's income (y_j)	**positively** with the average GDP in j
B) Link Specific Variables		
Trade Resistance Variables		
Transportation and transaction costs from i to j	Road distance from i to j (d_{ij})	**negatively** with increasing distance from i to j
Aggregate price level of exports from i to j	Unit-value index of exports from i to j (p_{ij})	**negatively** with increasing unit values
Dissimilarity of demand structures in i and j	Cross-country difference of incomes between i and j (Δy_{ij})	**negatively** with increasing dissimilarity
Trade Preference Variables		
Adjacency of trading countries	Adjacency dummy (d_{ij}^B)	**positively** with the existence of a common border
Trading preference areas	Trade preference dummy (d_{ij}^E)	**positively** with trade preferences, **negatively** with trade barriers

The model (4.1) – (4.2) has been estimated on cross-section product-category specific data referring to the year 1995 (see subsection 4.4). It is best thought of as providing a long-run equilibrium view of trading patterns. The model does not relate trade directly to prices. This exclusion stems from the general equilibrium nature of analysis, in which prices are endogenous and only adjust to equate supply and demand. Prices are assumed to adjust quickly, and demand and supply are assumed to be sufficiently responsive to prices to bring about an equilibrium rapidly. It is important to note that the model can not distinguish between trade creation and trade diversion (Hamilton and Winters 1992) so that one cannot generalise from the increases in intra-preference area trade to increases in national welfare.

4.3 The Multiregional General Equilibrium Model

This section describes the multiregional general equilibrium model that aims to estimate the effects of trade flows (as predicted by the bilateral trade model) on regional welfare and production in Austria. Industrialised economies (such as Austria) are characterised by highly interconnected home markets. Estimating welfare effects therefore requires a general equilibrium approach that is capable of dealing explicitly with the interrelationships between the different markets, sectors and regions of the economy (Dinwiddy and Teal 1988).

4.3.1 Fundamental Assumptions

In order to focus on the impact on regional welfare, some severe assumptions have to be made in terms of the representation of the domestic economy. First, despite some reservations, the study lies in the long tradition of representing regions as if they were dimensionless points. Second, the model presented here is a rather simple version of a multiregional general equilibrium model. It is based on a set of rather heroic assumptions, which are known in the literature as the Arrow-Debreu equilibrium. Production activities are characterised by constant returns to scale. Perfect competition prevails in all markets. The firms are assumed to take prices as given and maximise their profits. Due to the linear-homogeneity of their technologies, the zero profit condition applies. This means that at equilibrium prices equal the minimal unit-costs and no profits are left. The households are assumed to maximise their utility under a budget constraint (Shoven and Whalley 1984). Although commodities are traded between regions, production factors are postulated to be immobile between regions. In regard to the production activities of firms, we assume each firm of a sector produces only one characteristic commodity, i.e. the number of commodities equals the number of sectors.

4.3.2 Structure and Notation

Austria's economy is represented by $r=1,..,M$ regions. The following activities are carried out in each domestic region r:

- production of $k=1,...,N$ commodities by using $l=1,...,N$ intermediate goods and $t=1,..,T$ primary factors,
- consumption is carried out by a representative household,
- interregional trade with all domestic regions $s=1,...,M$, and
- international trade with $v=1,...,V$ foreign countries.

4.3.3 Behavioural Equations

In general terms, the basic structure of the model may be described as follows. *Production activities* in each region r are carried out by firms k that are assumed to produce an output z^k_r with intermediate inputs $l=1,...,N$, primary inputs $t=1,...,T$ and a certain production technology. The production technology may be described by the following unit cost function (see Varian 1994):

$$^1q_r^k = cf^k(\,^2q_r, ^3q_r) \qquad k=1,...,K, \quad r=1,...,M \qquad (4.3)$$

where $^1q_r^k$ denotes the price of commodity k in region r. The price is postulated to depend on the price vector of the intermediate inputs $^2q_r := (^2q_r^1,...,^2q_r^N)$ and on the factor price vector $^3q_r := (^3q_r^1,...,^3q_r^T)$.

Equations (4.4) and (4.5) serve to characterise the factor demand functions for intermediate inputs and primary factors, respectively. They describe the quantity of intermediate inputs $^1a_r^{kl}$ and primary factors $^2a_r^{tk}$ that is necessary to produce one unit of output, and can be derived from (4.3) as the first partial derivative with respect to the price of the respective input (see Varian 1994):

$$^1a_r^{kl} = \frac{\partial cf^k(\,^2q_r,^3q_r)}{\partial\,^2q_r^l} \qquad k,l=1,...,N, \quad r=1,...,M \qquad (4.4)$$

$$^2a_r^{tk} = \frac{\partial cf^k(\,^2q_r,^3q_r)}{\partial\,^3q_r^t} \qquad t=1,...,T,\; k=1,...,N,\; r=1,...,M \qquad (4.5)$$

Let us assume *consumption* activities in region r are represented by one representative and immobile household. The household earns income $\tilde{y}_r$ by selling primary factors g_r^t to firms in region r at factor prices $^3q_r^t$:

$$\tilde{y}_r = \sum_{t=1}^{T} g_r^t\,^3q_r^t \qquad r=1,...,M \qquad (4.6)$$

The income of the household in region r is spent on the consumption of commodities $l=1,...,L$ supplied by the pool of commodities in this region. The expenditure function describes the expenditures ch_r needed to reach one unit of utility depending on the vector ^2q_r of commodity prices:

$$ch_r = ch_r(\,^2q_r) \qquad r=1,...,M \qquad (4.7)$$

Total expenditures (eh_r) of the household depend on the expenditures per unit of utility ch_r and the level of utility u_r the household can reach. Thus:

$$eh_r = eh_r(ch_r(\,^2q_r), u_r) \qquad r=1,...,M \qquad (4.8)$$

At equilibrium total expenditures equal total income:

$$\tilde{y}_r = eh_r(ch_r(\,^2q_r), u_r) \qquad r=1,...,M \qquad (4.9)$$

The resulting demand function of the household in region r for commodity l is homogenous of degree zero in income and prices. This implies that multiplying the income and all prices by the same factor leaves demand unchanged (see Schumann 1984):

$$h_r^l = \frac{\partial eh_r(ch_r(^2q_r), u_r)}{\partial\, ^2q_r^l} \qquad l=1,...,N, \quad r=1,...,M \qquad (4.10)$$

The domestic regions are linked by *interregional commodity flows*, which are modelled using the pool concept (see Nijkamp *et al.,* 1986). This means that all commodities produced by sector l in region o with $o \in \{1,...,M,\, M+1,...,M+V =: O\}$ and delivered to region s are first merged into a pool of commodity l in s. This pool supplies all commodities used in that region for intermediate or final use. Transport activities aim to minimise transportation and transaction costs and are described by the following unit-cost function:

$$^2q_s^l = ct_s^l(\,^1q^l,\,^5q^l, \eta^l, d_{os} | o=1,...,O) \qquad l=1,...,N, \quad s=1,...,M \qquad (4.11)$$

where $^2q_s^l$ denotes the price of pool good l in region s, $^1q^l := (^1q_1^l,...,\,^1q_M^l)$ the price vector of the domestically produced commodities, $^5q^l := (^5q_1^l,...,\,^5q_V^l)$ the prices of the imported commodities. In order to keep the model as simple as possible, transportation and transaction costs are modelled according to the 'iceberg model' (Samuelson 1954). This means that a part of the commodities is used up during transportation. The quantity of commodities sent off from o exceeds that arriving in s by the factor exp ($d_{os}\, \eta^l$), where η^l denotes the transport rate and d_{os} the distance between o and s (see Bröcker 1998).

If commodities originating from different regions o are treated as homogenous, the cost minimisation principle would lead to a concentration of commodity flows from that region o with the lowest c.i.f prices. Since this behaviour contradicts empirical observations, one further assumption has to be made in order to let the model correspond better to reality. Thus it is assumed that commodities are distinguished by region of origin and treated as distinct from each other (Armington 1969). It is postulated that at the benchmark equilibrium the quantity of interregional commodities flows of sector l between domestic regions r and s depends on three types of variables. These are potential supply in region r, potential demand in region s and transportation and transaction costs. Potential supply for domestic regions r is represented by output z_r^l. Potential demand in s is represented by total demand for commodity l in this region (=intermediate plus final demand).

The derived demand functions state the quantity of domestic commodities $^3a_{rs}^l$ and imports $^4a_{vs}^l$ needed to produce one unit of pool good l in region s. They can be derived by the first partial derivatives of (4.11) with respect to the respective price as:

$$^3a_{rs}^l = \frac{\partial ct_s^l(\,^1q^l,\,^5q^l,\eta^l,d_{os}|o=1,...,O)}{\partial\, ^1q_r^k} \qquad k=1,...,N, \quad r,s=1,...,M \qquad (4.12)$$

$$^4a_{vs}^l = \frac{\partial ct_s^l(\,^1q^l,\,^5q^l,\eta^l,d_{os}|o=1,...,O)}{\partial\, ^5q_v^k} \qquad l=1,...,N,\, v=1,...,V\, s=1,...,M \qquad (4.13)$$

The model is linked with the CEEC-4 and the rest of the world by *international trade flows*. They are modelled by product-category specific export-demand and import-supply functions (see Whalley and Yeung 1984; Hirte and Wiegard 1988; Shoven and Whalley 1992). Foreign demand for Austrian *exports* is determined by two steps. First, aggregate foreign export demand e^k_v of country v for domestic commodities k is determined by the following export demand function:

$$e^k_v = \tau^k_v \left(\frac{{}^4q^k_v}{\xi}\right)^{-\delta^k_v} \qquad k=1,...,N, \quad v=1,...,V \qquad (4.14)$$

where ${}^4q^k_v$ denotes the c.i.f. price of exports k to foreign country v, divided by the exchange rate ξ, the parameter τ^k_v and the price elasticity δ^k_v. The c.i.f. price of exports depends on the vector of regional output prices ${}^1q^k := ({}^1q^k_1, ..., {}^1q^k_M)$ and transportation and transaction costs, represented as $\exp(d_v \eta^k)$, with $d_v := (d_{v1}, ..., d_{vM})$ denoting the vector of distances between domestic regions and foreign country v and η^k denoting the transport rate:

$$ {}^4q^k_v = ce^k_v({}^1q^k, d_{vr}, \eta^k) \qquad k=1,...,N, \quad v=1,...,V \qquad (4.15)$$

Second, this aggregate export demand is distributed to the domestic regions by regional export demand functions (4.16), which are derived from (4.15) as the first partial derivative:

$$ {}^5a^k_{rv} = \frac{\partial ce^k_v({}^1q^k, d_{vr}, \eta^k)}{\partial {}^1q^k_r} \qquad k=1,...,N, \; r=1,...,M, \; v=1,...,V \qquad (4.16)$$

where ${}^5a^k_{rv}$ denotes the share of national exports k to country v, which origins from region r, and ${}^1q^k_r$ the output price in region r. The quantity of foreign demand of country v for exports of region r e^k_{rv} can then be obtained by:

$$ e^k_{rv} = e^k_v \, {}^5a^k_{rv} \qquad k=1,...,N, \; r=1,...,M, \; v=1,...,V \qquad (4.17)$$

While exports are one-sided demand-driven, *imports* are determined by both import supply and import demand functions. This is an extension and adaptation of the external sector closing rule presented by Whalley and Yeung (1984). The import supply m^l_v of commodities of sector k from foreign country v is described by a constant elasticity function:

$$ m^l_v = \pi^l_v \left(\frac{{}^5q^l_v}{\xi}\right)^{\mu^l_v} \qquad l=1,...,N, \quad v=1,...,V \qquad (4.18)$$

where π_v^l denotes a parameter, $^5q_v^l$ the price of imports l from country v, ξ the exchange rate and μ_v^l the price elasticity of import supply. The import demand function is given by (4.13). The balance of payment for Austria's total trade is automatically fulfilled owing to the budget restrictions of the households. But it is not guaranteed that the balance of payment is fulfilled for each of the foreign countries, because they are not modelled as a general equilibrium system.

In addition to the above behavioural equations, three types of equilibrium conditions are needed. Equilibrium conditions for the commodities guarantee that total demand for commodity k in region r, given by the right hand side of (4.19), equals regional output z_r^k:

$$z_r^k = \sum_{s=1}^{M} {}^3a_{rs}^k (h_s^l + \sum_{k=1}^{N} {}^1a_s^{kl} z_s^k) + \sum_{v=1}^{V} {}^5a_{rv}^k e_v^k \quad k=1,...,N, \ r=1,...,M \quad (4.19)$$

where $^3a_{rs}^k$ denotes demand for commodity k in region r to produce one unit of pool good k in region s, h_s^l final demand for commodity l in region s, $^1a_s^{kl}$ the quantity of commodity k to produce one unit of commodity l in region s, z_s^k output of commodity k in region s, $^5a_{rv}^k$ the share of region r in total exports of commodity k to country v and e_v^k total exports of k to country v.

The factor market clearing condition that follows next, states that total demand for factor t in region r has to equal the supply g_r^t. This demand is given by summing up over the demand across all sectors k:

$$g_r^t = \sum_{k=1}^{N} {}^2a_r^{tk} z_r^k \quad t=1,...,T \ r=1,...,M \quad (4.20)$$

with $^2a_r^{tk}$ denoting the quantity of intermediate inputs needed to product one unit of output z_r^k.

The market clearing condition for the import markets ensures that the quantity of imports k supplied by country v (m_v^l) equals total demand for these imports:

$$m_v^l = \sum_{s=1}^{M} {}^4a_{vs}^l (h_s^l + \sum_{k=1}^{N} {}^1a_s^{kl} z_s^k) \quad l=1,...,N \ v=1,...,V \quad (4.21)$$

where $^4a_{vs}^l$ denotes the demand for imports l from country v needed to produce one unit of pool goods l in region s. Total demand for pool good l in s is given by the second term of the right hand side of (4.21). A price index equation is needed to obtain the relative price level:

$$w = \frac{\sum_{r=1}^{M} u_r ch({}^2q_r)}{\sum_{r=1}^{M} u_r} \quad (4.22)$$

where w denotes the price index, u_r the level of utility of the household in region r and $ch(^2q_r)$ the expenditure function of the household in region r.

4.4 Gravity Model Data, Estimation and Results

The standard approach to estimate the model (4.1)-(4.2) of bilateral trade flows is to assume that a normally distributed multiplicative error term holds. Then, OLS can be applied after a logarithmic transformation. The model is estimated with data from 1995. The sample utilized for estimation purposes includes seven Western countries (Austria, Belgium, France, Germany, Greece, the Netherlands, Switzerland) and twelve Central and Eastern European countries (Czech Republic, Slovak Republic, Hungary, Slovenia, and moreover, Croatia, Macedonia, the Rest of Yugoslavia, Bulgaria, Romania, Moldavia, Ukraine and Russia).

The data used in the study are expressed in \$US millions and refer to the merchandise for the following six broad product categories as defined in Fischer and Rammer (1993):

- Agricultural Products (SITC 00-09, 41-43),
- Raw Materials (SITC 21-25, 27-29, 32-33, 56),
- Labour Intensive Products (SITC 26, 61, 63-66, 69, 81-85, 89),
- Capital Intensive Products (SITC 11-12, 53, 55, 62, 67-68, 78),
- Low-End R&D Intensive Products (SITC 51-52, 54, 58-59, 75-76), High-End R&D Intensive Products (SITC 57, 71-74, 77, 79, 87-88).

The data were obtained from the UN world trade computer database. Export unit values were taken from the same source, while GDP data (measured in \$ US million) came from the World Bank database, probably the most reliable in the field. Additional data were observed from national statistics.

The estimates reported in table 4.2 support the hypotheses that have been put forward in section 4.3. Nearly all the coefficients have the expected signs, and most coefficients are statistically significant. The whole set of variables has considerable explanatory power, especially in the case of labour, capital and R&D intensive products, as evident by the coefficient of determination. We find strong income effects on trade with elasticities around unity, which accords well with other studies (Collins and Rodrik 1991; Johansson and Westin 1994; Hamilton and Winters 1992; Fischer and Johansson 1996). As in all gravity models applied to international trade data so far, distance is an important variable. Trade declines by about 45% to 50% per 1000 kilometres. The adjacency dummy variable has a positive coefficient sign in all product category specific cases. The preferential trade arrangement has an important effect on trade with agricultural products and capital intensive products.

Table 4.2. Estimates of the gravity trade model for six product classes in 1995 (t-statistics in parenthesis)

	Agricultural Products	Raw Materials	Labour Intensive Products	Capital Intensive Products	Low-End R&D Intensive Products	High-End R&D Intensive Products
Constant	-2.18	-4.61	-2.21	-2.37	-3.73	-3.48
	*(-8.83)**	*(-16.06)**	*(-12.06)**	*(-11.45)**	*(-17.22)**	*(-16.69)**
Country I's income	0.67	0.92	0.75	0.93	1.11	0.95
	*(9.62)**	*(11.68)**	*(14.20)**	*(15.53)**	*(17.41)**	*(15.24)**
Country j's income	0.72	0.79	0.83	0.76	0.81	0.84
	*(10.49)**	*(9.85)**	*(16.11)**	*(13.23)**	*(13.47)**	*(14.53)**
Cross-country difference of incomes	-0.07	-0.12	0.06	-0.07	-0.20	0.00
	(-0.96)	*(-1.47)*	*(1.11)*	*(-1.21)*	*(-3.26)**	*(-0.07)*
Unit-value index of exports from i to j	0.02	-0.81	0.02	-0.43	0.25	0.42
	(0.33)	*(-9.53)**	*(0.22)*	*(-4.57)**	*(4.66)**	*(4.66)**
Distance (dij10-3)	-0.62	-0.63	-0.77	-0.55	-0.75	-1.05
	*(-4.65)**	*(-4.07)**	*(-7.72)**	*(-4.89)**	*(-6.32)**	*(-9.25)**
Dummies						
Adjacency dummy	0.90	1.21	0.75	0.85	1.01	0.64
	*(3.67)**	*(4.22)**	*(4.08)**	*(4.12)**	*(4.65)**	*(3.06)**
Trade preference dummy	0.74	-0.40	0.40	0.51	0.31	0.07
	*(2.41)**	*(-1.14)*	*(1.76)*	*(1.99)**	*(1.16)*	*(0.25)*
R^2	0.58	0.63	0.76	0.72	0.73	0.75
F	69.3	85.9	162.5	126.0	132.9	151.8
Degrees of freedom	352	360	351	349	345	351

Notes: * Estimates are significant at 5% level

4.5 The Catch-up Scenario as a Pointer to Predict Size, Composition and Direction of Trade

This section gives some quantitative estimates of the impact which successful economic reforms in the four Central and Eastern European countries and their integration into the EU will have in the long run on the disaggregated trade flows between these countries and Austria. The (long term) estimates are naturally subject to a great range of uncertainty with respect to the likely developments in each of the countries concerned. The numbers presented might end up wide of the mark in the long term, but we think they are useful nonetheless in helping to frame the issues and focus thinking on the potential regional impact.

In order to quantify the impact we consider a catch-up long-term scenario which assumes that the four Central and Eastern European countries achieve a stable growth path. This assumption is based on the view that the transition from a planned to a well functioning market economy is associated with a specific development in output and employment; that is a J-curve in output and employment with a dip, a valley, or a deeper gorge, where output and employment will fall first and then begin to rise. The shape of this J-curve is not exactly known, but relevant for the political economy of transition and, thus, for the scenario. Clearly, the shape depends on the inefficiency of the existing firms, on the speed and methods of privatisation with which new firms come into existence, and on the conditions of the process of industrial restructuring including the opening up of the economy to the international division of labour.

All four Central and Eastern European countries have entered the growth stage by now. In the long-run catching-up scenario, the real gross domestic product of the CEEC-4 is assumed to grow by 3% p. a., while the real gross domestic product of Austria is assumed to grow by 1.5% p. a.. Moreover, it is assumed that the catching-up of incomes in the CEEC-4 countries will lead to an increase of prices. Thus, the unit values of the exports of the CEEC-4 are postulated to increase up to 75% of the EU average.

Table 4.3 reports Austria's long-term (2015) trade potential with the four Central and Eastern European countries, in terms of the six broad product categories. The changes predicted are dramatic for Central and Eastern Europe, but they are also important for Austria. First, and most relevant, the long-term increases in Austria's total trade are huge. Exports rise by 190% and imports by 160%. Such increases offer scope for new specialisation and economies of scale. Second, the model suggests that given the level of income assumed in the scenario, the Central and Eastern European countries should expect in equilibrium to have an excess of imports over exports, i.e. to run small trade deficits with Austria, caused primarily by capital-intensive and high-end R&D intensive products.

Table 4.3. Austria's long-run trade potential with the CEEC-4 for six product categories.

	Changes in quantities of trade flows between 1995 and 2015 measured in terms of			
	$US million		% change	
	Exports	Imports	Exports	Imports
Agricultural Products	659	842	290%	300%
Raw Materials	324	37	40%	10%
Labour Intensive Products	530	1,362	220%	220%
Capital Intensive Products	1,423	333	250%	140%
Low-End R&D Intensive Products	1,212	1,545	200%	250%
High-End R&D Intensive Products	5,523	1,111	140%	200%
Total	9,670	5,231	190%	160%

The figures suggest strong adjustment pressures in agriculture and labour-intensive products as well as in low-end R&D intensive production in Austria. But the model offers only broad and trend patterns, not precise indicators of year-by-year developments. Third, a weakness of the gravity model is that the increased trade predicted between, say, Austria and the Czech Republic, has no implications in the model for Austria's other trade. It apparently neither diverts imports from other sources nor absorbs exports destined for elsewhere. This implies that the new imports displace only domestic sales, while the new exports are met by curtailing domestic sales or increasing output. Given this approximation, one should expect some spillovers from the growth of Central and Eastern European trade in terms of declining intra-industrial interregional flows. That is, Austria may find that it faces extra competition in export as well as home markets. It can not be overemphasised that the opportunities for buying and selling cannot be decoupled. Austria as well as other Western economies must offer market access to Central and Eastern Europe if they themselves are to benefit from the gains that the increased volume of trade permits. This entails not only opening markets to sensitive sectors such as agriculture, but also accepting large volumes of impacts on products. The consequent changes in the sources of Austrian production and consumption will demand a high degree of flexibility and adjustment in both exportable and importable industries, although they will be spread over longer periods of time. It seems unlikely that the Central and Eastern European countries could realise their trading potential in much less than 15-20 years.

4.6 The Impact on Regional Welfare and Production in Austria

The commercial map of Europe is changing as Eastern Europe moves towards freer access to the Western European market. What effect will the integration of the former Central and Eastern European countries have on regional welfare and production in Austria?

If these countries catch-up to the required specification of the functions in (4.3)-(4.22) (see Schneider 1998 and Bröcker and Schneider 1998 for more details on this issue) then the model may provide some interesting insights into the impact on regional welfare and production in Austria even though the underlying assumptions are admittedly rather heroic in nature. The model has been calibrated for $M=9$ domestic NUTS-II regions, $N=10$ economic sectors, $V=2$ groups of foreign countries (CEEC-4 and the rest of the world) and $T=2$ production factors, i.e. capital and labour. The results obtained at the national level are summarised in table 4.4. The analyses are based on three types of indicators: first, indicators representing trade, output and final demand quantities; second, indicators representing prices such as expected import prices, production factor prices and commodity prices; and, finally, indicators measuring real income and trade volume.

74 M. Schneider and M.M. Fischer

Table 4.4. Impact of the catching-up scenario on Austria at the national level (% changes compared to 1995).

Type of Indicator	Indicator	% Change
Quantities	Total exports	12.4
	Exports to CEEC-4	135.2
	Exports to the rest of the world	-6.9
	Total imports	11.5
	Imports from CEEC-4	87.7
	Imports from the rest of the world	7.7
	Output	0.3
	Final demand	1.2
Prices*	Export prices	0.1
	Import prices	-3.3
	Wage rate	1.3
	Factor price of capital	0.9
	Commodity prices	1.2
	Price index	2.6
Values*	Real income	1.2
	Trade volume	7.8
Other variables	Gini index of regional real income	0.5

*) measured in terms of the price index

It is interesting to note that the catching-up scenario of the four Central and Eastern European countries does not only lead to a likely expansion in the overall volume of trade with these countries as discussed above, but also to an increase in national welfare in terms of real income. This is due to two major reasons. First, the additional trade volume induces a more efficient allocation of domestic resources according to changes in relative prices. Second, the pressure of declining import prices from the CEEC-4 increases real income in Austria.

The effects on sectoral production are reported in table 4.5. In a neoclassical general equilibrium model, changes in sectoral output are caused by the reallocation of resources between economic activities, whereas aggregate output remains almost constant. The highest increases in output can be observed for low-end R&D intensive products and high-end R&D intensive products, mainly on the cost of raw materials, which shrinks most. This result is in line with the Heckscher-Ohlin model, which states that each country should specialize in the production of commodities in which it has a comparative advantage.

The welfare gains will not necessarily be shared equally by all the nine Austrian regions. To assess the likely implications of the catching-up process of the CEEC-4 we will focus on the consequences for regional welfare gains, measured in terms of percentage changes of regional real income relative to 1995. The results are displayed in figure 4.1. It is interesting to note that regional income gains range from 0.95 % in

Tyrol to 1.27% in Vienna. There is some evidence of a clear East-West pattern. The Eastern part of the country (with the capital region of Vienna, Lower Austria and Burgenland) gains most, followed by Carinthia, then by Upper Austria and Styria, while the regions further away from the CEEC-4 profit least. Vienna itself is not only very close to the capital city of Slovakia, but also exhibits a higher share of low-end and high-end R&D intensive production. The question of whether disparities between regions will be affected can be answered by taking a look at the Gini index of regional real income. The Gini index rises by 0.5%, which represents a situation where the distribution becomes more unequal, whereas the magnitude of this increase is rather low (see table 4.4). Thus, one cannot expect a dramatic impact of the catching-up scenario on regional disparities in Austria.

Table 4.5. Impact of the long-run trade scenario (2015) on sectoral production in Austria (% changes compared to 1995).

Sector	% change
Agricultural Industries	-1.0%
Manufacturing of Raw Materials	-4.3%
Manufacturing of Labour Intensive Products	1.6%
Manufacturing of Capital Intensive Products	1.2%
Manufacturing of Low-End R&D Intensive Products	5.7%
Manufacturing of High-End R&D Intensive Products	8.2%
Construction	0.3%
Trade & Accommodation	-0.1%
Production Services	-1.2%
Other Services	0.2%
Total	0.3%

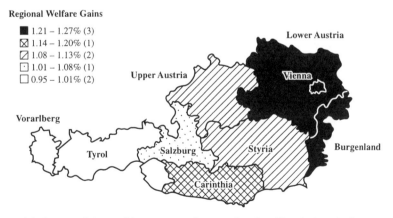

Figure 4.1. Impact of the catching-up scenario on regional welfare in Austria (measured in terms of percentage changes compared to 1995)

Sensitivity analysis has been performed in order to test the sensitivity of the simulation results to the values of the elasticities of the model. The main finding is that the magnitude of the welfare gains depends mainly on the values of the trade elasticities, especially the export demand elasticity δ^k (see equation (4.14)). Doubling the values of all export demand elasticities δ^k decreases national welfare gains by 29%. This confirms the well-known property of applied general equilibrium trade models that trade elasticities play a crucial role in determining the magnitude of the results (see van der Mensbrugghe et al. 1990). The values of the elasticities of substitution in production and consumption have only a very small impact on the magnitude of the aggregate welfare gains.

4.7 Summary and Conclusions

In this chapter an attempt has been made to illustrate how two distinct modelling traditions at the centre of regional science – spatial interaction modelling and multiregional computable general equilibrium modelling – can be combined to tackle some interesting real world problems such as the question of what changes will be brought about by the integration of Central and Eastern European countries into the European common market. Emphasis has been laid on the consequences for trade patterns and the economies of Austrian regions. The chapter has presented estimates based upon a two stage conceptual framework feeding the results of product category specific trade models into a multiregional computable general equilibrium model. The approach in estimating the trade and regional economic implications started from what has been called the catching-up scenario: a long-run scenario in which the four neighbouring Central and Eastern European countries have been completely integrated into the Western European economy. This assumption helps us to focus the analysis and to provide an outer bound on the magnitude of the potential impact on welfare.

The empirical results indicate that there is still a huge trade potential between Austria and its neighbouring countries. The realisation of this potential may lead to an increase in the trade deficit of the Central and Eastern European countries with Austria. Thus, Austria's trade balance should improve. In addition to the opportunities this trade potential offers, it must not be forgotten that some sectors may experience strong adjustment pressures. National real income in Austria's regions can be expected to rise, although the regions will be affected differently. The results indicate a clear East-West pattern with the Eastern regions of Austria gaining most.

References

Armington P. (1969) A theory of demand for products distinguished by place of production, *International Monetary Stuff Papers*, 16, 159-176.

Balassa B. and Bauwens B. (1988) The determinants of intra-European trade: A temporal cross-section analysis, *American Economic Review*, 69, 881-892.

Bröcker J. (1998) Operational spatial computable general equilibrium modelling, *Annals of Regional Science*, 32(3), 367–387.

Bröcker J. and Jäger-Roschko O. (1996) Eastern reforms, trade, and spatial change in the EU, *Papers in Regional Science*, 75(1), 23-40.

Bröcker J. and Schneider M. (1998) *How does economic development in Eastern Europe affect Austria's regions? A multiregional general equilibrium framework*. Diskussionsbeiträge aus dem Institut für Wirtschaft und Verkehr der Technischen Universität Dresden, Nr. 1/99.

Collins S.M. and Rodrik D. (1991) *Eastern Europe and the Soviet Union in the world economy*. Institute for International Economics, Washington DC.

Dinwiddy C.L. and Teal F.J. (1988) *The Two-Sector general equilibrium model: a new approach,*. Philip Allan, Oxford.

Fischer M.M. and Johansson B. (1996) Opening up international trade in Eastern European countries. Consequences for aggregate trade flows in the Rhine-Main-Danube area. *Papers in Regional Science*, 1, 65-78

Fischer M.M. and Rammer C. (1993) *Zur Entwicklung des grenzüberschreitenden Warenverkehrs im Einzugsbereich der RMD-Wasserstraße und der Mengenpotentiale der Binnenschiffahrt*. WSG Research Report 4, Department of Economic and Social Geography, Vienna University of Economics and Business Administration

Fischer M.M. and Rammer C. (1995) Trade Effects of the Emerging Market Economies: A Study of the Transport Potential of the Rhine-Main-Danube Waterway. In: Coccossis, H., Nijkamp P. (eds.) *Overcoming Isolation - Information and Transportation Networks in Development Strategies for Peripheral Areas*. Berlin and Heidelberg: Springer, 179-194

Hamilton C.B. and Winters L.A. (1992) Opening up international trade in Eastern Europe. *Economic Policy*, 7, 78-116.

Hirte G. and Wiegard W. (1988) An Introduction to Applied General Equilibrium Tax Modeling. In Bös D., Rose M. and Seidl C. (eds.), *Welfare and Efficiency in Public Economics*, Heidelberg: Springer, 167-203.

Johansson B. and Westin L. (1994) Revealing network properties of Sweden's trade with Europe. in Johansson B., Karlsson C. and Westin L. (eds.), *Patterns of a Network Economy*. Heidelberg: Springer, 125-141.

King A. (1993) A Note on Export Unit Value Indices in Competitiveness Variables. *Bulletin of Economic Research*, 45:1, 69-77.

Marsh I.W., Tokarick S.P. (1996) An Assessment of Three Measures of Competitiveness. *Weltwirtschaftliches Archiv*, 132:4, 700-722.

Nijkamp P., Rietveld P. and Snickars F. (1986) Regional and multiregional economic models: A survey. In Nijkamp P. (ed.), *Handbook of regional and urban economics, Volume 1 - Regional Economics*, Amsterdam: North-Holland, 257-294.

Samuelson P.A. (1954) The transfer problem and transport cost, II. analysis of effects of trade impediments. *Economic Journal*, 64, 264-289.

Schneider M. (1998) *Modelling the Effects of Future East-West Trade on Austria's Regions: A Multiregional General Equilibrium Approach*, PhD Dissertation, University of Vienna.

Schumann J. (1984) *Grundzüge der mikroökonomischen Theorie*. 4. ed., Berlin: Springer.

Shoven J.B. and Whalley J. (1984) Applied General-Equilibrium Models of Taxation and Trade. An Introduction and Survey. *Journal of Economic Literature*, 22, 1007-1051.

Shoven J.B. and Whalley J. (1992) *Applying General Equilibrium*. Cambridge, NY: Cambridge University Press.

van der Mensbrugghe D., Martin J.P. and Burniaux J.-M. (1990) How robust are Walras results? in *OECD Economic Studies, Modelling the effects of Agriciultural Policies*, No. 13/1989-1990, Paris, 174-204.

Varian H. (1994) *Mikroökonomie*. 3. ed., München and Vienna: Oldenbourg.

Whalley J. and Yeung B. (1984) External Sector 'Closing' Rules in Applied General Equilibrium Models. *Journal of International Economics*, 16, 123-138.

5 The Regional Cambridge Multisectoral Dynamic Model of the UK Economy

Terry Barker, Bernie Fingleton, K.Homenidou and R.Lewney

5.1 Introduction

This chapter reports on a regionalised, fully specified and coherent model of the various UK regional economies. It has a clear economic structure allowing incorporation of incomplete and partial data in a similar manner to the procedure followed in general equilibrium modelling, but at the same time validating the model's projections against the available data for employment and output. The regionalised model is a development of the Cambridge Multisectoral Dynamic Model of the UK economy (MDM); see Barker and Peterson (1987) for an account of version 6 of the model. This is a time-series, cross-section (input-output) model distinguishing, inter alia, 49 industries and 68 categories of consumers' expenditure. The standard UK regions are treated as one of several classifications in the model, with several commodity, industry and employment variables regionalised according to the availability of data. The current version of the model (MDM94) has been re-estimated on 1997 National Accounts and consistent Regional Accounts data (on the 1990 price base) and incorporates the 1990 input-output tables for the UK.

5.2 Regional Input-Output Models

Stone (1961) provides a useful analysis of the development of regional models. The possibilities of regional applications were recognised early in input-output analysis (Isard, 1951). Isard's model envisaged a full set of input-output tables for all the regions and a full set of data on inter-regional trade, ie trade was to be disaggregated for each region with every other region. Trade between the regions was in fixed proportions to output. This was soon followed by two variants, both limiting the data requirements and altering the restrictive assumption of proportionality. The first by Leontief (1953), elaborated further by Leontief and Strout (1963), assumed that each industry had the same cost structure in each region, that commodities were either entirely locally produced and consumed or traded, and that the traded commodities were traded in fixed proportions to total output. The second variant (Moses, 1955) allowed cost structures to vary, but fixed the trading patterns. The choice between these models depends on the data available and the likely importance of distance and differences in technology between regions. If distance is less important and the regions can be expected to share common technologies, then Leontief's model seems more appropriate.

More recent surveys and research (Nijkamp *et al.*, 1986; Hewings and Jensen, 1986; Harrigan and McGregor, 1988) demonstrate the large scale of research activity in the last 30 years in the theory and application of regional input-output models. Two developments are worth noting here. The first is the extension of the Social Accounting Matrix framework (the basis of the accounting system used for MDM) to Regional Accounts (Round, 1988). The second is the integration of regional input-output tables with national General Equilibrium Models, such as the ORANI model of Australia (Dixon *et al.*, 1982) and with time-series econometric models (Joun and Conway, 1983). The regionalised MDM is an example of this second type of development (see also chapters 2-4).

In the absence of inter-regional trade data and with the generally poor quality of regional data, a model of the regional economies of the UK with significant economic content inevitably entails a substantial exercise in data construction. Some considerable reduction of data requirements can be made by adopting Leontief's approach, in which each region with all the rest as a group rather than with each other region separately, but there remains a need for data on each region's exports and imports in total. (The data requirements involved in a full inter-regional model are of a higher order of magnitude – see Polenske, 1980). However the approach does allow important economic relationships to be embedded in the model. For example, in the input-output model regional output can be determined from regional exports and domestic demand depending on tradeability; and regional employment can be determined from output.

The construction of such a model has appeared until recently to be too ambitious in relation to the available data and the resources required. The new approach has been to build up a regional econometric input-output model and database as an integral part of the MDM model and MDM database. The model has a clear economic and accounting structure, uses incomplete and partial data, and applies techniques drawn from general equilibrium modelling. The forecasts and projections for the recent past are calibrated so as to reproduce the available data for employment and output. A sensible direction of economic causation for employment is an inherent feature of the model. There are several advantages of the MDM approach:

- *improved economic explanation*

Regional output responds to changes in *regional* final demand. Thus, for example, the link between a slump in consumer spending in the South East and output in the same region, or other regions, is explicitly identified.

- *regional accounts*

The treatment allows the calculation of full regional accounting balances for commodity supply and demand, exactly corresponding to the balances for the whole UK including the Continental Shelf. These balances cover inter-regional trade and an allocation of the UK commodity imbalances across the regions. They are in constant and in current prices, under the assumption that annual changes in prices for each commodity are the same for all the regions. These commodity balances provide an important consistency check on any forecasts of regional output and the components of regional final demand.

- *modelling regional markets*

The approach allows full feedback from the regional economies to the UK economy. This facility is especially important in modelling those areas of economic life where markets are restricted by costs of travel or other costs associated with distance. For example, it is clear from the data that there are distinct regional differences in patterns of saving and consumption; the new approach allows total consumption by region to be estimated and solved and UK consumers' expenditure to be formed as the sum of the regional expenditures. Another example is in the operations of the labour market which tend to be restricted to travel-to-work areas; here it has been possible to estimate regional employment and wage rate equations to reflect different conditions in each of the regional labour markets. UK employment and the UK rate of wage inflation can then be found from the regional rates.

Such feedbacks, however, are an option in the software and in the current version of the model they are only operative in the case of employment. The regional forecasts depend on the UK forecasts and if necessary they are scaled to add up to the UK forecasts.

- *distance and location effects*

In the regionalised MDM, distance and location have three main influences.

1. Economic distance determines the regional export activity indices, such that the closer one region is to another in economic distance, the more its domestic demand affects the other region's exports.

2. For certain location-based activities, such as transport and distribution, the location of the infrastructure in the form of transport links and warehousing determines the regional supply.

3. The location of large new investment projects, eg tidal barrages, are introduced directly into the regional investment projections.

- *use of regional information*

It is much easier to incorporate partial and incomplete information into a fully specified economic model than into a reduced form model. For example, estimates are available for costs and impacts of infrastructure projects such as the Channel Tunnel rail link or the Severn Barrage. These will have strong regional effects. If investment is fully specified in the model, such exogenous increases can be introduced explicitly into the forecast. Similarly, estimates of the local multiplier effects of the Toyota car plant in Derbyshire can be directly implemented in the model.

- *inter-industry links*

One of the great strengths of I-O models is their simulation of inter-industry links, allowing the calculation of industrial multiplier effects. These multipliers show the effects on the industrial structure of changes in exogenous variables, or in behaviour, for example an increase in the propensity to save in one particular region. They show, under simplifying assumptions, how extra demand is transmitted from one industry to another – for example the large increases in output of cars from the Nissan plant in Sunderland will have effects on suppliers of parts, and in turn, effects on steel and glass production and imports.

5.3 Clients' Requirements

The model is applied to provide information for clients in two broad ways. Firstly, clients who subscribe to Cambridge Econometrics' UK Regional Forecasting Service receive base forecasts and scenarios for regional economic growth, represented in a variety of macroeconomic and sectoral indicators at regional level. Secondly, clients with a specific (sometimes commercially confidential) requirement commission special scenarios tailored to their needs.

5.3.1 Regular Forecasts and Scenarios

The organisations who subscribe to CE's UK Regional Service have varied requirements, and so to encompass these the service has to incorporate a wide range of indicators. All are typically interested in overall GDP and employment growth in the regions, in the short and medium term. Utilities and organisations responsible for or advising on regional planning (and particularly housing) also require a long-term perspective, stretching out to ten years and beyond, and have a particular interest in demographic forecasts. Consultants and organisations responsible for training and development require a detailed labour market analysis, including employment (by sector and occupation), participation and unemployment rates. Those whose interest is in commercial property require forecasts for indicators that are related to the demand for retail, office, factory and warehouse space. Typically they use a measure of activity (either output or employment) for relevant sectors.

5.3.2 Special Scenarios

The requirement for special scenarios typically involves the development of projections to flesh out a view of the future that the client wants to specify and which differs from the base case available in the regular forecasts. This includes, for example, testing the sensitivity of the base forecast to a more severe recession or a sharper recovery than projected in the base case.

Often the client's view is initially expressed only in terms of very broad indicators (eg the outturn for GDP growth). But the sensitivity of the regions to faster or slower growth in the UK economy as a whole depends critically on the nature of the scenario: for example, an upturn led by a faster recovery in continental Europe would have different implications to one led by a lower value for sterling, and this in turn would have a different regional impact from a recovery in consumer confidence and spending. Hence, work on a special scenario involves a careful debriefing of the client to translate their view into a specification that is operational for applications of the model.

5.4 Data Issues

The reason why this approach has not been adopted before, despite the obvious advantages, is the poor quality (if not the absence) of much of the required regional data. In developing a regional model for the UK, the data problems have been tackled as follows. Where the data are of suspect quality, three checks have been carried out. First, all sources have been used to cross-check data where possible. Second, the UK totals have been used to control the regional data in as much detail as possible. Third, the views of regional experts are sought. Where no data exist, they have been imputed from other regional data, as in the case of trade flows, or from UK data, as in the case of the input-output coefficients. The methods used are described in Section 5.5. The methods have been applied systematically for all the standard regions and this provides a further check in the case of Scotland, Wales and Northern Ireland, where more detailed data are available.

The classifications adopted in the model are as far as possible those of MDM, and the regional variables and data are consistent with the UK variables and data. This means that a 49-industry classification has been adopted for the commodity and industry variables in the regionalised MDM (such as gross output, GDP, employment, regional exports and imports). A list of industries distinguished in the model is included in Appendix 5. Most of the regional data are provided by the UK Office for National Statistics (ONS). The ONS publishes annually a series of Regional Accounts consistent with the UK National Accounts. These include data on nominal GDP, consumers' expenditure, personal incomes and GDFCF for the 11 standard planning regions of the UK and the two South East subregions, Greater London and the rest of South East England. Most of the data are available since 1971 but some disaggregated series are available only since 1978. Some data at disaggregated level exist for 1971-78 but these are based on the 1968 SIC and much other data is on the 1980 SIC; in the process of creating long-term series these were all translated to 1992 SIC categories. Total consumers' expenditure is disaggregated into 68 categories, using information from the Regional Accounts and making the most of the disaggregated data on expenditure available by region from the Family Expenditure Survey.

The source for employment and unemployment data is also the ONS. Employment is defined as the total of employees in employment, self-employment and HM Forces and is the June count seasonally unadjusted. Unemployment is defined by the annual average, seasonally adjusted, of benefit claimants aged 18 and over. Other data such as regional population, working-age population and migration are obtained from the ONS and the Registrars General for Scotland and Northern Ireland.

5.5 Development of the Regional Econometric Input-Output Model

This account of the model emphasises the accounting balances adopted. In general the regional equations follow their UK counterparts in terms of the explanatory variables, with the UK totals added as a further variable. The model comprises: the accounting balances; the various fixed allocations to the regions of UK government expenditure, stock levels and commodity residuals; and 7 sets of time-series econometric equations (commodity exports, total consumers' expenditure, disaggregated consumers' expenditure, industrial fixed investment, industrial employment, industrial wage rates and population change). In the rest of this section the subscript r is used to denote a regional variable while the subscript i is used to denote an industry.

5.5.1 Commodity Balances and Imbalances

For each region r, in constant prices:

$$Q_r + QM0_r = QY0_r + QC0_r + QG0_r + QK0_r + QS0_r \tag{5.1}$$

where each variable is a vector as follows:

The Regional Cambridge Multisectoral Dynamic Model of the UK Economy 85

Q output by 49 commodities (goods and services)
$QM0$ imports
$QY0$ absorption of commodities by industries
$QC0$ consumers' expenditure classified by commodity
$QG0$ government final current expenditure classified by commodity
$QK0$ investment in fixed assets classified by commodity
$QS0$ stockbuilding by source commodity
$QX0$ exports
QER residual imbalance

and in current prices:

$$VQ_r + VQM0_r = VQY0_r + VQC0_r + VQG0_r + VQK0_r \\ + VQS0_r + VQX0_r + VQER_r \qquad (5.2)$$

where the variables of equation (5.1) are now in current prices.

The UK commodity imbalances (vector $QERUK$) are defined as:

$$QER\ UK = Q\ UK + QM0\ UK - QY0\ UK - QC0UK\ UK \qquad (5.3)$$

where the right-hand-side variables are as defined in equation (5.1) but for the UK economy.

This residual for each commodity (i), given by the ith element at the vector $QERUK$ and denoted by $QER_i UK$, is allocated across the regions (r) according to their share in total supplies:

$$QER_{ir} = QER_i UK \times \\ \frac{Q_{ir} + QM0_{ir}}{\Sigma(Q_{ir} + QM0_{ir})} \qquad (5.4)$$

where Σr denotes summation over all regions of the UK.

5.5.2 The Inter-Regional Export-Import Balance

There is an accounting balance for net regional trade and UK trade with the rest of the world. If the regional commodity balances in equation (5.1) are summed across the regions, and the UK totals for each variable are subtracted from each side, the fundamental inter-regional export-import identity can be found:

$$\Sigma_r Q_r - Q\ UK + \Sigma_r QM0_r - QM0\ UK = \\ \Sigma_r QY0_r - QY0\ UK + \Sigma_r QC0_r - QC0\ UK$$

$$+ \sum_r QG0_r - QG0\ UK + \sum_r QK0_r - QK0\ UK$$
$$+ \sum_r QS0_r - QS0\ UK + \sum_r QX0_r - QX0\ UK \tag{5.5}$$

Since regional gross output and each component of domestic demand (ie excluding exports) adds up to the corresponding UK variable, in constant prices and for each commodity, these variables cancel out in the identity, leaving UK exports and imports and inter-regional exports and imports. The identity becomes:

$$\sum_r QM0_r + QX0\ UK = \sum_r QX0_r + QM0\ UK \tag{5.6}$$

In words, taking all UK regions together, regional imports plus imports of the rest of the world (ie UK exports) are equal to regional exports plus exports of the rest of the world (ie UK imports). This balance is enforced in the projections of the model as an adding-up constraint on regional exports.

5.5.3 Regional Trade Data

An important part of the modelling is the imputation of trade data which is required to form commodity balances for each region. The basic assumption here is that, for most commodities, there is a national 'pool' into which each region supplies its production and from which each region satisfies its demands. This might be compared to electricity supply and demand: each power plant supplies to the national grid and each user draws power from the grid and it is not possible or necessary to link a particular supply to a particular demand. This permits the estimation of the volume of regional exports from data on regional output:

$$QX0_r = Q_r * XP_r \tag{5.7}$$

where XP is the proportion exported into the national pool.

For English regions, where there are no trade data, XP is given by assumption and depends on inter-regional tradeability of the commodity in question. For Scotland and Wales, XP is derived from the trade flows given in the available input-output tables.

The imputation of data on imports is more straightforward. Given the regional commodity balances and assumed allocation of UK imbalances, imports can be found as a residual from equation (5.1).

5.5.4 Regional Export Equations

Equation (5.7) does not determine the demand for exports, being only a means of constructing data. In the model, the demand for a region's exports of a commodity is related to domestic demand for the commodity in all the UK regions, weighted by their economic distance from the region in question, and to activity in UK export markets as measured by UK exports, again weighted by economic distance, this time by the distance of the region from the main UK export markets for the commodity. The economic distance variable is normalised with a weight of 1 being given to activity in the home region; the weights for the other regions are inversely proportional to the economic distances of the other regions from the exporting region.

Figure 5.1. Regional consumption (money flows)

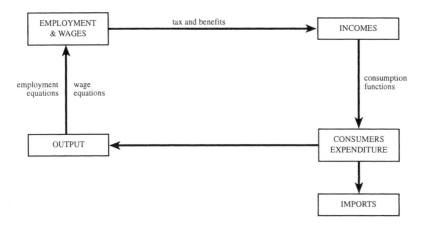

5.5.5 Intermediate Demand

$QY0_r$ is a vector comprising the row sums of the industrial absorption matrix, QY_r, for region r.

$$QY_r = matp(QYC_r, diag(Q_r)) \qquad (5.8)$$

where

QYC a matrix of input-output coefficients
Q a vector of output for 49 commodities
matp a matrix multiplier operator
diag an operator converting from a vector to a diagonal matrix

In the current version of the model, the UK input-output tables 1975, 1979, 1985 and 1990 are interpolated and used for all the regions, ie $QYC_r = QYCUK$ for all r. The coefficients are calculated as inputs of commodities from whatever source, including imports, per unit of gross industrial output; they are therefore applicable to the regional economies on the assumption that the regional technology and mix are the same as those for the UK for each industry. This assumption will be tested for Scotland, where independent estimates of the input-output coefficients are available.

5.5.6 Income and Expenditure

Figure 5.1 shows money flows for regional consumption. Personal disposable incomes are derived from income from employment, government transfers and taxes. Expenditure is related to incomes. Output generates employment and income, using regional wage rates. Total consumer spending by region is derived from consumption functions estimated from time-series data. These equations relate consumption to regional personal disposable income, wealth and demographic characteristics. Regional per capita consumers' expenditure is modelled in an error-correction specification, distinguishing short and long-term effects.

5.5.7 Long-term Equation

$$rsc_{rt} = const_r + \alpha_r^1 \, rpdi_{rt} + \alpha_r^2 \, rhw_{rt} + \alpha_r^3 \, roap_{rt} + \alpha_r^4 \, dpce_t \tag{5.9}$$

where

rsc_{rt} (log) consumers' expenditure in region r at time t
$rpdi_{rt}$ (log) real personal disposable income in region r at time t
rhw_{rt} (log) housing wealth in region r at time t
$roap_{rt}$ (log) retired population in region r at time t
$dpce_t$ change in (log) consumer price index at time t

5.5.8 Short-term Equation

$$drsc_{rt} = const_r + \alpha_r^1 \, drpdi_{rt} + \alpha_r^2 \, dhp_{rt} + \alpha_r^3 \, dunr_{rt} + \alpha_r^4 \, drsc_{rt-1} + \alpha_r^5 \, ECT_{t-1} \tag{5.10}$$

where

$drsc_{rt}$ change in (log) consumers' expenditure in region r at time t
$drpdi_{rt}$ change in (log) real personal disposable income in region r at time t
$dunr_{rt}$ change in (log) unemployment rate in region r at time t
dhp_{rt} change in (log) real house prices in region r at time t
ECT_{rt-1} residual from long-term equation for region r at time t-1

Furthermore, for each region, a set of cointegrating equations, viz a long-term equation and a dynamic error-correction equation, are estimated for each of the 68 categories of expenditure used in the model. Explanatory variables in both the dynamic and long-term regional cointegrating equations include total regional spending, the implicit (UK) price deflator of the spending category relative to the overall (UK) price deflator for consumers' expenditure, and some demographic characteristics of the region. In the long run a positive relationship is imposed between spending by disaggregated category and total spending. Restrictions are also imposed such that movements in relative prices result in an increase in spending on the relatively cheaper categories of expenditure. However, the data are allowed to determine the size of the relative elasticities in each region. Demographic characteristics are modelled by the inclusion of the proportions of pensioners and of children in the total population of a region. The composition of spending is found to be influenced by these demographic characteristics, for example, a larger proportion of pensioners in the population has a large positive effect on spending on gas.

5.5.9 Government Expenditure

This is exogenous, being calculated by applying regional shares to UK totals by the five functional categories shown in the ONS's National Accounts.

5.5.10 Fixed Investment

Regional investment by investing industry is determined in the model by regional industrial output, and UK investment in total by the industry. This treatment follows the accelerator equations in MDM for national investment. In order to find the demands on the industries supplying the investment goods, these investments have to be converted using UK converters and assuming that each region's asset structures for each industry correspond to those of the UK.

5.5.11 Stockbuilding

Stock levels are assumed to be allocated across the regions in the same proportions as output. Stockbuilding is calculated as the changes in these stock levels.

5.5.12 Gross Output

In the data construction exercise, gross industrial output, Y_r, is grossed up from net output, YVA_r, as follows:

$$Y_r = matp(inv(I-diag(csum(QYC_r))),YVA_r) \qquad (5.11)$$

Where

QYC a matrix of input-output coefficients
I an identity matrix
$matp$ a matrix multiplier operator

inv denotes that the matrix following the operator is inverted
matp a matrix multiplier operator
diag an operator converting from a vector to a diagonal matrix
csum an operator forming columnμ sums from a matrix

Gross commodity output, Q_r, for region r is found from gross industrial output of the region:

$$Q_r = matp(YQC_r, diag(Y_r)) \qquad (5.12)$$

where

YQC a matrix m of commodity output per unit of industry outputs (the 'mix' matrix)

When the model is used for projections, gross commodity output is divided into two components, exported output which is identical to the regions exports and output sold to meet domestic demand within the region. Output for domestic demand is directly related to that demand.

5.5.13 Regional Imports

In both the data construction exercise and the projections, imports are residual supplies, calculated from the commodity balances.

5.5.14 Employment

Employment is treated as a demand for labour derived from the regional demand for goods and services. Regional employment equations are estimated relating industrial employment in each industry to its output in the region, to wage rates in the region relative to output prices and to national variables such as average hours worked. Long-run cointegrating relationships are identified and estimated and dynamic error-correction equations estimated to allow for short-run effects. In general the equations are well-determined and the parameters are of the expected sign and magnitude.

5.5.15 Average Wage Rates

In general, wage rates in the UK are formed as the outcome of a bargaining process between employers, sometimes organised into employer organisations, and employees, sometimes organised into trade unions. The government affects the process as a major employer and as a source of legislation affecting all aspects of the process: the legal standing of the parties; the taxes imposed at various stages in the earning and spending of wages; the provision of benefits to the unemployed and the non-employed; and the direct legislation of permitted or recommended wage rates via income policies or minimum wages. The dependent variable in the wage equation in MDM94 is the gross nominal wage, that is, the contractual wage which is a common component of real wages for the main players in the labour market. The external influences on wage bargaining in an industry are divided into those from other industries in the same region, and those from the same industry in other regions. Regional average wage rates by industry are also determined by national price inflation, benefit rates and regional unemployment rates.

5.5.16 Migration and Population

The migration and population submodel consists of six parts, which estimate gross in-migration, gross out-migration, retired population, child population, the workforce and the numbers of households. In each set of equations, the same set of explanatory variables and functional form is used for all of the 12 regions. Once estimated, the residuals from the long-run equations were lagged and incorporated as part of an error-correction model to model the short-term dynamics in line with the Engle-Granger two-stage estimation procedure. A brief description of how each series was modelled follows.

Regional in and out-migration of working-age population are both assumed to be affected by the same economic factors. The migration is modelled as occurring from the region to the outside world and vice versa. The explanatory variables used are a measure of relative house prices, a measure of regional surplus labour relative to the UK, the mortgage rate, relative wages and a linear time trend. The equation is estimated in levels, as the surplus labour measure can be either positive or negative. Various restrictions are placed on the independent variables in line with a priori expectations. Except for the mortgage rate, which is assumed to slow down labour mobility in general, the sign restrictions switch sign between in and out-migration. Those for in-migration are as follows.

Relative house prices are expected to have a negative coefficient, as the higher a region's house price relative to the UK, the less likely people are to leave the region. When the surplus labour variable is positive it indicates the region has a loose labour market relative to the UK, and as one of the main reasons for migration is job search, the coefficient should be negative. The mortgage rate raises the borrowing cost and so impedes movement between regions (in either direction), while a region with higher than average labour payments attracts people and so a positive sign on relative wages is expected. Net migration is then derived as the difference between the forecasts for the gross flows.

The old-age population equations are estimated in logs and include regional working-age population, the UK old-age population, UK working-age population, the mortgage rate and a linear time trend. The only restriction is that the coefficient on the working-age population be positive. The children equations are also estimated in logs and the regressors are regional working-age population, the UK child population and UK working-age population. The signs on the latter two variables are imposed as being equal and opposite which for every region means that they combine to the UK child population as a proportion of the UK working-age population.

The workforce equations have as explanatory variables the UK working-age population, the UK workforce, the regional working-age population and a linear time trend. Restrictions are imposed such that the equation is in effect modelling the regional participation rate as a function of the UK participation rate. Using estimates of the workforce and of employment, unemployment is simply the difference of the two series.

Working-age population is calculated as the sum of assumptions for the natural increase and net migration. The number of households is calculated as the region's population divided by assumptions for average household size in each region, based on DETR projections.

5.6 Results and Concluding Comments

A key uncertainty for the UK over recent years has been how long sterling would remain at a relatively high level and what impact this would have on competitiveness and hence output and employment at the regional level. In July 1997 the model was used to identify the impact of sterling remaining strong over the following three years, relative to a base case in which sterling fell back to a more sustainable level.

In the base forecast it was assumed that an orderly depreciation would bring the DM/£ rate down to average DM2.55 in 1998 and DM2.35 in 1999. This was expected to be encouraged by a tighter UK fiscal policy, sending the signal that interest rates were unlikely to rise much higher, and by an upturn in activity in continental Europe (suggesting that interest rates there would begin to rise). By contrast, in the scenario sterling was assumed to remain at DM3.00 (a view which turned out to be closer to the eventual outturn).

5.6.1 Effects of High Sterling on Macroeconomic and Sectoral Performance

Under the scenario, consumers' expenditure growth was stronger than in the base case in 1998 but then much weaker in 1999, reflecting the temporary impact of more favourable terms of trade on the real purchasing power of domestic incomes. Initially UK GDP growth was boosted slightly, but thereafter it was lower than in the base case as the deterioration in the trade balance offset the boost from higher consumer spending. A recession would have been precipitated by 1999.

Because MDM distinguishes industries in considerable detail and models their trade performance separately, it allows the different effects by sector to be captured, and this can be important for understanding the regional impacts because of the different specialisation of the various regions. Strong sterling produces price signals that discourage production in the traded sectors. Industries competing in 'commodity' markets respond by cutting margins and profitability to maintain prices in foreign currency terms. This is true of crude and refined oil products and for metals and metal goods. Industries dominated by international companies for which production is integrated across Europe also tend to see prices maintained in foreign currency terms (e.g., cars, aerospace). The longer such low profitability is sustained, the greater the extent to which production is scaled back in the UK.

Industries where trade is most price-sensitive naturally see the worst deterioration in trade performance. How much this affects output depends on the importance of trade to the industry compared with domestic demand. Thus textiles, paper, most of the engineering industries and tourism see the largest reduction in output, whereas retailing and personal services see less of an impact, mainly coming from second-round multiplier effects. As sterling's strength persists, the exposed industries accelerate moves to shift into less price-sensitive, higher value-added parts of the business.

Table 5.1. High sterling scenario – Differences in GDP from base

	2000 (per cent)
Greater London	-2.1
Rest of South East	-2.4
East Anglia	-2.1
South West	-2.2
West Midlands	-2.3
East Midlands	-2.2
Yorks & Humberside	-1.9
North West	-2.0
North	-2.0
Wales	-2.5
Scotland	-2.6
Northern Ireland	-1.8
UK	-2.2

5.6.2 Effects of High Sterling on Regional Performance

The effects on the regions depend partly on each region's industry mix, but also on the sensitivity of a given industry in a given region to weaker UK activity in that industry, as modelled in the regional export equations.

Table 5.1 shows the impact of a strong sterling on the regions' output by the year 2000. UK GDP was expected to be about 2% below the level forecast in the base case. All the regions suffer, but Scotland and Wales see the sharpest slowdown, with the South East and the West Midlands also vulnerable. This reflects the impact of a worse trade performance on manufacturing industries; in Scotland manufacturing output in 2000 is 7% lower than in the base case, cutting output growth in Scottish manufacturing to less than 0.5% pa up to 2000. The next most-affected manufacturing sectors are in the South East and the West Midlands.

In the case of tourism, the sensitivity of the regions to strong sterling is more uniform although the East Midlands, the North West and the North see the least impact.

In the event, sterling did remain strong and so the model's predictions can be compared with the outturn, at least in broad terms (the outturn for the model's other assumptions has not necessarily followed the same path as assumed in the scenario). At the macroeconomic level, consumer spending did remain stronger for longer than expected in 1998 and was subsequently followed by a sharp reduction in growth, as the scenario suggested. However, consumer price inflation has not slowed by as much as the model predicted. At the level of UK industries the scenario results have been borne out in that output in textiles, clothing and leather saw a sharp fall in 1998 with somewhat smaller falls also seen in the output of metals and mineral products, mechanical engineering and in the hotels and catering sector. Regional output data are not yet available but survey evidence (BCC, 1998) has indicated some correspondence with the model results in that manufacturing industry has performed badly in the West Midlands, Wales and in Scotland with the exception of the electronics sector.

References

Barker T. and Peterson W. (eds, 1987) *The Cambridge Multisectoral Dynamic Model of the British Economy*, Cambridge University Press, Cambridge, Great Britain.

British Chambers of Commerce (1998) *Quarterly Economic Survey*, The British Chambers of Commerce.

Dixon P.B. *et al.,* (1982) *ORANI: A Multisectoral Model of the Australian Economy*, North-Holland, Amsterdam.

Harrigan F. and McGregor P.G. (1988) (editors) *Recent Advances in Regional Economic Modelling*, Pion, London.

Hewings G.J.D. and Jensen R.C. (1986) 'Regional, interregional and multiregional input-output analysis', pp. 295-355 in Nijkamp P. (ed 1986) *Handbook of Regional and Urban Economics Volume 1 Regional Economics*, North-Holland, Amsterdam.

Isard W. (1951) 'Interregional and regional input-output analysis: a model of a space economy', *The Review of Economics and Statistics*, Vol. 33, no. 4, pp. 318-328.

Joun R.Y.P. and Conway Jr R.S. (1983) 'Regional economic demographic models: a case study for the Washington and Hawaii models', *Socio-Economic Planning*, Vol. 17, pp. 345-353.

Leontief W. (1953) 'Interregional theory', pp. 93-115 in *Studies in the Structure of the American Economy*, Oxford University Press, New York.

Leontief W. and Strout A. (1963) 'Multiregional input-output analysis', pp. 243-259 in T Barna (ed) *Structural Interdependence and Economic Development*, St Martins Press, London.

Moses L.N. (1955) 'The stability of interregional trading patterns and input-output analysis', *The American Economic Review*, Vol. 65, no. 5, pp. 803-832.

Nijkamp P., Rietveld P. and Snickars F. (1986) 'Regional and multiregional economic models: a survey', pp. 257-294 in Nijkamp P. (ed) *Handbook of Regional and Urban Economics Volume 1 Regional Economics*, North-Holland, Amsterdam.

Polenske K. (1980) *The US Multiregional Input-Output Accounts and Model*, Lexington Books, Lexington, Mass. US.

Round J.I. (1988) 'Incorporating the international, regional, and spatial dimension into a SAM: some methods and applications', pp. 24-45 in Harrigan F. and McGregor P. G. (editors) *Recent Advances in Regional Economic Modelling*, Pion, London.

Stone R. (1961) 'Social accounts at the regional level: a survey', pp. 263-295 in Isard W. and Cumberland J. (eds) *Economic Planning: Techniques for Analysis for Less Developed Areas*, OEEC, Paris.

MDM9 Industries defined in terms of the 1992 and 1980 standard industrial classification

	MDM9	SIC 92	SIC80
1	Agriculture	01, 02, 05	0
2	Coal	10	11
3	Oil and Gas etc.	11, 12	13
4	Other Mining	3, 14	21, 23
5	Food	15.1-15.8	41-423
6	Drink	15.9	424-428
7	Tobacco	16	429
8	Textiles	17	43
9	Clothing and Leather	18, 19	44, 45
10	Wood and Wood Products	20	461-466
11	Paper, Printing and Publishing	21, 22	471-475
12	Manufactured Fuels	23	12, 14, 152
13	Pharmaceuticals	24.4	257
14	Chemicals	24(ex 24.4)	25(ex 257), 26
15	Rubber and Plastics	25	48
16	Non-Metallic Mineral Products	26	24
17	Basic Metals	27	221-224, 311
18	Metal Goods	28	312-319, 320
19	Mechanical Engineering	29	321-329, 346
20	Electronics	30, 32	33, 344, 345
21	Electrical Engineering	31	341-343, 347, 348
22	Instruments	33	37
23	Motor Vehicles	34	35
24	Aerospace	35.3	364
25	Other transport Equipment	35 (ex 35.3)	36 (ex 364)
26	Manufacturing and Recycling	36, 37	467, 49
27	Electricity	40.1, 40.3	161-163
28	Gas Supply	40.2	162
29	Water Supply	41	17
30	Construction	45	5
31	Retailing	52	64/65 (ex 651/652)
32	Distribution	50, 51	61-63, 651/652, 67
33	Hotels and Catering	55	66
34	Rail Transport	60.1	71
35	Other Land Transport	60.2, 60.3	72
36	Water Transport	61	74
37	Air Transport	62	75
38	Other transport Services	63	76, 77
39	Communications	64	79
40	Banking and Finance	65	81
41	Insurance	66	82
42	Professional Services	67, 73, 74.1-74.4	831, 832, 835-838, 94
43	Computing Services	72	8394
44	Other Business Services	70, 71, 74.5-74.8	834, 8395-6, 84-5, 923
45	Public administration & defence	75	91
46	Education	80	93
47	Health and social work	85	95, 961
48	Waste Treatment	90	921
49	Miscellaneous Services	91-99	9 (rest of)
50	Unallocated		

6 Discrete Choice Modelling: Basic Principles and Application to Parking Policy Assessment

Harmen Oppewal and Harry Timmermans

6.1 Introduction

The prediction of the spatial distribution of consumer demand constitutes an important step in the planning and strategic decision making processes of businesses and government alike. For example, the planning of service facilities is largely determined by how many facilities are required as a function of population growth or other indicators of demand. Similarly, when entering new spatial markets, businesses will assess the market potential as reflected by the potential demand in the market and the strength and position of their competitors. Hence, information concerning spatial demand allows businesses and governments to assess their plans in terms of indicators such as market shares, turnover, and feasibility.

The total demand in any given market area will be allocated to particular businesses or zones as a function of individual choice behaviour. Consumers will compare the various attributes of products and businesses and then decide which one to choose. When attempting to predict the distribution of demand across space, the task of the modeller then is to predict the probability that a consumer will choose the various choice alternatives (stores, transport modes, recreational areas, products, etc.) as a function of their locational and non-locational attributes and possibly as a function of consumer characteristics.

Over the years, many different modelling approaches have been suggested to address this problem. Perhaps the best known of these are the spatial interaction models, which are discussed by Clarke and Clarke in chapter 8. In this chapter, the discrete choice modelling approach will be outlined. It was introduced in the literature as an alternative to spatial interaction models in the mid 1970s. Spatial interaction models are based on a division of the study area in terms of zones and predict interaction (traffic flows, shopping expenditures, etc.) between these zones. Spatial interaction models have been criticised for their lack of behavioural content. Progress in the statistical literature on categorical data analysis in the 1970s offered the opportunity to combine economic and psychological theories of consumption and individual choice behaviour and the estimation of individual choice models based on categorical data (for introductory textbooks, see e.g., Ben-Akiva and Lerman, 1985; Wrigley,

1985). The development of discrete choice models took place first in the transportation literature, but was soon followed by regional science and quantitative geography (and a little later by marketing). It seems fair to say that in the 1980s and early 1990s, discrete choice models largely replaced spatial interaction models in transportation, but their popularity in regional science has been much less, reflecting the dominance of meso-scale applications in the spatial sciences.

The purpose of this chapter is to outline the principles of discrete choice models and to illustrate their use in an application in the context of predicting spatial shopping behaviour based on the existing transport network and changes to that network. To that end, the chapter is organised as follows. First we explore the development of discrete choice models. Then we examine our client's needs. Next, we will discuss the application, focusing on the policy problem, data collection, model specification, estimation results and policy simulation. This chapter will be concluded by a discussion of the prospects of discrete choice modelling in the spatial sciences.

6.2 Discrete Choice Models

In general, discrete choice models allow the prediction of the probability of choices between discrete choice outcomes. Examples include the probability of choosing the car versus public transport, or the probability of choosing product A versus products B, C, D, and E. Discrete choice models serve to predict the probability that an individual will choose a particular discrete choice alternative as a function of its locational and non-locational attributes and socio-economic characteristics. It results in predictions of market share, consumer demand, and when linked with external criteria, provides indicators of feasibility and competitive impact.

Discrete choice models can be derived from different theories, such as Lancaster's consumption theory, random utility theory and Luce's psychological choice theory (see e.g., Ben-Akiva and Lerman, 1985). We will not pursue all these theories, but rather focus on some common elements. This is consistent with the typical discussion of these models in the spatial sciences literature, but one should realise that these theories differ considerably, and to be consistent one has to be very cautious when interpreting discrete choice models.

Assume that an individual i is faced with J choice alternatives in his/her choice set A_i. Each choice alternative j can be described by a bundle of attributes $\mathbf{x}_j$. Assume that individuals derive some utility for each attribute, and that these attribute utilities are combined according to some algebraic equation to arrive at an overall utility for choice alternative j. Hence,

$$U_{ij} = f(\mathbf{x}_j) \qquad \forall j \in A_i \qquad (6.1)$$

Discrete Choice Modelling: Basic Principles and Application to Parking Policy Assessment

Several functional forms can be chosen for f, but in most cases a linear additive function is assumed to describe this integration process. The linear additive function represents a compensatory decision making process in the sense that lower values on some attributes may at least partially be compensated by higher evaluation scores on one or more of the remaining attributes. Thus:

$$U_{ij} = \sum_k \beta_k x_{jk} \qquad (6.2)$$

where, x_{jk} describes the level (or value) of attribute k for choice alternative j and the β's are the parameters to estimate. If these beta parameters can be derived, the above set of assumptions allows one to rank the choice alternative in terms of preference. In order to predict choice behaviour, one needs an additional assumption relating preference to choice. For example, Luce (1959) assumed that the choice probabilities are proportional to the utilities. Hence, (dropping the subscript for the individual) the probability of choosing j over j' is equal to:

$$p_{j/j'|j,j' \in A_i} = \frac{U_{ij}}{U_{ij'}} \qquad (6.3)$$

and the choice model reads as:

$$p_{j|A_i} = \frac{U_{ij}}{\sum_{j' \in A_i} U_{ij'}} \qquad (6.4)$$

A problem with this set of assumptions is that inconsistency in consumer choice behaviour is not accounted for. If one assumes that preferences are not fixed but stochastic, and in addition that the modeller is faced with measurement error, then the utility function should be expressed as a random utility in terms of the sum of a systematic utility component (V_{ij}) and an error component (ε_{ij}) reflecting stochastic preferences, measurement error and the like. Thus,

$$U_{ij} = V_{ij} + \varepsilon_{ij} \qquad (6.5)$$

The choice model now differs. The probability of any alternative j being selected by utility-maximising individual i from choice set A_i is now given by the following equation:

$$P(j | A_i) = \Pr(U_{ij} \geq U_{ij'}, \forall j \in A_i) \qquad (6.6)$$
$$= \Pr(V_{ij} + \varepsilon_{ij} \geq U_{ij'} + \varepsilon_{ij'} \; \forall j \in A_i)$$

The above equation shows that the actual choice probabilities depend on the assumptions made with respect to the distribution of the error terms of the utility function.

Over the years, many different choice models have been derived from the above expressions. We will not discuss most of these here: interested readers are referred to Timmermans and Golledge (1992) for further details. Perhaps the most appealing and certainly most flexible model is the multinomial probit model, which can be derived from the above expression by assuming that the errors are normally distributed and correlated across choice alternatives. A problem associated with this model, however, is that no closed-form expression for the choice probabilities can be derived, limiting the application value of the model. Recently, other simulation methods have been developed, but the probit model has still not found wide application.

A much more rigorous set of operational assumptions would be that the error terms are identically and independently distributed. If, in addition, we assume that they follow a Gumbel distribution, then a close-form expression can be derived. This has become known as the *multinomial logit model*, which has the following form:

$$p_{j|A_i} = \frac{\exp(U_{ij})}{\sum_{j' \in A_i} \exp(U_{ij'})} \tag{6.7}$$

$$= \frac{\exp(\sum_k \beta_k x_{ij})}{\sum_{j' \in A_i} \exp(\sum_k \beta_k x_{ij'})} \tag{6.8}$$

Note that due to these additional assumptions, the relationship between choice probabilities and attribute values has become exponential. In the case of spatial applications, one of the attributes might be distance (*d*), resulting in the following spatial choice model (χ being the parameter for distance):

$$p_{j|A_i} = \frac{\exp(\sum_k \beta_k x_{ij} - \chi d_{ij})}{\sum_{j' \in A_i} \exp(\sum_k \beta_k x_{ij'} - \chi d_{ij'})} \tag{6.9}$$

The above set of rigorous assumptions results in an easy-to-use model, but at the cost of ending up with a model specification which can be rather unrealistic under particular circumstances. Consider the following example: suppose that an individual originally is faced with the choice between two alternatives C and D. Suppose that a third choice alternative D', which is exactly equal to D, is added to his choice set. What is to be expected in terms of shifting choice probabilities? Because D and D'

Discrete Choice Modelling: Basic Principles and Application to Parking Policy Assessment 101

are exactly the same, it seems reasonable to expect that the probability of choosing C will not be affected at all, whereas the original market share of D will now be equally split between D and D'. Now suppose that $\exp(U_C) = 0.4$, and $\exp(U_D) = 1.6$. Then,

$$P_{C|C,D} = \frac{0.4}{0.4 + 1.6} = 2/10 \tag{6.10}$$

$$P_{D|C,D} = \frac{1.6}{0.4 + 1.6} = 8/10 \tag{6.11}$$

The introduction of D' leads to the following choice probabilities:

$$P_{C|C,D,D'} = \frac{0.4}{0.4 + 1.6 + 1.6} = 1/9 \tag{6.12}$$

$$P_{D|C,D,D'} = \frac{1.6}{0.4 + 1.6 + 1.6} = 4/9 \tag{6.13}$$

$$P_{D'|C,D,D'} = \frac{1.6}{0.4 + 1.6 + 1.6} = 4/9 \tag{6.14}$$

Thus, the introduction of D' has increased the market share of the D's to 8/9, whereas the market share of C has been reduced from 2/10 to 1/9. Note that this is in direct proportion (2:8) to their original shares before the introduction of D'. Thus, the multinomial logit has the property that the introduction of any new choice alternative is predicted to attract a market share from existing choice alternatives in direct proportion to their utility, which as the simple example shows may be unrealistic. This property is directed related to the so-called *Independence from Irrelevant Alternatives* property, which can be easily derived:

$$p_{j/j'|j,j' \in A_i} = \frac{U_{ij} \Big/ \sum_{j'' \in A_i} U_{ij''}}{U_{ij'} \Big/ \sum_{j'' \in A_i} U_{ij''}} = \frac{U_{ij}}{U_{ij'}} \tag{6.15}$$

Thus, the odds of choosing any two choice alternatives from the choice set is dependent only on these two alternatives and is independent of the characteristics of any other alternatives that may exist. Non-availability and similarity among choice alternatives do not have any impact at all on the choice probabilities predicted by the model. This is also true for relative locations. The location of choice alternatives vis-à-vis competing alternatives does not influence the predictions, and hence competition and agglomeration effects are not accounted for by the conventional multinomial logit model.

There are several alternatives to avoid the IIA-property of the multinomial logit model. The assumption of identically and independently distributed error terms may be relaxed. Alternatively, the existence and/or attributes of competing choice alternatives may be included in the specification of the utility function of a particular choice alternative, leading to universal logit model (e.g., Timmermans, Borgers and van der Waerden, 1991). However, the best-known way of developing non-IIA choice models is to specify a nested logit model. The idea behind the nested logit model is that the modeller constructs a hierarchical structure representing the order in which alternatives are assumed to share unobserved attributes. By placing similar alternatives in the same nest, the expected shifts in market shares for our hypothetical example can be approximated. The nested logit choice model can be expressed as a product of marginal and conditional choice probabilities:

$$P_i(x_1, x_2, x_3) = P_i(x_3 \mid x_1, x_2) P_i(x_2 \mid x_1) P_i(x_1) \tag{6.16}$$

where,

$$P_i(x_3 \mid x_2 x_1) = \frac{\exp(V_{x_3} + V_{x_2 x_3} + V_{x_1 x_3} + V_{x_1 x_2 x_3}) \mu_{x_3}}{\sum_{x'_3} \exp(V_{x'_3} + V_{x_2 x'_3} + V_{x_1 x'_3} + V_{x_1 x_2 x'_3}) \mu_{x_3}} \tag{6.17}$$

$$P_i(x_2 \mid x_1) = \frac{\exp(V_{x_2} + V_{x_1 x_2} + V'_{x_1 x_2}) \mu_{x_2}}{\sum_{x'_2} \exp(V_{x'_2} + V_{x_1 x'_2} + V'_{x_1 x'_2}) \mu_{x_2}} \tag{6.18}$$

$$P_i(x_1) = \frac{\exp(V_{x_1} + V'_{x_1}) \mu_{x_1}}{\sum_{x'_1} \exp(V_{x'_1} + V'_{x'_1}) \mu_{x_1}} \tag{6.19}$$

and where we define

$$V'_{x_1 x_2} = \frac{1}{\mu_{x_3}} \ln \sum_{x_3} \exp(V_{x_3} + V_{x_2 x_3} + V_{x_1 x_3} + V_{x_1 x_2 x_3}) \mu_{x_3} \tag{6.20}$$

$$V'_{x_1} = \frac{1}{\mu_{x_2}} \ln \sum_{x_2} \exp(V_{x_2} + V_{x_1 x_2} + V'_{x_1 x_2}) \mu_{x_2} \tag{6.21}$$

Discrete Choice Modelling: Basic Principles and Application to Parking Policy Assessment 103

The V' terms are called inclusive values. To be consistent with the principle of utility-maximising behaviour, the following conditions must hold:

$$0 > \frac{\mu^{x_2}}{\mu^{x_3}}; \quad \frac{\mu^{x_1}}{\mu^{x_2}} \leq 1 \qquad (6.22)$$

and

$$\mu^{x_1} \leq \mu^{x_2} \leq \mu^{x_3} \qquad (6.23)$$

The μ's are inversely proportional to the error variance of the model. Therefore, the variance of the random utilities is the smallest at the lowest level of the tree, and it cannot decrease as we move from a low to a higher level. The same principles can be applied to develop models of multidimensional choice, such as for example the choice of transport and destination choice (see, e.g., Ben-Akiva and Lerman, 1985).

The dependent variable of these choice models is the probability that consumers will choose a particular alternative. Hence, the data collection effort involved gathering data on the frequencies with which the various choice alternatives are chosen as a function of their attributes. In general, two means of data collection are described in the literature. *Revealed choice models* are based on choices as observed in the real world. An example would be to collect data on actual shopping behaviour by asking consumers how they allocate their expenditures among the shopping alternatives in the study area. An alternative means of data collection is asking respondents to report their most likely choice among a set of alternatives that are defined by the researcher. This is known as *stated choice modelling*. The latter approach involves the design of experiments that vary the attributes of interest such that the necessary and sufficient conditions to estimate the choice model of interest are satisfied. The main advantage of the approach is that consumer choice behaviour can be related to alternatives or attribute values that do not yet exist. The approach has become very popular in marketing and is rapidly gaining popularity in transportation research, but has still not become a main modelling approach in the spatial sciences. For more details see Louviere (1988), Louviere and Timmermans (1990), or Hensher (1994).

The explanatory variables of the choice model consist of (perceived) attribute values and distance or travel time (in the case of a spatial choice model). Hence, the collection of data about consumer choice, typically collected by means of a survey, is supplemented by fieldwork to collect the positioning of the choice alternatives on the explanatory variables. Sometimes, these data already exist as part of generally available statistical data sources, provided by governmental agencies or commercial vendors.

Several methods can be used to estimate the parameters of the discussed choice models. The most commonly used method is to construct the likelihood function and then use an algorithm to find the parameter values that maximise this function. Most commercial software packages for estimating discrete choice models are based on this method. The nested logit model was originally estimated by using a method that is called sequential estimation. It involves first estimating the conditional choice model or models for the lowest level. The estimated parameters are then used to calculate the inclusive value for the next level. Given this included value, the marginal choice model corresponding with the higher level is estimated, including the inclusive value parameter(s). This process is repeated until all levels have been processed. The estimates provided by this method are consistent, but they are asymptotically not efficient. Thus, the reported standard errors tend to be too small, implying that one may falsely conclude a parameter to be significant, where in fact it is not. More recent software includes full information estimation procedures, which are not hampered by this property. The parameters of the various nests are estimated simultaneously in this case.

A widely used indicator of model fit is a measure called Rho-square. The interpretation of this measure is similar to the R-square measure that is used in ordinary regression, and as with the R-square, its value lies between zero and unity. Its size is however typically smaller than the R-square. One should be careful however that, like R-square, the value of the measure goes up if the data are aggregated. This happens for example when individual choice represented as 0/1 scores are aggregated into choice frequencies. A rule of thumb for totally disaggregated data is that a Rho-square value of .20 is good.

Validation of discrete choice models is not significantly different from the validation of other models. Additional data is required to assess how the model is performing outside of the sample. Tests that are specific for discrete choice models are tests to show that the data are consistent with the IIA-property of the multinomial logit model (e.g., McFadden *et al.*, 1977).

Finally, what about simulation? If the choice model is capable of reproducing the observations within acceptable limits, it can be used for policy simulation. This implies that policies are translated into the explanatory variables of the model, and that the estimated parameters are used to predict choice behaviour. By comparing the estimated choice probabilities and related market shares to the original shares or the shares obtained for different policy options, the choice model allows decision makers to assess the likely consequences of their decisions on consumer choice behaviour, and hence on market shares. The simulation results can thus be used to assess competitive effects and obtain feasibility and similar performance indicators.

6.3 Client Needs

Discrete choice models allow one to predict the probability that a consumer will choose a particular choice alternative as a function of variables that may be manipulable by policy-makers or managers. Hence, these models are potentially valuable to all those clients wishing to predict and/or assess the likely consequences of their policies or marketing or strategic decisions on consumer choice behavior. Typical clients include planning authorities, businesses, non-profit organisations, etc. The predicted choice probabilities are typically not used as the ultimate piece of information but linked to more relevant indicators. For example, choice probabilities may be multiplied by expenditures, resulting in predicted sales, which can be compared to anticipated costs to assess the financial feasibility of a project. Other typical indicators relate to market share, equity and efficacy.

The application described in the rest of this chapter uses a nested logit model applied within the concept of sustainability, which is discussed in many different disciplines. To contribute to the notion of sustainability, many planning authorities in the Netherlands have formulated a policy to reduce car use by a series of actions, including limited parking duration, higher parking fees, a reduced number of parking lots and specific routing in inner-city areas. These policies are not particularly appreciated by retailers located in inner-city areas. Many of these retailers are anxious that consumers will not patronage their stores if they cannot come by car or if parking becomes more difficult. Municipalities are also concerned about the possible parking policies. They do not like their shopping centres to loose market share to centres in neighbouring municipalities. In a similar vein, there is concern about the competition between city centres vis-à-vis the regional shopping centres.

To get more insight into this problem, the larger cities of the Province of Brabant, the Chamber of Commerce and transport authorities joined forces and commissioned a study on the impact of parking policies on spatial shopping behaviour. The study was conducted by EIRASS (the European Institute of Retailing and Services Studies, where the authors are based) and completed in November 1998. This chapter summarises some results of the study. Interested readers should consult the research reports (Oppewal *et al.*, 1997; Arentze *et al.*, 1998a, 1998b) for more details. The purpose of the study was two-fold. First, the actual effects of parking policy measures on sales levels and consumer shopping behaviour should be monitored. Secondly, a model should be developed to predict the likely consequences of such policy decisions. If such a model could be built successfully, it would give the various parties a tool to a priori assess their policies. The goal of the model building process was to develop a model of joint shopping centre and transport mode choice, primarily as a function of parking variables and travel time. Thus, the idea was not to build a full-blown shopping model, but to investigate to what extent parking variables plus travel time influence shopping choice behaviour.

6.4 Data Issues

As the aim of the application described above was to monitor change, a panel survey was administered. A random sample, consisting of 1704 households, was selected to provide information about their shopping behaviour. Respondents were contacted by phone in the period December 1995 - February 1996. They were asked to report how often they buy a pre-selected list of goods, where they mostly buy these goods (choice categories were 25 shopping centres plus a category 'elsewhere'), the travel time involved, whether they typically visit the centre involved on a weekday or on a Saturday, and the transport mode they use when shopping. In addition, they were asked to describe their last shopping trip. It was important to collect reliable information for the 25 shopping centres that would be monitored and, to guarantee sufficient spread across the days of the week and across the centres that the respondents visit more and less often. Therefore, the following procedure was adopted. During the interview by phone, the computer selected a shopping centre that was mentioned by the respondent such that in 50 percent of the cases the selected centre was one of the 25 monitored centres and in 50 percent of the cases it was not. A second shopping centre was selected using the same principle. Weekdays versus Saturdays were also selected in a balanced manner. Respondents were then required (for every selected shopping centre) to provide detailed information about their last trip to this centre. This included the amount spent, if they used the car, where they parked the car and how they evaluated the parking situation. They were also asked if the shopping trip was linked to any other activity, and how they evaluated the shopping centre in terms of parking opportunities, parking fees and quality of signage.

Two years later, the sample respondents were approached again to collect information about their current shopping behaviour and measure whether there have been any shifts in their behaviour. The focus was on the extent to which these shifts resulted from the implemented parking policies. Respondents were also asked to evaluate a set of attributes related to shopping and parking, on a 10-point evaluation scale. These scores were used to identify any shifts in the evaluation of the various shopping centres.

These surveys were used to collect information about the choice of shopping centre and transport mode. In terms of the explanatory variables, for each of the 366 shopping areas in the study area, plus 20 shopping areas outside of the study that were frequently visited by the respondents, data on the number of stores and total floorspace were collected. In principle, such data are available in the Netherlands on a national basis (Retail Information System), but this database has not been well maintained in some cities (which recently led to debates whether or not to continue this information system). Some additional fieldwork was thus required to update the system. The most important explanatory variables concerned the parking situation. Municipalities were requested to provide information regarding the number of paid parking lots, the number of lots where one can park for free, the number of special parking lots (for permit holders), parking fees, the number of parking lots with maximum parking duration, the number of parking lots where one needs to pay up front, and the walking distance to the shops in meters. Occupancy rates at selected times across the day were collected through additional field work. All this information seems easy to collect, but it turned out that not all municipalities have this information to hand, and if they had, the operational definitions were often different. Hence, considerable time and effort was required to design and implement a uniform and consistent data recording system.

Finally, the model required data on travel times. To keep open the possibility that the model can be linked to the regional transportation model, it was decided to use the zoning system and travel times by mode that underlies this transportation system. Hence, data on travel times were provided by the Provincial Transport Planning Authority for car, bicycle and public transport. The data on travel time by car were based on the road network, updated in 1994. Travel times for public transport and bicycles were based on the situation of 1990. There were no substantial changes in the bicycle and public transport networks in the 1990-1995 time period. Travel times were provided for a system, consisting of 436 zones, plus 100 external zones. Travel times were linked to residences and shopping centres on the basis of the official zip code system for the Netherlands.

6.5 Model Development and Refinement

Since the major requirement of the application is the need to predict the combined choice of shopping centre and transport mode, and because different travel modes to the same shopping centre share the attributes of this centre, it was decided to construct a nested logit model. Let s be a subscript to denote the choice of shopping centre, m represent the choice of transport mode, and A represent the choice set, then the estimated model can be formulated as follows:

$$p(s,m \mid A) = p(s \mid m) * p(m \mid A) \tag{6.25}$$

$$= \frac{\exp(V_{s\mid m})}{\sum_{s'} \exp(V_{s'\mid m})} * \frac{\exp(V_m)}{\sum_{m'} \exp(V_{m'})} \tag{6.26}$$

where,

$$V_{s\mid m} = \beta_0 + \beta_1 x_{1s} + \beta_2 x_{2s} + \ldots + \beta\, x_{Ks} \tag{6.27}$$

$$V_m = \alpha_0 + \alpha_1 x_{1m} + \alpha_2 x_{2m} + \ldots + \alpha_K x_{Km} + \theta I_m \tag{6.28}$$

$$I_m = \ln \sum \exp(V_{s\mid m}) \tag{6.29}$$

The model was estimated separately for groceries, clothing and shoes, and household appliances. For illustration, we will focus on the model estimated for clothing and shoes, which are assumed to be indicative of non-grocery shopping. The nested structure assumed that consumers will choose between the car and any other means of transport, and between shopping centres. The following explanatory variables were selected: minimum travel time by transport mode, number of shops in the shopping centre in the category of interest, average floorspace of shops, and the subjective evaluation of the parking situation. The latter variable was explained in a separate modelling step in terms of the set of physical descriptors of the parking situation.

A full information maximum likelihood procedure was used to estimate the model. The estimation required destination sampling. To that effect, for each (zip coded) origin zone the travel times to the frequently visited shopping centres by category of goods and transport mode was derived. For each of these distributions, the travel time within which 95 percent of the shopping trips was made was identified and this travel time was then used to delineate a feasible set of potential destinations. Choice sets of up to a maximum of 20 centres were constructed by including all visited centres, plus randomly selected centres from this feasible set.

6.6 Results

The estimated equations were as follows:

$$V_{car} = -2.071 + I \tag{6.30}$$

$$V_{non-car} = 0.710 * I \tag{6.31}$$

$$V_{sl car} = 1.169 * NumberOfStores / 100 + 0.287 * AverageSizeStores \\ - 0.156 TravelTime + 0.202 EvaluationParking \tag{6.32}$$

$$V_{sl non-car} = 1.169 * NumberofStores / 100 + 0.287 AverageSizeStores \\ - 0.132 TravelTime \tag{6.33}$$

These estimated equations jointly determine the probability that a combination of transport mode and shopping centre will be chosen to buy clothing and/or shoes. All parameters were significant, hence the assumed explanatory variables were all related to the joint choice of shopping centre and transport mode beyond chance. The estimated parameter for the inclusive value suggests that consumers are more inclined to shift to another shopping centre than to another transport mode when changes in the parking situation occur. The results also indicate that the utility of a shopping centre, regardless of transport mode, increases with an increasing number of stores, and with the average size of the stores in the centre, but decreases with increasing travel time. Utility is also positively correlated with the evaluation of the parking situation.

The subjective evaluations of the parking situation were regressed with a series of descriptors, representing the parking attributes. The estimated equation was:

$$Evaluation = 7.111 + 0.465 * AverageSizeStores - 1.261 * OccupancyLevel \\ - 0.256 ParkingFee - 0.315 MaximumDuration \tag{6.34}$$

This equation shows that the evaluation of the parking situation increased with the average size of the stores, and was negatively related to higher occupancy levels, higher parking fees, and a higher percentage of parking lots with a maximised duration. The overall goodness-of-fit of the nested logit model was quite good, as indicated by a Rho-Square value of 0.570.

6.7 Policy Simulations

The nested logit model, described in the previous section, was developed as a tool for policy makers to assess the likely consequences of their parking policy decisions on consumer spatial shopping behaviour and hence market shares and turnover levels of the 25 monitored shopping centres. Before the municipalities implemented their policies, several simulations were conducted to assess the impact of these alternative policies. These policies were formulated by the advisory board of the research project, which consisted of planners from the 4 majors cities in the Province, representatives of the local and regional Chambers of Commerce, representatives of regional transport planning authorities, planners from the Province of Noord-Brabant, and representatives of the Ministry of Economic Affairs and the Ministry of Transport, Public Works and Water Management. Four scenarios were formulated:

6.7.1. A *Consumer Evaluation* Scenario

In this scenario, it is assumed that a change is driven by a changed consumer evaluation of the parking situation in shopping centres that have parking. The simulation reported in this section is based on the assumption that the evaluation of the parking situation in Tilburg (one of the main cities in the study area) will increase by 0.5 points and that the evaluation of the parking situation of one of its regional competitors (Etten Leur) will decrease by 0.5 points on the 10-point scale that was used to measure consumer evaluations of parking situations.

6.7.2. A *Local* Scenario

This scenario assumes that a parking fee of NLG 1,00/hour will be introduced in one of the smaller shopping centres, and that the parking capacity in the non-adjacent large shopping centre will be reduced. It reflects a policy discussion in the area about whether or not to introduce a parking fee in the smaller shopping centres. At the time of the investigation, parking was free in these centres. The advisory board was interested to evaluate whether the effects of policy measures to reduce car-use to the major shopping centres would be counterbalanced by such local policy measures.

The choice of choosing non-adjacent shopping centres was motivated by the desire to learn about cross-regional effects. The simulation reported in this chapter is based on the assumption that the parking capacity of Breda, (another main city in the study area) will be reduced by 15%, resulting in an increased parking occupancy rate of 74 percent on Saturdays, and that paid parking (NLG 1,00/hour) will be introduced in Rosmalen, a medium sized centre in the region, dominated by s'Hertogenbosch. To compensate for the introduction of a parking fee, it is assumed that the parking duration restrictions in Rosmalen will be lifted.

6.7.3. A *City* Scenario

During the project, it became evident that the four major cities in the Province were all concerned that unpopular policy measures taking in a particular city would be a vehicle for competing cities to take measures that would improve their position vis-à-vis their competitors. Under this scenario, it was therefore assumed that all parking variables would be exactly the same in all four major cities, reflecting a rather strict parking policy. In particular, it was assumed that (i) the average parking fee would be NLG 2,50/hour in the zone of 0-100 m from the core of city centre, and NLG 1,50/hour in the zone of 101-500 m from the core of city centre, resulting in changes in the percentage of paid parking lots; (ii) the parking capacity will be reduced in the 0-100 m zone such that for the current number of shoppers who visit the city centre by car the average parking occupancy rate on a Saturday would be 75 percent of its former size.

6.7.4. A *Brabant* Scenario

The final scenario consisted of the combination of all other scenarios plus the assumption of the same parking policy in all regional shopping centres in the Province. The latter assumption was formulated in terms of a parking fee of NLG1,00/hour, compensated by a discontinuation of parking duration restrictions. The assumption were made for the following shopping centres: Berkel-Ensschot, Goirle, Kaatsheuvel, Tilburg-Wagnerplein, Tilburg-Westermarkt, Etten-Leur, Rosmalen, s'Hertogenbosch-Rompert, Boxtel, Geldrop and Veldhoven.

The policies were translated into the explanatory variables of the estimated nested logit model. By keeping the parameters of the model fixed, the effects of the parking policies on transport mode/shopping centre combinations were calculated, assuming that the nature of the relationships between the explanatory variables and the dependent variable remained invariant over time. Predictions for each product class were then aggregated and multiplied by average expenditures to derive sales levels per class per shopping centre. These predictions were then compared to the original situation to provide information about percentage shifts in sales levels. Table 6.1 summarises the changes that result from the implementation of the various parking scenarios.

A detailed discussion of these results goes beyond the purpose of this chapter. We will confine the discussion to some main results. First, Table 6.1 demonstrates that the consumer evaluation scenario does not have a substantial impact on transport modal choice and the choice of shopping centre. The probability of choosing the non-car (bicycle or public transport) combination is slightly reduced in the Tilburg scenario (6.7.1 above). This is expected as the scenario assumed that the evaluation of the parking situation in Tilburg-Centre would improve. The results also indicate that for the non-car users, this policy scenario does not have an impact beyond the shopping market area of Tilburg, suggesting that cross-market area shopping by bicycle or public transport is rather the exception than the rule.

Table 6.1. Effects of the parking policy scenarios

Shopping centre	Consumer evaluation scenario		Local scenario		City scenario		Brabant scenario	
	non-car	car	non-car	car	non-car	car	non-car	car
Eindhoven	-0.04	-0.03	0.00	0.00	1.78	-5.90	1.82	-5.77
Woensel	-0.01	-0.01	0.00	0.00	0.70	1.64	0.72	1.75
Geldrop	0.00	0.00	0.00	0.00	0.93	0.79	1.07	-0.95
Helmond	0.00	0.00	0.00	0.00	0.43	0.87	0.44	0.88
Valkenswaard	0.00	-0.01	0.00	0.00	1.62	2.13	1.79	2.31
Veldhoven	0.00	0.00	0.00	0.00	0.88	1.28	1.49	-3.37
Tilburg-centre	-0.36	10.13	0.03	0.09	0.41	-6.62	0.68	-5.92
Tilburg-Wagnerplein	-0.21	-0.48	0.01	0.01	0.17	0.41	0.47	-4.24
Tilburg-Westermark	-0.38	-1.01	0.01	0.02	0.34	0.77	0.81	-4.45
Berkel-Ensschot	-0.52	-0.46	0.02	0.01	0.53	0.47	1.41	-3.24
Dongen	-0.39	-0.37	0.29	0.30	0.64	0.64	0.71	0.67
Goirle	-0.83	-0.96	0.06	0.04	0.77	0.83	1.28	-4.02
Kaatsheuvel	-0.17	-0.14	0.03	0.05	0.22	0.24	0.86	-2.80
Loon op Zand	-0.20	-0.66	0.00	0.02	0.19	0.63	0.37	1.44
Oisterwijk	-0.58	-0.65	0.02	0.02	0.72	0.78	0.82	1.23
Gilze-Rijen	-0.16	-0.08	0.10	0.10	0.25	0.22	0.26	0.27
Waalwijk	-0.18	-0.27	0.04	0.12	0.32	0.59	0.39	0.90
Breda	0.03	0.11	0.54	-2.15	0.67	-2.34	0.72	-2.20
Breda-Heksenwiel	0.02	0.10	0.21	0.61	0.25	0.74	0.26	0.82
Etten Leur	1.57	-9.40	0.40	0.47	0.47	0.56	1.17	-3.59
Oosterhout	0.00	-0.07	0.27	0.35	0.35	0.50	0.36	0.57
S'Hertogenbosch	-0.04	-0.12	0.02	0.06	0.61	-4.95	0.73	-4.74
s'Hertogenbosch-Rompert	0.00	-0.02	0.03	0.20	0.09	0.47	0.25	-5.59
Boxtel	-0.08	-0.06	0.00	0.00	0.29	0.28	0.75	-2.32
Rosmalen	0.00	-0.01	0.17	-1.53	0.59	0.59	0.87	0.25

The effects of this parking policy for the combination of car and shopping centre choice are listed in the third column of Table 6.1. The improved evaluation of the parking situation in Tilburg is predicted to result in a 10 percent increase in expenditure for this segment. As a result, the other shopping centres loose market share. Etten Leur is hit hardest due to the assumed lowest parking evaluation score. Because Etten Leur is also competing with the city of Breda, the net result of Breda is still positive: the sales lost to Tilburg are more than compensated by the sales won from Etten-Leur.

As for the local scenario, Table 6.1 demonstrates that, as the name of this scenario suggests, the effects are mainly localised. Breda is predicted to loose some market share, and this loss is distributed among the other shopping centres in the Breda region, and to Tilburg, the closest major city in the study area. There is also evidence of a minor shift from car to other means of transport.

The nested logit model predicts a reduction in sales levels between 2.3 and 6.6 percent by shoppers using the car as a means to go shopping when the city scenario would be implemented. Remember that this scenario involved a very strict parking policy in the four major cities in the study area. Table 6.1 however also demonstrates that this loss is at least partially recuperated by shoppers switching to another means of transport, but still going to the same city centres. On balance then, the impact of this very strict parking policy on the sales level of the relevant shopping centres is predicted to be modest, ranging from 1.03 to 3.66 percent. It seems that retailers should be able to easily remedy this loss by appropriate measures to improve service provision in these centres.

The last two columns of Table 6.1 report the predicted impact of the *Brabant* scenario, which combined all three previous scenarios, plus a more strict parking policy in the regional shopping centres. The predicted effects show some interesting patterns. First, as compared to the city scenario, the regional shopping centres should now also expect to loose sales levels. Secondly, because under this scenario the regional centres are less competitive, the negative impact of strict parking policies in the four main shopping centres is less, although still negative. Thirdly, the negative effects of stricter parking policies in the regional shopping centres for the segment of car users is partly compensated by higher sales levels for non-car users, indicating an expected shift in the choice of transport mode if stricter parking policies are implemented. Note that the counterbalancing effects are stronger for the regional centres than for the city centres, suggesting that people are willing to switch from car to bicycle or public transport for shopping when smaller distances or travel times are involved. Fourthly, ultimately the smaller shopping centres are the winners as indicated by the positive percentages for both car users and non-car users. Thus, the Brabant scenario is predicted to result in a general shift from shopping in larger shopping centres to shopping in smaller shopping centres, plus a shift from car use to other means of transport. This is the kind of result that many planners would like to see as it is consistent with their objective of reducing car use. On the other hand, the size of the effects is not to be ignored, especially for retail categories with low margins. Most retailers in the larger shopping centres should be able to take appropriate measures to reduce the predicted effects, but the prediction is only an average for each centre; the variation among retailers within centres may be large and hence the parking measures may affect some retailers more than others.

6.8 Conclusion and Discussion

The purpose of this chapter was to outline the principles underlying discrete choice modelling and illustrate their use in a study on parking policies and spatial shopping behaviour. The principles of discrete choice modelling were discussed and it was shown that discrete choice models allow one to predict consumer choice among discrete choice alternatives, such as shopping centres, transport modes, work zones, recreation facilities, and the like. Compared to their major alternative approach (spatial interaction models), discrete choice models are directly derived from a theory of individual consumer choice behaviour and rely on data and assumptions about individual choice behaviour as opposed to statistical relations between zonal attributes. However, this potential theoretical advantage comes at a cost as one has to collect data about individual choice behaviour, which can often be expensive and time-consuming.

We have also indicated that the earlier discrete choice models, such as the multinomial logit model, are based on rather rigorous assumptions that often seem unrealistic from a theoretical perspective. Although we did not discuss these in any detail, the further development of discrete choice models can therefore be best described as an attempt to relax these rigorous assumptions. Increasingly realistic but also increasingly more complex models of consumer choice behaviour have been built over the years. More advanced models allow one to incorporate spatial agglomeration and substitution effects, effects of choice set composition, constraints and aspects of dynamic choice behaviour. Most of this work, however, has typically remained academic.

In terms of predicting shopping behaviour, the field has also moved forward. Most of the models, including the one presented in this chapter, are implicitly based on the assumption of single-purpose, single-stop behaviour. However, increased time pressure, a growing participating of women in the work force, and changes in task allocation in the household, to name a few, have resulted in increasingly complex behaviour. Although there is evidence that single shopping trips are still dominant, an increasing share of shopping trips involve multi-stops and combination trips, where shopping is combined with other activities such as work, recreation and social visits. This has led to new model types that try to incorporate the mechanisms underlying such behaviour. To the extent that a discrete choice modelling framework has been used in this regard, the trip as the unit of observation has been replaced first by journeys, then by tours, and lately by activity patterns. Although the principles of the model building process remain the same, the structure of the model has become considerably more complex with more nests. It goes without saying that the estimation effort, data requirements and estimation problems have also rapidly increased.

Whether this increased sophistication is worthwhile from an applied perspective is open to debate. It seems that the additional effort and costs can only be motivated by improved predictive success. Unfortunately, there is a fundamental lack of comparative research that would allow one to draw any concrete conclusions in this regard.

References

Arentze T., Oppewal H., Timmermans H. and van der Waerden P. (1998a) Economische Effecten van Parkeermaatregelen in Noord-Brabant: Nameting 1997. European Institute of Retailing and Services Studies, Eindhoven

Arentze T., Oppewal H., Timmermans H. and van der Waerden P. (1998b) Economische Effecten van Parkeermaatregelen in Noord-Brabant. European Institute of Retailing and Services Studies, Eindhoven.

Ben-Akiva M. and Lerman S. (1985), *Discrete Choice Analysis: Theory and Application to Travel Demand*. Cambridge, Mass., MIT Press.

Hensher D.A. (1994), Stated Preference Analysis of Travel Choices: The State of Practice, *Transportation*, 21 (2), 107-134.

McFadden D., Train K. and Tyne W.B. (1977), An Application of Diagnostic Tests for the Independence of Irrelevant Alternatives Property of the Multinomial Logit Model, *Transportation Research Record*, 637, 39-46.

Louviere J.J. (1988), *Analyzing Decision Making: Metric Conjoint Analysis*. Sage University Paper Series on Quantitative Applications in the Social Sciences, series no. 07-067. Beverly Hills: Sage Publications.

Louviere J.J. and Timmermans H. (1990), A Review of Recent Advances in Decompositional Preference and Choice Modeling, *Tijdschrift voor Sociale en Economische Geografie*, 81, 214-224.

Luce R.D. (1959). *Individual Choice Behaviour: A Theoretical Analysis*. New York: Wiley and Sons.

Oppewal H., Arentze T., Timmermans H. and van der Waerden P. (1997) Economische Effecten van Parkeermaatregelen in Noord-Brabant: Winkelcentrum- en vervoerwijzekeuze van consumenten in Noord-Brabant. European Institute of Retailing and Services Studies, Eindhoven, 81 pp.

Timmermans H.J.P. and Golledge R.G. (1992) Applications of Behavioural Research on Spatial Choice Problems II: Preference and Choice, *Progress in Human Geography*, 14, 311-354.

Timmermans H.J.P., Borgers A. and van der Waerden P. (1991), Mother Logit Analysis of Substitution Effects in Consumer Shopping Destination Choice, *Journal of Business Research*, 23 (4), 311-323.

Wrigley N. (1985) *Categorial Data Analysis for Geographers and Environmental Scientists*. London: Longman, 392 pp.

7 Applied Population Projection for Regional and Local Planning

John Stillwell and Phil Rees

7.1 Introduction

Academics are increasingly required to use their skills and expertise in applied contexts. Population geography is perhaps not regarded as the most obvious academic sub-discipline for applied research, but the techniques of spatial demographic analysis have become important to the commercial world as businesses have sought to improve their market share and develop their strategies for providing goods and services to maximum effect. Major commercial organisations have recognized the necessity to identify and target sub-sections of the population most likely to be their primary customers or to site their retail outlets at locations likely to generate maximum revenue (see chapters 8 and 12). Financial organisations selling life insurance have required increasingly detailed geographical information about life expectancy and illness in order to fix insurance premiums as the population ages. Whilst geodemographics has been a growth industry of the 1990s in the private sector (see chapter 14), there has also been a trend towards closer collaboration between academic demographers and public sector organisations such as local authorities and government departments in the last decade, as the latter have recognized the expertise available in universities and as the former have become more conscious of the need to respond to government emphasis on more policy relevant research.

Both the authors have been involved in research consultancy projects in the 1990s to develop new methods of population projection to support local authorities in their planning activities either at a broad regional scale or at a more detailed intra-urban or local scale. In this chapter, we report on our experience using two case studies in which applied research was undertaken for local authorities in Yorkshire and Humberside. The first case study involves the development of a methodology for updating population and household projections for local authority areas as part of the formal regional planning process. The second case study involves the development of a small area population projection model. The work associated with both projects has been undertaken by teams of researchers whose individual contributions deserve acknowledgement. The first project was undertaken by the Yorkshire and Humberside Regional Research Observatory (ReRO), a network of associated research units in universities across the region, and involved staff from the School of Geography at the University of Leeds and the Centre for Regional Economic and Social Research (CRESR) at Sheffield Hallam University. The second project was carried out in collaboration with GMAP Ltd, a university owned company, and involved technical and managerial staff employed by GMAP, working with researchers from the School of Geography, University of Leeds, with whom GMAP has a close relationship.

Although the case studies are contrasting in their aims and objectives, they illustrate the demands made upon the consultants to develop innovative projection methods, to assemble large databases and estimate unknown information, to create alternative scenarios of what might happen in the future and thus to generate plausible results that are acceptable to professional planners. The structure of the chapter follows this sequence. The requirements of the client are specified in each case in section 7.3, the core methodologies are summarized in 7.4, data estimation issues are reviewed in 7.5 and selected results from each case study are illustrated in 7.6. In the final section, we reflect on our experiences and on the lessons we have learnt. Before embarking on this review, however, we focus attention in section 7.2 on the national context of population and household projection in which both these case studies are located, provide a short synopsis of the state of play, and identify the key reasons why the applied work was required.

7.2 Population and Household Projections

Sub-national population projections are produced for each part of the United Kingdom (UK) separately. The General Register Office Scotland (GROS) and the Welsh Office (WO) use a standard cohort-survival model with net migration terms for both international and internal migration (reviewed by Van Imhoff et al.,1994). The Northern Ireland Statistics and Resource Agency (NISRA) are currently preparing district projections for Northern Ireland using the same methods (NISRA 1998). Population and household projections for sub-national areas in England have been produced periodically by the official statistical agencies since the 1960s, with the first population projections generated by the Office of Population Censuses and Surveys (OPCS) in 1965 and the first household projections produced by the Department of the Environment (DoE) and the Building Research Establishment (BRE) in 1969. Champion et al., (1998) have provided a timely review of developments since then. The last round of sub-national population projections in England was produced in 1998 by OPCS' successor, the Office of National Statistics (ONS) using 1996 as the baseline. Previous rounds of projections in the last ten years have been 1989-based (OPCS 1991a) and 1993-based (OPCS 1995). The latest set of household projections are those based on 1992 (DoE 1995) and these replaced the 1989-based projections (DoE 1991). A new set of household projections will be published in 1999 by the Department of the Environment, Transport and the Regions

(DETR) based on the 1996 round of national projections. The sub-national population projections provide numbers by age, sex and marital status for London boroughs, metropolitan districts and shire counties of England and Wales for the next 25 years, whilst the household projections provide numbers in different household type categories (married couple, cohabiting couple, lone parent, other multi-person and one person households in the 1992-based projections) over a similar time period for the same set of geographical areas in England. Projections of concealed couples and concealed lone parents are also contained in the household projections.

Although distinct from the official mid-year population *estimates* whose derivation is explained in OPCS (1991b), the population *projections* are conceptually similar in that they start from the base year for which all information is known and are then rolled forward on an annual basis by adding projections of births, subtracting projections of deaths and adjusting for projections of migration. The methodology for producing the 1989-based projections, summarised by Capron and Corner (1990), has been maintained, although modifications were introduced in the 1993-based projections to treat students and armed forces personnel separately from the rest of the population. The model used for projecting the internal migration component has been critically reviewed by Boden *et al.,* (1992) and the validity of different sets of population projections have been evaluated by Capron (1994).

The 1996-based projections for sub-national areas in England use the new local government areas defined over the 1996-98 period (see Wilson and Rees 1998 for details of the reorganisation) and a substantially revised model (MVA 1996, DETR 1998). The intensities of inter-area migration are based on National Health Service (NHS) register patient transfer data (supplemented by 1991 Census migration data to fill the gaps in the origin-destination matrix) averaged for 1992-96. The 1993-based projections used 1991 Census migration data for 1990-91, whilst previous projections in the 1980s used 1980-81 migration data from the 1981 Census. No attempt as yet has been made to forecast forward from the benchmark period the intensities of inter-area migration by UK national statistical offices. However, Van de Gaag *et al.,* (1997) have recently carried out regional projections for the European Union which include high and low scenarios for the European Union, corresponding to forecasts of increasing and decreasing net transfers of migrants.

The procedure used for household projection, in simple terms, involves taking the OPCS or ONS population projections for each area and converting them to household numbers by applying projections of marital status and household formation to the non-institutional figures. This methodology, explained in a technical internal BRE working paper (Corner 1992), is an intricate and complex process, involving reference to diverse historical data sets and the adoption of some fairly limiting assumptions.

Guidance notes from the national agencies indicate that the 'official' population and household projections described above should not be seen by local authorities as targets for planning purposes but represent a view of the most likely future on the basis of past trends. They provide a starting point for planning and policy-making for local authorities, who might accept them as they are, adjust them, or produce their own independent projections. The results of a questionnaire survey of the eleven local authorities in Yorkshire and Humberside (ReRO 1993) indicate local authority reaction to the 1989-based official projections:

- four local authorities found the population projections very useful, six found them quite useful and one found them not useful;
- three local authorities found the household projections useful, four found them useful and three found them not useful;
- seven local authorities did not understand the population projection methodology;
- eight local authorities did not understand the household projection methodology;
- all local authorities asked for a clearer explanation of each method;
- ten local authorities produced their own independent population projections;
- eight local authorities produced their own independent household projections; and
- only two local authorities were satisfied with the current practice of consultation with the DoE.

These results demonstrate that local authorities treat the official sub-national projections with some scepticism. Very few authorities accepted either set of projections and individual authorities from this sample objected to particular assumptions about different components of each method.

Whilst this short, non-technical review of projection methods gives an indication of the background associated with the two case studies, there are two specific issues with the 'official' government projections that were identified by all local authorities as being problematic. The first issue relates to the timing of the projections in relation to the availability of the results of the 1991 Census. The 1989-based projections were published in 1991 and the subsequent 1993-based projections that incorporated the results of the 1991 Census were published in 1995. This meant that local authorities in 1993 were confronted with 'official' population and household projections in need of updating to reflect socio-demographic changes measured by the census in the knowledge that the next round of projections was still two years away. This updating requirement provided part of the rationale for the first case study described in the chapter.

The second issue relates to the geographical scale of the sub-national projections. Local authority planners have long been concerned to develop a fuller understanding of demographic and household structures and processes in geographical areas smaller than the London boroughs, metropolitan districts and shire counties for which 'official' projections are made. Population and household dynamics in large urban areas are of particular importance because of the implications of change for service provision, and yet very few local authorities throughout the country have developed their own small area projection systems. The development of a population and household projection system for small areas was the focus of the second project and is discussed in more detail in later sections of the chapter.

7.3 Clients Requirements

7.3.1 Updating Projections for Regional Planning Guidance

Regional Planning Guidance (RPG) provides a broad framework for the development of a region over a 20 year planning period. One important aspect of RPG relates to the provision which needs to be made for new housing in local authority development plans. At the present time, this is a matter of national debate because the 1992-based household projections forecast an increase of 4.4 million new households between 1991 and 2016, while it is anticipated that in the 1996-based projections the forecast increase will rise to 5 million households. The 1992-based household projections have prompted a regional inquiry by the Town and Country Planning Association into housing need and provision (Breheny and Hall 1996) and focused local authority attention on the difficult task of matching the growth in demand for dwelling space with an increased provision of land and housing. It is widely recognized that 'official' population projections are intended to be planning guidance tools and that *"the availability of a consistent set of official projections provides a framework within which producers of strategies, structure plans and other proposals for particular parts of the country may be required to justify their use of different figures"* (OPCS 1991a).

In the early 1990s, the availability of the results of the 1991 Census gave local authorities, regional planning agencies and government departments in the UK a unique opportunity to assess the accuracy of the 'official' 1989-based population and household projections for 1991 and to utilize the findings in the formulation of a range of alternative population and household projections which could then be used to estimate future dwelling requirements. It was expected that those local authorities pursuing policies of development constraint would opt for lower projections of population and household growth whilst those with interests in house-building would be likely to favour higher projected totals. These were the circumstances in which the local authorities in Yorkshire and Humberside, drawn together by the Yorkshire and Humberside Regional Planning Conference (YHRPC), in collaboration with the Regional Office of the DoE, commissioned a study by the Yorkshire and Humberside Regional Research Observatory. The local authorities included the five metropolitan districts of West Yorkshire, the four metropolitan districts of South Yorkshire and the two former two-tier shire counties of North Yorkshire and Humberside whose boundaries are illustrated in Figure 7.1. Since 1996, local government reform has meant that Humberside has been replaced by four new unitary authorities and in North Yorkshire, York has been designated as a unitary authority whilst the remainder of the county remains two-tier.

Figure 7.1. Boundaries of the local authorities of Yorkshire and Humberside, 1993

The project was undertaken as part of the preparation of RPG under the provisions for Planning Policy Guidance (PPG) Note 12 (DoE 1992). Its objectives included (a) an assessment of the existing projections in the light of the 1991 Census results and information from other relevant data sources and (b) the generation of a series of alternative sets of population and household projections for 2001 and for 2006 based on differing assumptions about migration, marital status and headship rates. In this chapter, we outline the methodology used to adjust the projections, report on some of the data issues, explain the procedure used to generate different scenarios and present a suite of revised household projections. The detailed evaluation of the 1989-based population projections for 1991 when compared against the revised final mid-1991 population estimates for local authority areas (OPCS 1993) is presented elsewhere (Stillwell and Gore 1997).

7.3.2 Projecting Populations for Small Areas

Monitoring and understanding population change is at the heart of urban and regional planning since it is the people who require the physical infrastructure and functional services that make cities what they are. Population structures and dynamics vary across any city and reflect the characteristics of the residents of each neighbourhood or suburb. Inner city areas with a higher proportion of younger, single people are distinguishable from outer suburban areas where families or the elderly are predominant, for example. The process of suburbanisation in which people migrate as their housing needs and aspirations change is very familiar whilst the process of gentrification changes the socio-demographic character of established urban neighbourhoods. It is therefore important that local authorities, with the responsibilities for physical planning (e.g., housing provision) and socio-economic planning (e.g., services provision), are able to look ahead and make decisions about the future based on historical information about population stocks and current trends in fertility, mortality and migration. In order to do this, it is necessary to construct a population projection model.

At the beginning of the 1990s, the metropolitan district authorities in West Yorkshire (Bradford, Calderdale, Kirklees, Leeds and Wakefield) (Figure 7.1) recognised the importance of projecting the populations of urban communities and commissioned GMAP Ltd to develop a projection model as a tool that they could use to carry out their own projections of the populations of wards within their districts. The model was designed to generate detailed annual projections of the age and sex structure of ward populations, to allow users to explore the implications of different demographic scenarios, and to have the flexibility of incorporating the latest information as it became available. The initial concept was to build a core model that would project populations and households and then to develop routines to estimate more disaggregated measures such as populations and households by ethnic group, marital status and household structure, in the first instance, and socio-economic groups, jobs and the demand for service facilities thereafter. In this chapter, we outline the main stages in the projection model, discuss some of the data estimation requirements that necessitated the development of an information system, and illustrate the results of alternative scenarios using the wards of Bradford. A more detailed account of the methodology and of the evaluation of the estimation and projection method for ward populations using 1991 Census data can be found in Rees (1994).

7.4 Adjustment and Modelling Methods

7.4.1 Adjustment Methodology for the Local Authority Projections

The methodology developed for updating the population and household projections for local authorities in Yorkshire and Humberside was one of adjustment of existing 1989-based projections rather than the construction of a new projection model. The population projections were adjusted initially to take into account the differences revealed by a comparison of the projected age-specific totals for 1991 with the revised final mid-1991 estimates which, in aggregate, underprojected the population of Yorkshire and Humberside by 0.5% of its 4.9 million. Thereafter, adjustments were made for net migration because the most contentious issue for local authorities in the region at this time was the set of assumptions made in the official projections about net migration. A comparison of net movement counts recorded by the National Health Service Central Register (NHSCR) with the 'official' net migration assumptions for mid-1989 to mid-1992 had revealed significant differences for the local authorities at a time when the region as a whole was experiencing a quite dramatic change of fortune (Figure 7.2) in its net migration balance.

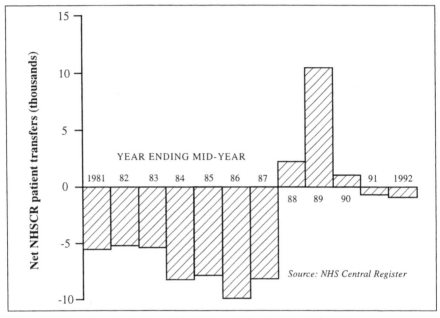

Figure 7.2. Net migration balances, Yorkshire and Humberside, 1981-92

The adjustment procedure for population projections for each local authority area can therefore be summarized in two stages as follows:

(1) application of revised population base factors (mid-1991 estimate divided by mid-1991 projection) to existing projections for 2001 and 2006 by age and sex, on the assumption that the error in the population base remains constant over the period; and

(2) adjustment of the net migration assumptions by:
  (a) summing the all age net migration assumptions for males and females for individual years over the projection period;

(b) summing the recorded NHSCR net migration balances over the first three years of the projection period, 1989-92;
(c) recalculating net migration assumptions for the remaining years of the projection period (1992-2001 and 1992-2006) using different assumptions about the range of possible futures; and
(d) subtracting the new revised assumptions from the original assumptions to give a set of all age adjustments to the population totals in 2001 and 2006.

The methodology then used to convert populations to households and to derive a revised set of household projections involved three stages as follows:

(1) subtraction of the institutional populations from the total population projections to produce a private household population projection for each quinary age group and sex;
(2) application of marital status factors – the proportion of a given population group that are either single, married or widowed /divorced – to male and female population projections by quinary age group; and
(3) application of headship rates – the proportion of members of a population group defined by age, sex and marital status that are heads of specific types of households – to corresponding projected population groups.

The detailed methodology for updating the population projections is spelt out in Stillwell and Gore (1993) and for household projections in Gore and Stillwell (1994).

7.4.2 Modelling Framework for Small Area Projections

A large range of different techniques for multi-regional population projection exists (Rees 1989) and the West Yorkshire model brought together a multi-regional cohort survival model with a set of spatial interaction models that link migration to changes in the housing stock so that possible impacts of housing developments can be assessed. As with every projection model, it is necessary to specify the spatial units, age groups, sexes, the projection time interval and the components to be included. In the West Yorkshire model, wards were the principal geographic units because the maximum number of reliable data sets existing at this scale. Projections are generated for males and females but also for total persons. A single year of age classification was adopted because of the importance of detailed age specification, particularly amongst younger age groups and because it was essential for the age group interval to match the projection period interval of one year. The accounting model has the following mathematical form for each gender and age cohort:

$$P_{ag}^w(t+1) = P_{ag}^w(t) - D_{ag}^w(t,t+1) - E_{ag}^w(t,t+1) + I_{ag}^w(t,t+1) \\ - Out_{ag}^w(t,t+1) + In_{ag}^w(t,t+1) + B_{0g}^w(t,t+1) \tag{7.1}$$

where:

$P_{ag}^w(t+1)$ is the population of ward w of age a and gender g at the end of the projection interval;

$P_{ag}^w(t)$ is the population at the start of the interval;

$D_{ag}^w(t,t+1)$ are the deaths estimated to occur during the time interval;

$E_{ag}^w(t,t+1)$ and $I_{ag}^w(t,t+1)$ are the emigrants and surviving immigrants;

$Out_{ag}^w(t,t+1)$ and $In_{ag}^w(t,t+1)$ are the internal out-migrants and surviving in-migrants; and

$B_{0g}^w(t,t+1)$ are the surviving births occurring during the time period.

The components of population change incorporated in the model therefore included those produced by mortality, emigration to and immigration from areas outside the UK, migration to and from elsewhere in the UK and fertility, with the internal migration components requiring a separate estimation model. The seven steps in the ward-based projection model are as follows:

(1) multiply start-of-projection-period populations for each ward disaggregated by sex and single year of age by non-survival probabilities to generate the number of deaths which are then subtracted from the start population;

(2) multiply the start populations by emigration and survival probabilities to produce the number of surviving emigrants which are then subtracted from the start population;

(3) multiply the start populations by immigration and survival ratios (previous immigration flows divided by destination populations) which are added to the start populations;

(4) apply inter-district migration probabilities to the district start populations to yield the number of inter-district migrants which are shared out by wards using proportions derived from the census for both in-migration and out-migration to give the numbers of surviving in-migrants and out-migrants;

(5) use spatial interaction models to project the 'turnover' migration between wards (within the existing housing stock), the migration to wards with new housing and the migration out of wards where houses have been demolished. To exemplify, turnover migration, TM^{wz}, from ward w to ward z is projected on the basis of an origin-constrained spatial interaction model, calibrated using 1991 Census Special Migration Statistics ward data, as follows:

$$TM^{wz} = A^w O^w W^z \exp(-\beta^v d^{wz}) \qquad (7.2)$$

where

A^w is a balancing factor for ward w;

O^w is the total turnover migration originating from ward w;

W^z is the attraction of ward z to in-migrants, measured by the number of housing vacancies and the rate at which they are taken up;

d^{wz} is the distance (in km) between the centres of wards w and z; and
β^w is the ward origin-specific distance decay parameter.

Whereas a production-constrained model is also used for demolition migration, migration to wards with new housing requires a destination-constrained model. The projected total migration flows are computed as the number of surviving migrants in each case and the surviving out-migrants from and in-migrants to each ward are summed and then disaggregated by age and gender using model age profiles computed from the census. The number of surviving out-migrants are subtracted from the ward start populations and the number of surviving in-migrants are added;

(6) compute the births during the projection period and add these to the start populations; and finally
(7) allow the final populations of the first projection period to become the start populations of the next projection period.

Each of these stages required a substantial amount of historical data, some of which were available from standard sources and some of which were estimated. Data estimation issues are considered in section 7.5.2. The three spatial interaction models used at stage (5) of the modelling procedure to project intra-district, inter-ward migration flows were based on the family of spatial interaction models developed by Wilson (1974). The turnover and demolition migration streams were modelled using origin (out-migration) and destination (in-migration) constrained models whilst the new housing migration stream was modelled by a destination (in-migration) constrained model (Rees 1994) using a software package called IMP (Stillwell 1984).

7.5 Data Issues

7.5.1 Assembling Data for Local Authorities in Yorkshire and Humberside

Data collection is one of the most time consuming and critical aspects of applied population research. In the case of the regional study, it was necessary to assemble data from various sources for the eleven local authorities that constituted the system of interest. Rebased mid-year population estimates for 1991 and 1989-based population projections to mid-2011 were provided by the DoE. The mid-1991 estimates were derived from the 1991 Census population counts and adjusted for underenumeration, for the definitional differences between the census count and the mid-year population estimate of the resident population, and for the natural change and net migration occurring between 21 April and 30 June 1991. The mid-year estimates for local authority areas thus include all members of the resident population, whatever their nationality, together with members of HM and non-UK armed forces stationed within the area as well as students assumed to be resident at their term-time address.

Particular problems in the adjustment methodology involved how to generate projections of institutional populations and students. The procedure adopted to estimate institutional populations was based on the assumption that the proportion of people living in medical and care establishments, defence establishments and prisons at Census 1991 would remain the same over the projection period. However, additional adjustments were included for North Yorkshire on the basis of additional information provided by the County Council, and in recognition of the area's special position with regard to armed forces personnel. The method of student projection, outlined in more detail in Gore and Stillwell (1994), was based on a breakdown of the student population into categories of *imported* (those from elsewhere studying in the region), *locally based* (those living and studying in the region) and *visiting* (those who neither live or study in the region but who happened to be enumerated there on census night). Different growth rates for each category of students were used based on admission forecasts from several of the higher education institutions in the region.

The 1989-based net migration assumptions involving projected totals of out-migration, in-migration and net-migration for each local authority and the rest of England by broad age group, together with all age assumptions of net migration for each local authority and the rest of the world, were also provided by DoE. It was fortuitous that the School of Geography at the University of Leeds had been purchasing from OPCS for several years data on patient transfers between Family Health Service Authorities (FHSAs) recorded in the NHSCR. The project therefore had access to a consistent set of patient reregistration data from 1975 to 1992 which had been carefully analysed by the authors in previous research projects (Stillwell *et al.*, 1992).

Whilst data on marital status and headship rates for 1991 was extracted directly from the 1991 Census, the Building Research Establishment (BRE) supplied the headship rates from the previous four censuses, the mid-year headship rate estimates from 1981 to 1990, and the 1989-based marital status and controlled headship rate projections for all the local authorities concerned. The BRE also supplied the mid-year household estimates from 1981 to 1990 and the 1989-based household projections to mid-2011 by household type. All the data was held and manipulated in Excel spreadsheets.

7.5.2 Building a Population Information System for West Yorkshire

The conceptual structure of the West Yorkshire 'model' (Figure 7.3) indicates that the projection procedures outlined in section 7.4.2 required significant amounts of information. Four databases made up the information system and contained data that could be viewed by the user in table, graph or map form. These databases were referred to as the raw, estimated, scenario and projection databases. The software programs that constituted the projection model took data from the first three databases and output data to the last database containing the projections of population and other variables.

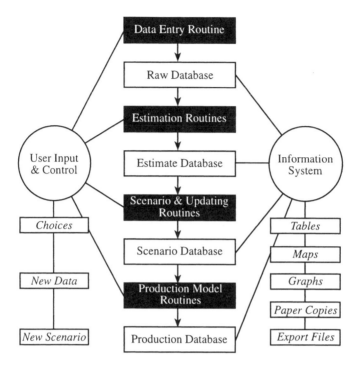

Figure 7.3. The structure of the West Yorkshire model

The *raw database* contained the demographic and other statistics about wards or districts in their original form. In order to project small area populations, it was necessary to estimate the values of the component variables for a base year. These estimates were then held in the *estimate database*. One example component is the base populations for wards. Prior to the release of 1991 Census data, it was necessary to use the 1981 Census populations for electoral wards as the raw data and to roll forward these values to give estimates for any particular base year. The methodology for doing this is outlined in Rees (1994) where an evaluation of the rolled forward estimates against mid-1991 estimates is also reported. A general review of small area estimation methods is provided by Simpson *et al.*, (1997). The West Yorkshire model results constitute the single 'cohort-survival' case in this analysis. The updating procedure involved generating population estimates by single year of age and sex for each ward, summing up the ward populations over each district and then adjusting the ward populations using iterative proportional fitting (IPF) techniques to ensure that the sum of the populations of each ward was consistent with the 'official' population estimate for the corresponding district and an estimate of the total ward population based on the electoral register. Raw ward population totals from the 1991 Census were incorporated with the same updating procedure as soon as the data became available.

The *estimate database* also contained other information required by the projection routines: information about fertility, mortality, and both emigration to and immigration from abroad. In the case of fertility, for example, the estimate database contained ward-based total fertility rates (TFRs) summarizing age-specific fertility and indicating the number of children that a woman would have were she to experience current age-specific fertility rates throughout her life. Life expectancy was the indicator used to report mortality in each ward, and as with the TFR, appreciable differences were observed between wards, particularly in certain inner city areas compared with outer suburban and rural areas. The estimation of ward level emigration and immigration rates was problematic because of the unreliability of available data and involved using the national age-specific migration statistics derived from the International Passenger Survey (IPS), national mid-year populations by single year of age and sex, and model migration rates from the standard model migration age function computed by Rogers and Castro (1981).

The third database was the *scenario database* which included key summary indicators used to lead the population projection. The definition of these indicators was left partly in the hands of the user since it is the user who must design the scenario for which the projections are to be generated. Thus, it was possible to make assumptions that the fertility rates would increase, mortality would decline and migration would continue at the same rate as in the past, for example. Alternatively, a 'no change' scenario can be used in which historical rates and probabilities are assumed to remain the same. Finally there was the *projection database* that contained projected age-specific populations for up to 25 years into the future, together with statistics that allow the user quick and easy comparison with the base year figures to see what changes are projected to occur under the assumptions used in the scenario. An example is provided in section 7.5.2.

7.6 Projections under Alternative Scenarios

7.6.1 Revised Regional Population and Household Projections

From the questionnaire survey and follow-up interviews with the local authorities in Yorkshire and Humberside, it became apparent that most had much less concern about the natural change assumptions contained in the official projections than with the net migration assumptions. Some local authorities expressed views that the external net migration assumptions, based on a very small sample, were unacceptable and their derivation needed to be clarified in future rounds of projections, but all the local authorities found considerable problems with the internal net migration assumptions. Thus, our adjustment methodology was designed to produce three new sets of population projections for 2001 and 2006 based on assumptions about net migration derived by calculating averages of historical net migration over the last three years (1989-92), five years (1987-92) and ten years (1982-92). These sets of assumptions allowed three alternative scenarios to be identified: under the assumption of future net migration at 1989-92 levels (scenario A), an additional 66,600 persons would be added to the regional population by 2001 whereas a further 77,800 would be added using the five year net migration levels (scenario B), and if the ten year net migration assumptions (scenario C) are adopted, the increment to the population would be much lower at around 28,500 people (Stillwell and Gore 1993). Under all three net migration scenarios, the population of the region would increase (Table 7.1), indicating that the official assumptions were too low.

Table 7.1. Population projections for 2001, local authority areas of Yorkshire and Humberside

1989-based	Scenario projection (000)	A (000)	B (000)	C (000)
Bradford	489	503	505	498
Calderdale	205	204	205	206
Kirklees	373	381	383	380
Leeds	-711	731	728	724
Wakefield	320	328	330	324
West Yorkshire	2097	2147	2151	2131
Barnsley	223	226	227	222
Doncaster	301	308	309	304
Rotherham	259	259	261	258
Sheffield	510	522	514	512
South Yorkshire	1294	1315	1311	1293
Humberside	875	884	889	877
North Yorkshire	782	768	776	776
Yorkshire and Humberside	5048	5115	5126	5077

Note: Figures rounded to nearest thousand
Source: Stillwell and Gore (1993)

In order to convert these population projections into household projections, the institutional and student populations must be projected and subtracted from the population base in order to apply revised marital status factors and headship rates to the remaining population. Separate projections of student households were then added back into the household projections at a later stage. This procedure is different from the 'official' methodology which treats students as though they were part of the resident population. It was then necessary to estimate the numbers of single, married and widowed/divorced people, recognizing the social changes in living arrangements that had been occurring. The adjustment procedure makes use of the 1989-based marital status factors for males and females in quinary age groups except where there was a considerable difference between the values in the 1991 Census and the 1989-based projections across several local authorities. Differences were found to be significant for males aged 20-24, 25-29, 30-34, 75-79, 80-84 and 85+, and for females aged 20-24, 25-29, 75-79 and 85+. For these groups, the marital status proportions were projected on the basis of extrapolation of past trends as illustrated by the examples of males aged 3-34 and females aged 75-79 in Bradford (Fig. 7.4). The probability of males aged 30-34 being single or widowed/divorced increases to 2006 whilst the probability of being married declines. The probability of women aged 75-79 being married increases as the probability of being single reduces.

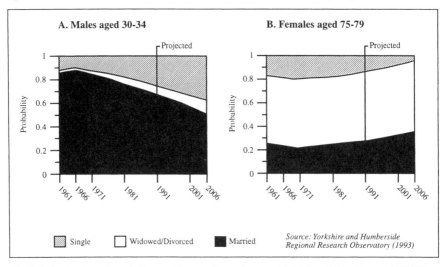

Fig 7.4. Projection of marital status factors for males aged 30-34 and females aged 75-79 in Bradford

A similar comparison of headship rates by broad age group from the 1991 Census with those used in the 1989-based household projections indicated that the most important differences were for married and widowed/divorced males and females aged 15-19, married males and females aged 20-24, widowed/divorced males aged 20-24 and 25-29, single males aged 85+ and married females aged 85+. The projected headship rates for these ten categories were revised in line with extrapolated trends whilst the official headship rates were used for the remaining categories.

The overall effect of using these revised marital status factors and headship rates was to bring the projections of households downwards and Table 7.2 presents the household projections for mid-2006 using the revised assumptions for each of the three net migration scenarios. In each case, the scenario projections A, B and C for the region as a whole were lower that the 1989-based household projections whereas our revised projections of population suggested that the official population projections were underestimated. These rather paradoxical results occurred primarily for two reasons: firstly, the separate treatment of students which, as reported elsewhere (Gore and Stillwell 1994), generated far fewer households than when such people remain in the private household population; and secondly, the effect of revised marital status and headship rates, which suggested that the trend to smaller households would not occur at as fast a pace as officially projected.

7.6.2 Alternative Ward Population Projections

Scenario design is a difficult task and frequently requires a good deal of local knowledge about the populations within the system. Will migration rates for the wards of one district all remain the same in future? Will ward migration rates all change by the same amount? What might be the changes in migration rates experienced by different wards? The answers to these types of questions focused only on the migration component are frequently not straightforward but the existence of a computerised system facilitates the process by allowing different scenarios to be run and the results evaluated and compared. Furthermore, the existence of a model-based projection system also makes it possible to assess scenarios that are policy-related.

Table 7.2. Household projections for 2006, local authority areas of Yorkshire and Humberside

	1989-based projection (000)	Scenario A (000)	Scenario B (000)	Scenario C (000)
Bradford	206	207	209	203
Calderdale	87	85	87	87
Kirklees	159	156	157	155
Leeds	313	313	311	308
Wakefield	137	138	139	135
West Yorkshire	901	900	904	888
Barnsley	93	90	90	87
Doncaster	127	130	131	127
Rotherham	109	104	105	103
Sheffield	227	223	514	215
South Yorkshire	555	547	544	531
Humberside	378	365	368	359
North Yorkshire	343	331	333	333
Yorkshire and Humberside	2158	2142	2148	2112

Note: Figures rounded to nearest thousand
Source: Stillwell and Gore (1993)

Two example scenarios from the West Yorkshire model are identified here for illustration, based on Bradford wards and taking 1990-91 as the base year. The first (scenario 1) assumed that benchmark rates continue unchanged to 2000-01 and that no new housing would appear in the city over the decade. The second (scenario 2) assumed that fertility rates in the wards with above-replacement fertility would decrease steadily and that fertility would stay constant elsewhere, mortality would decline at 2% per year and out-migration from Bradford would double over the decade. In addition to these demographic trends, there would be a major new suburban housing programme which added 500 new houses per year in each of the wards of Craven, Ilkley, Rombalds and Worth Valley, thereby increasing Bradford's housing stock by 10% over the projection period.

The results of the scenarios are presented in Figure 7.5 in terms of the percentage population change projected for each ward in Bradford between 1991 and 2001. The scenario 1 projections, with no policy 'driver', resulted in quite appreciable population gains in the wards of central Bradford and declines in some of the outer suburban and semi- rural wards. This contrasts with the projections of scenario 2, where three of the outer wards with substantial amounts of new housing were among the wards with most population change. Both these sets of projections suggest that Bradford's inner wards were destined to experience substantial population growth when the demographic rates are assumed to continue unchanged. Whilst this may be a realistic assumption, the end result implied that further housing would be required to cater for the population change and would necessitate a response from the Housing Department to avoid chronic overcrowding of dwellings in these areas, several of which are characterised by having concentrations of non-white ethnic populations.

7.7 Conclusions

In this chapter, we have outlined two case studies of applied population geography 'in action' for public sector authorities. What conclusions can be drawn and what lessons have been learnt from our experiences? Firstly, there are a lot of positive dimensions to undertaking applied research of this type. The stimulus of providing information and tools to support decision-making in the real world of local authority planning at urban and regional scales is important and the development of products which are acceptable to professional planners is rewarding. The challenge of producing solutions to difficult problems where data sets are incomplete, or where existing methods are not appropriate, demands original and innovative thinking. The development of procedures that are not necessarily conventional methods found in textbooks on spatial demography or population geography is creative. The data sets that are assembled in the course of conducting the project have much potential for further research that would provide further insights into what is happening in the local region.

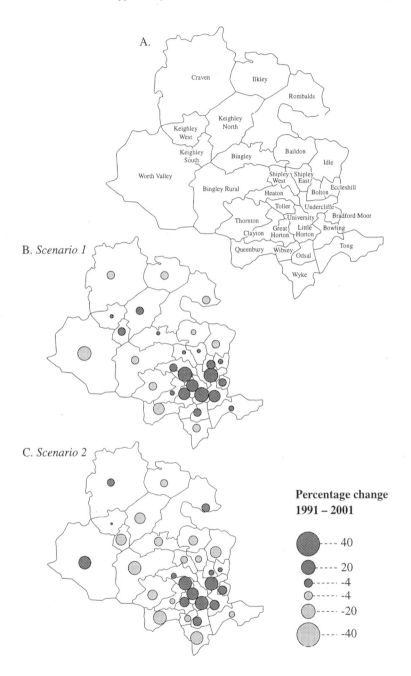

Figure 7.5. Projected changes in population of Bradford wards under different scenarios, 1991-2001

On the other hand, contract research does have its difficulties. The managerial expertise and time required to organise and conduct projects with a number of different clients is considerable. Most academics have to fit their consultancy work around their teaching, administrative and other research activities, which means a lot of work has to be done out of normal working hours. It is very important to ensure that there is clear understanding and agreement between the research team and the clients, from the outset, about what will be the outputs of the research. Our experience in both the projects described above suggests that local authority officers can be very demanding in terms of what they want the consultants to achieve, sometimes assuming that they can change the original specification without recognition of the implications. Inevitably there are problems of language communication between consultants and clients when the project involves complex modelling techniques and the interpretation of mathematical formulae, but there is also the added problem that, in the case of the West Yorkshire model, for example, local authority officers need quite extensive training in order to use and update the information system with confidence.

'Doing' applied population geography therefore has its benefits and its shortcomings. It also has its frustrations. One example of this is that the time and resources that are available for this type of work are rarely sufficient to do the job properly. In the case of population projections for RPG, it became very clear that what was really required by the local authorities in Yorkshire and Humberside was not a projection updating procedure but a full multi-regional population projection model linked to a regional model for household projections which would generate outputs that could be converted into projected demand for dwellings. Similarly, the West Yorkshire population projection model had enormous potential for improvement and further development as a decision support system for land use, transport or employment planning, for example.

As a final conclusion, it is perhaps worth reminding ourselves of the context in which the studies undertaken and the tools developed by consultants are used. Planning or policy decisions at the regional or local scale, particularly on matters relating to land allocation for housing, are frequently made for political reasons. It is therefore very important to acknowledge that population and household projections systems of the type we have described in this chapter should not attempt to produce a single set of 'optimum' projections since agreement between local authorities is unlikely to be forthcoming; instead, the important contribution of the projection exercise is to generate a suite of alternatives, thereby giving the politicians the opportunity for debate and compromise.

References

Boden P., Stillwell J.C.H. and Rees P.H. (1992) Internal migration projection in England: the OPCS/DoE model examined, Chapter 14 in Stillwell J.C.H. and Congdon P. (eds.) *Migration Models: Macro and Micro Approaches*, Belhaven Press, London: 262-286.

Breheny M. and Hall P. (eds). (1996) *The People – Where Will They Go?*, Town and Country Planning Association, London.

Capron D. (1994). How good are the subnational projections as forecasts?, *Population Trends*, 76:46-49.

Capron D. and Corner I. (1990) Subnational population an household projections by Central Government, in *OPCS Occasional Paper 38*, BSPS Conference Papers 1990, HMSO, London: 55-63.

Corner I. (1992) A technical summary of the DoE household projection model, *Building Research Establishment Report N57/92*, Watford.

Champion T., Fotheringham S., Rees P.H., Boyle P. and Stillwell J.C.H. (1998) *The Determinants of Migration Flows in England: A Review of Existing Data and Evidence*, Report for the Department of Environment, Transport and the Regions, Department of Geography, University of Newcastle, pp. 200.

DETR. (1998) Website: http://www.planning.detr.gov.uk/consult/subnatproj/index.htm

DoE. (1991) *Household Projections England 1989-2011: 1989-based Estimates of the Numbers of Households for Regions, Counties, Metropolitan Districts and London Boroughs*, Housing Data and Statistics Division, DoE, London.

DoE. (1992) *PPG 12, Development Plans and Regional Planning Guidance*, DoE, London.

DoE. (1995) *Projections of Households in England to 2016*, HMSO, London.

Gore T. and Stillwell J. (1994) Yorkshire and Humberside household composition and projections to 2006, *Working Paper 94/1*, Yorkshire and Humberside Regional Research Observatory, Leeds, pp. 68.

MVA. (1996) *England Sub-national Projections: Model Dynamics*, Report for the Office for National Statistics, MVA Systematica in association with London Research Centre.

OPCS. (1991a) *Subnational Population Projections: Population Projections by Sex and Age for Standard Regions, Counties, London Boroughs, Metropolitan Districts and Regional and District Health Authorities of England from Mid-1989*, Series PP3 no 8, HMSO, London.

OPCS. (1991b) Making a population estimate in England and Wales, *OPCS Occasional Paper 37*, OPCS, London.

OPCS. (1993) *Revised Final Mid-1991 Population Estimates for England and Wales and Constituent Local and Health Authorities*, OPCS Monitor PP1 93/2, HMSO, London.

OPCS. (1995) *1993-based Subnational Population Projections, England*, OPCS Series PP3 9, OPCS, London.

Rees P.H. (1989) Old model faces new challenge: a review of the state of the art in multistate population modelling, *Working Paper 531*, School of Geography, University of Leeds.

Rees P.H. (1994) Estimating and projecting the populations of urban communities, *Environment and Planning A*, 26: 1671-1697.

Rogers A. and Castro L. (1981) *Model Migration Schedules*, RR-8130, International Institute of Applied Systems Analysis, Laxenburg.

Simpson S., Middleton L., Diamond I. and Lunn D. (1997) Small area population estimates: a review of methods in Britain in the 1990s, *International Journal of Population Geography*, 3(3): 265-280.

Stillwell J.C.H. (1984) IMP2: A computer program for inter-area migration analysis and projection - user's manual, *Computer Manual 12*, School of Geography, University of Leeds.

Stillwell J.C.H. and Gore T. (1993) Yorkshire and Humberside mid-year population estimates and projections to 2006, *Working Paper 93/4*, Yorkshire and Humberside Regional Research Observatory, Leeds, pp. 70.

Stillwell J.C.H. and Gore T. (1997) The adjustment of population and household projections for regional planning in Yorkshire and Humberside, in Borgegård L-R., Findlay A. and Sondell E. (eds.) Population, Planning and Policies, CERUM Report No 5, Umeå University, Umeå:223-241.

Stillwell J.C.H., Rees P.H. and Boden P. (1992) *Migration Processes and Patterns Volume 2 Population Redistribution in the United Kingdom*, Belhaven Press, London.

Van de Gaag N., Van Imhoff E. and Van Wissen L. (1997) Long term internal migration scenarios for the countries of the European Union, *EUROSTAT Working Paper 97/1*, Statistical Office of the European Communities, Luxembourg.

Van Imhoff E., Van Wissen L. and Spiess K. (1994) *Regional Population Projections in the Countries of the European Economic Area,* Swets and Zeitlinger, Lisse.

Wilson A.G. (1974) *Urban and Regional Models in Geography and Planning,* Wiley, London.

Wilson T. and Rees P.H. (1998) Look up tables to link 1991 population statistics to the 1998 local government areas, *Working Paper 98/5*, School of Geography, University of Leeds, Leeds.

Yorkshire and Humberside Regional Research Observatory. (1993) *An Analysis of the 1991 Census Results for Population and Household Projections in Yorkshire and Humberside,* Final Report for the Yorkshire and Humberside Regional Planning Conference and the Department of the Environment, ReRO, Leeds, pp. 259.

8 Applied Spatial Interaction Modelling

Graham Clarke and Martin Clarke

8.1 Introduction

The aim of this chapter is to reflect on progress with spatial interaction modelling in a wide range of commercial environments and to focus on how applied work has fed back some important challenges to make spatial interaction models work more effectively. Such applied models build on thirty years of academic model development beginning with Wilson's derivation of the model through entropy maximisation (Wilson 1967, 1970). We briefly review theoretical progress with this model in section 8.2. In section 8.3 we turn our attention to applied modelling and discuss why such models have been so successful in a commercial environment and how they are able to address a wide variety of client requirements. In section 8.4 we discuss the increasing availability of better quality data which enables calibration to be undertaken more effectively. In turn, the availability of better data can shape the form of the models themselves (through greater disaggregation), although it quickly becomes apparent when dealing with applied problems that models are 'context' dependent, usually requiring fine tuning to work in specific sectors. We thus also explore the theoretical extensions to the models based on commercial applications and the data they bring. A particular interest of ours, and essential in a commercial context, is the ability to demonstrate that spatial interaction models 'work'. This is a subject that has been largely ignored by the academic community except in the narrow technical sense of calibration. We shall discuss this issue in section 8.5. In section 8.6 we go through a typical application of the models, showing how a user would interface with the information system to analyse the impacts of the changes they simulate on the computer. Concluding comments are given in section 8.7.

Although applied work has produced models with are undoubtedly richer, as we mention above such work also demonstrates that spatial modelling is highly context specific. There is no simple or standard approach, which has implications for the development of 'off-the-shelf' products. Many vendors of computer packages (such as GIS – see Hirschfield *et al.,* chapter 17, and Fotheringham *et al.,* chapter 15) have now introduced interaction models as part of their analysis routines. However, as we discuss elsewhere (Benoit and Clarke 1997) these provide very simple aggregate models with only one or two parameters. These can be of little practical use in dealing with the complexity apparent in most applied situations. We hope to demonstrate some of this complexity in the remainder of the chapter.

8.2 The Spatial Interaction Model

Our starting point for the review of spatial interaction models is Wilson's entropy-maximisation derivation at the end of the 1960s (Wilson 1967, 1970). His work produced the mathematical form of the model still in common usage today. The model can be written as:

$$S_{ij} = A_i e_i P_i W_j^\alpha e^{-\beta c_{ij}} \tag{8.1}$$

where

$$A_i = 1 / \sum_k W_k e^{-\beta c_{ik}} \tag{8.2}$$

to ensure that

$$\sum_j S_{ij} = e_i P_i \tag{8.3}$$

The model can be used to estimate any type of flows in a city or region. For example, in a retailing context S_{ij} is the flow of expenditure from residential zone i to shops (shopping centres or a store of a particular kind) in j; e_i is the per capita expenditure of residents of i; P_i, the population of i; W_j, a measure of the 'attractiveness' of j, usually measured via floorspace as a proxy; c_{ij}, a measure of travel distance or cost between i and j. α and β are parameters.

A typical product of the flow model for retailing is revenue, D_j:

$$D_j = \sum_i S_{ij} \tag{8.4}$$

These models have been adapted to a wide range of application areas. They have been used to examine flows of people to shops, offices, work, schools, hospitals and even pubs and dry ski slopes! Indeed, Fischer and Getis (1999) remark that models of spatial interaction have been fundamental in regional science. A detailed history of the development of these models takes us well beyond the scope of this chapter. Interested readers can follow this development chronologically through key texts such as Wilson (1974), Fotheringham and O'Kelly (1989), Sen and Smith (1995), and Fischer and Getis (1999). We attempt below to summarise some key developments of the models but concentrate our efforts in the remainder of the paper on developments largely driven by commercial applications.

First, the retail model has been widely used as a test-bed to explore a wide range of theoretical ideas. These include the extension of the models to include dynamics and different methods for testing of alternative methods of determining equilibrium solutions (Harris and Wilson 1978, Clarke and Wilson 1983, 1985, Phiri 1980, Rijk and Vorst 1983a,b, Fotheringham and Knudsen 1986, Crouchley 1987 and Kantorvich 1992). Clarke *et al.,* (1998) provide a useful summary of progress with theoretical and empirical work on dynamic models.

Second, extensions have been made to the theoretical properties of the models. These include relatively straightforward experiments to find more realistic formulations of the major variables. Some of these will be detailed in section 8.4. In a few cases, however, new versions of the basic models were presented. Fotheringham (1983, 1986) for example, experimented with new variables in the model to address the issues of agglomeration and spatial competition. These are particularly important issues if individual stores rather than shopping centres represent the supply side. The agglomeration or *competing destination* variable allows consumer's utility to increase for comparison shopping by selecting a retail outlet in close proximity to other outlets. He also explored new cost terms in the model, correctly arguing that the simple assumption that costs of supply are solely a function of floorspace was unrealistic and that variables such as location rents were also important (see section 8.4 for more discussion on this and also refer to Fik and Mulligan 1990). Others have followed this line of research. Pooler (1994a, b) looked at similar ways of developing interaction models with new 'extended' variables, whilst Guldmann (1999) develops and tests a model containing both competing destination and intervening opportunity factors (see also Roy 1999).

The final broad set of developments are labelled 'technical' by Fischer and Getis (1999). They refer to work on designing new models of interaction based largely on inductive modelling. The idea here is to find models that produce perfect fits between model outputs and real interaction data. New tools such as genetic algorithms and neural nets have been extensively used to search for these new models. Key works include Openshaw (1993), Diplock and Openshaw (1996), Fischer *et al.,* (1999). The difficulty with these models is they become less reliable for predictive work. That is, although they fit existing data well the same form of the models may not fit new data very well. The proof that these models work as well as traditional interaction models in predictive work remains to be presented.

In the rest of the paper we focus on commercial applications of the retail interaction model, although we use retail in the broadest sense of the word. Examples include high street retailing, out-of-town retailing, financial services, cars, restaurants and petrol stations. We build on the experience gained through GMAP – a company that grew out of the School of Geography, University of Leeds to exploit the commercial applications of this sort of modelling. By 1997 GMAP employed over 110 graduates and enjoyed a turnover of £5-6 million.

8.3 Client Needs

Many of the chapters in this book refer to a single client and provide a detailed case study of that application. In this chapter, we wish to draw on the requirements of many clients of the University of Leeds and GMAP. These include ASDA, Sainsburys, Esso, Toyota, Ford, Halifax, Barclays, Thorn EMI, Storehouse and W.H.Smith. Although there are slight variations in what these individual organisations want from this collaboration, we can generally group these requirements together. Thus for banks, oil companies, car manufacturers and so on channel management and network planning are key strategic and operational issues. The relationship between sales and store (or dealer) location are well-known (Birkin *et al.*, 1996). Therefore, any changes to the store network (investment in new outlets, rationalisation, mergers and acquisitions) need to demonstrate acceptable rates of return. The crucial variable in the equation is the ability to forecast or estimate the impacts on revenues of any strategic investment or disinvestment. The inability to forecast accurately leads to increased risk and potentially missed opportunities.

Spatial interaction models estimate revenues and flow patterns between origin (demand) zones and shopping centres or outlets. Both sets of information are useful for retail organisations, given the lack of high quality information on consumer activities. The first requirement most organisations have is one of benchmarking actual revenues against 'potential'. That is, a retailer may know how much a store is taking but is that a good or bad performance given the characteristics of the catchment area (number of affluent customers, level of competition etc). We know of one organisation that changed the way it rewarded its top performers based on this type of analysis. The rules for winning the trip to Kenya at Toyota changed from the measurement of annual total sales improvement to one based on sales performance against model estimates. This allowed the retailer in say Truro, Cornwall to compete fairly with the top London dealer. Although the Truro dealer sold only 5-6 new unit sales per year this was more than predicted whereas the Croydon dealer with 20-30 new unit sales per year should actually have sold more!

Another important indicator produced by the models is small-area market shares. Most organisations know their national market share, and many will be able to disaggregate this to the regional level (and there are often market intelligence reports to help here). However, very few know their market shares at the local level (Birkin *et al.*, 1996). Mapping local market shares provides an immediate snapshot of retail performance. It shows areas where the organisation has few customers. These spatial gaps in the market become possible new store locations for any firm that is contemplating expansion. Having understood the existing performance of individual branches or outlets the models are then used in 'what-if?' fashion. Each one of the variables given in the models can be modified and the models re-run to estimate the impacts. An example of this will be provided in section 8.6.

Applied Spatial Interaction Modelling   141

8.4 Data Issues and Model Developments

8.4.1 Introduction

It is fair to say that lack of high quality real-world data in the past has been a major barrier that has prevented more commercial applications of spatial interaction models. The proliferation of retail data has been an important contributor to the new era of applied modelling (Clarke and Wilson 1987). On the demand side there are now many geodemographic products (Birkin 1995, Openshaw 1995, Openshaw and See, this volume) which allow the specification of very detailed demand portraits. Data on the supply side is still perhaps the weakest area. Directories are often incomplete and in most circumstances some form of data audit on store locations is necessary. Flow data (needed for calibrating models) is increasing in availability. Some markets, such as car retailing and financial services, have always had good interaction data. Other retailers are increasing their knowledge of customer flows through store loyalty cards. We shall explore data issues in relation to the theoretical development of the models below.

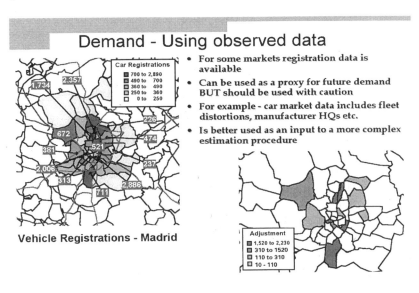

Figure 8.1. Car data in Madrid   **Figure 8.2.** Adjustments to the car data

8.4.2 Demand

Measuring small area demand for products or services serves a number of useful purposes for retailers. First, it allows a quantification of available expenditure by small area, which can be aggregated to produce total demand estimates for store catchments or regions. Estimating demand at the small area level usually requires combining demographic lifestyle market research and client data. Some markets are easier than others, particularly where organisations collectively decided to syndicate their sales or registration data to produce full market size estimates. For example, in most European countries, car manufacturers combine their vehicle registration data by small area and this historical data can be used as proxy for future demand. Figure 8.1 maps out the historical car registrations for the city of Madrid in Spain for 1997 by commune. However, in the car market it is often the case that these historical registrations contain distortions in areas where there are large fleet operators, manufacturers' headquarters or daily rental companies. To adjust the registration statistics to reflect the 'true' retail market requires the elimination of these distortions. Figure 8.2 shows the adjustments that had to be made to the Madrid data to create a more realistic picture of available demand. Other sectors where sharing of sales data is common place include the retail financial service industry. For example, in the UK, CACI operate a syndicated market sized service for financial service companies covering both mortgage and savings products. Most UK financial services organisations participate in this service supplying CACI with their sales data by postal sector. CACI then acts as a broker to produce market size estimates for each postal sector which are then supplied back to the participating organisations.

Figure 8.3. Demand estimations in NW England

Where no direct data is available some form of estimation of market size is required. Most particularly this involves applying market research data directly or indirectly to small area population data to derive demand estimates. Many companies collect data on the social class and age profiles of their customers either through market research or through 'point of sale' data. In every European country census data is available at the small area level which will easily yield similar population profiles for small areas. The two sets of data can then be combined to yield demand estimates. For example, if we were attempting to generate estimates of food expenditure by small area in the UK the following approach could be used. From the Family Expenditure Survey it is possible to derive the average weekly expenditure by households in different social class groups (A, B, C1, C2, D, E: the so-called Jictnar classification). From the 1991 census data sets we can obtain the number of households by social class for each postal sector. A simple multiplication of average weekly expenditure by population total for each social class group will then generate an estimate of available demand. Figure 8.3 shows the results of this exercise for the North West of England and in this case we have overlaid the locations of the principal supermarkets in this region (the small squares on the map). It is obviously possible to refine this analysis to take account of other factors such as household size, number of children and so on but the principles remain the same.

Thus, it is apparent that data rich systems provide as many challenges as data poor ones! The challenge comes not from the need for estimation but from the need to assimilate. Figure 8.4 shows the typical complexity of the alignment process on the demand side.

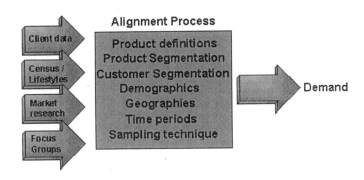

Figure 8.4. The Alignment process

In some markets the demand for goods or services might be related to availability of supply outlets providing that product or service. For example, Figure 8.5 shows the average number of ATM transactions per account for postal sectors within north and west Yorkshire. We also overlay on this map the location of branches without ATMs, branches with ATMs and remote ATMs. It can be clearly seen from this figure that ATM usage varies considerably across the region but is strongly influenced by the availability of ATM machines. Other sectors that show significant amounts of elasticity of demand include recreation, leisure and sport.

As supply clearly drives demand in this instance, opening more outlets could genuinely increase demand not simply redistribute it. The same could be said of many markets in the entertainment industry (it has even been proved for health – see Clarke and Clarke 1997). This has been discussed in the literature, but often in a theoretical sense (see Pooler 1994a,b, Ottensmann 1997). Demand elasticity can be included but should be done in a constrained rather than unconstrained way.

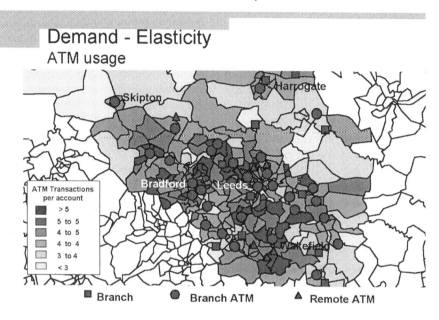

Figure 8.5. Observed demand at ATM branches in West Yorkshire

8.4.3 Supply

As we saw in section 8.2, attractiveness is most often represented as size in interaction models. Applied research shows that the real world is often far more complicated. There have been numerous studies since the 1970s to investigate alternative attractiveness terms. Spencer (1978) introduced qualitative variables to capture centre or store design in attractiveness, whilst Pacione (1974) attempted to quantify retail structure which he labelled as the number of supermarkets, department stores and national banks. Harry Timmerman's group of researchers in the Netherlands have also investigated many more behavioural components in shopping models (see chapter 6).

Our own investigations have also offered some alternative attractiveness terms. The first relates to brand preference. A key question in any application is what effect does the name over the door have (the retail marque)? i.e. how does a square foot of Sainsburys compare with a square foot of Tesco? Data suggests that the answer is partly driven by presence of store but it also relates to quality of the brand – some retail names are simply more attractive to consumers (consider Harvey Nichols and other department stores of the same size for example). Given that brand preference relates partly to store presence it is not surprising that there is a regional variation to this brand preference (see Figure 8.6).

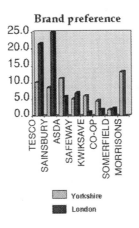

Figure 8.6. Brand preference for groceries in Yorkshire and London.

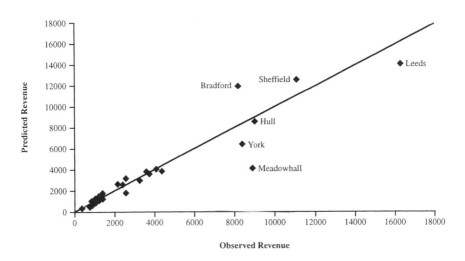

Figure 8.7. Scatter plot between observed and predicted centre revenues for the Yorkshire centres containing WH Smith Group stores

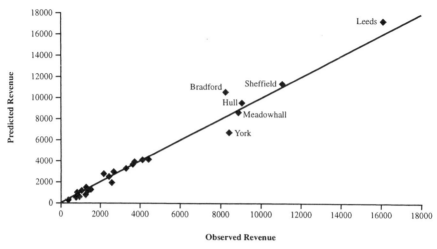

Figure 8.8. Scatter plot between observed and predicted centre revenues (using new attractiveness term), for the Yorkshire centres containing WH Smith Group stores

Eyre *et al.,* (2000) show the results of an investigation of results obtained within Yorkshire for a major UK high street retailer. Figure 8.7 shows that a conventional model (with size and brand preference as the major attractiveness terms) produced good results generally but showed some major outliers. The models seriously underpredicted the regional shopping centres such as Meadowhall and more 'attractive' centres such as Leeds and York. At the same time the models over-predicted the performance of older centres such as Bradford. By building an attractiveness term made up of a combination of size, quality of environment and presence of other major stores, they show that it is possible to build these factors into the model. Figure 8.8 shows the marked improvement on the fit of the model, especially in relation to the outliers.

Similarly, Fig 8.9 shows the variables that were found to be important in detailed GMAP modelling of the petrol station market.

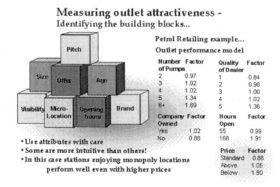

Figure 8.9. Outlet attractiveness for the petrol market

These kinds of study show that the supply-side variable is more likely, in reality, to take the following form:

$$W_j^{k\,\ln} = \omega_j^{k\,\ln} \sum_{j \in J} R_j^{kl} Z_j^k \gamma_{\Psi(j)}^{(j)kl\phi(j)n} \mu_j^{kl} \chi_j^{kl} \quad (8.5)$$

- $\omega_j^{k\,\ln}$ — Centre Performance
- R_j^{kl} — Size, Z_j^k — Brand → Store Performance
- $\gamma^{(j)kl\phi(j)n}_{\Psi(j)}$ — Store maturity
- μ_j^{kl} — Store agglomeration

Again, this demonstrates the difficulty of using very simple aggregate models in most applied situations (cf. Benoit and Clarke 1997).

8.4.4 Distance Deterrence

The distance deterrence term in spatial interaction models has often been measured as straight-line distance. The most widely debated question has been whether to use the exponential distance decay parameter or the negative power function (although see Taylor 1975 for many hybrid forms). Applications in the real world quickly highlight the problem of straight-line distance, especially where railway lines and rivers make accessibility difficult. Most commercial applications now use travel times that are provided by a number of private companies.

Another general issue is the definition of study regions. In real life customers travel in continuous geographic space and the imposition of study regions means some customers are always on the boundaries which leads to difficulties with cross-boundary flows. The power of modern computers means that boundary-free modelling is now possible and the problems of cross-boundary flows are limited to international borders.

Dealing with real clients usually means having to deal with a whole range of different products. The willingness of consumers to travel long distances will clearly vary by product. The usual solution to this problem is to disaggregate the deterrence term by product type. However, in some markets the data on interactions is so good that models can be calibrated by person type, product type etc. In the car market this leads to the calibration of a very detailed, highly parametised model (compare this to equation 8.1):

$$S_{ij}^{st} = A_i^{st} Q_{mi}^{st} O_i^{st} R_j^{st} W_{mj}^{st} e^{-B_i^{st} C_{ij}} \quad (8.6)$$

where:

- s - Segment
- t - Sales type
- m - Manufacturer
- A_i - Balancing factor
- Q_m - Observed registrations per head of population

W_j - Relative dealer attractiveness
O_i - Population
R_j - Floorspace
C_{ij} - Weighted drive time from zone i to dealer j
β - distance deterrence parameter

The complexity of modern trip making behaviour also means that models have to be flexible to handle new types of interaction.

8.5 Calibration and Model Accuracy

The success of this sort of modelling lies in its predictive power. In most markets the models need to be within 10% of real turnovers (when known) 80% of the time. Undertaking predictive experiments is probably the only way to 'prove' that models work. It is possible to use historical data to try and predict the present, although care has to be taken that all the dynamics in the background (competitor locations etc) are taken into account. More common is for clients to test the models by using contemporary data with 'missing data' (i.e. withholding sales data for a number of outlets and examining the model predictions for these outlets). Table 8.1 shows the results for nine 'missing' petrol stations withheld by the client in two regions of Italy. This is typical of model performance – six outlets are predicted very well, one reasonably and two are clearly outliers. The question now is whether there are local problems with these outlets (poor management, competitor sales campaigns etc) which would explain these results. In order to judge the quality of these results the client also needs to compare them with those done by alternative methods (see the discussion in Clarke and Clarke 1995).

8.6. Using the Model

In this section we describe how one major UK retailer uses a variety of business models to access the impact of potential changes to both it and its competitors' store configurations in terms of locations, floor space and store characteristics. The effects of simple or complex network changes may be evaluated, for example opening of new outlets, closure of existing outlets, outlet reinvestment and outlet re-branding. The model operates by simulating how customer patterns change for the network reconfigurations. The impacts upon sales and market share are evaluated and provide valuable information to the investment appraisal process. The business model allows the impacts of various scenarios to be viewed: by outlet, by geographical area, by brand and by product line.

We now proceed to demonstrate how a business scenario could be evaluated using this particular model. As an example we have chosen the Teeside region of NorthEast

England with suitably anonymised brand and outlet names. The first task is to assign the scenario a name and a description and this is shown in figure 8.10.

Table 8.1. Observed versus predicted sales volumes at selected petrol stations in Sicily

Petrol Station Location	Actual Volume	Predicted Volume	Perf Index
Corso Tukory	1,510	1,663	0.91
Piazza Giacchery	1,860	1,600	1.16
Via Uditore	1,002	1,050	0.95
Viale Regina Margherita	1,424	1,000	1.42
Corso Dei Mille	956	900	1.06

Petrol Station Location	Actual Volume	Predicted Volume	Perf Index
Viale Michelangelo	1,089	1,150	0.95
Via Aurispa	763	600	1.27
Via Sciuti	1,980	2,141	0.92
Niscemi	1,250	1,350	0.93

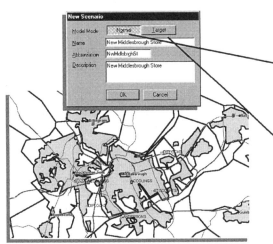

Figure 8.10.

We propose opening a new outlet under the Stacey's brand. The model requires us to assign a specific geographical location to the store (OSX and OSY in figure 8.11). We can then input details about the store including its size, the number of car parking spaces, the opening hours of the store, whether the store provides a petrol station facility, whether there are other retail adjacencies and so on. We can also close an existing store in the region. As shown in figure 8.12, we simply select a brand and store which is to be closed. We can also change the attributes of an existing outlet. For example, modelling the effect of a change in the physical attributes such as size and opening hours can be assessed and the effects of refurbishment can be simulated through alterations in the store attractiveness rating. The 'amends store' window from the business model is shown in figure 8.13. The next effect the business model can address is to simulate the impact of re-branding. The attractiveness of brands within the model may be altered to evaluate the effects of local marketing and any price promotions. This re-branding scenario may be applied regionally or to a complete chain across the country (see figure 8.14).

We now present a specific example in which we open a new supermarket in the Middlesborough Metropolitan Market. Figure 8.15 provides a summary of the scenario that we are about to examine. Figure 8.16 presents the model results showing the impact of the new Middlesborough store and projects a weekly sales revenue of £525,000 also indicating where the impacts of this new store opening will be felt. We can see that the biggest weekly reduction in revenues is at the Taylor's store in Thornaby but the biggest percentage reduction is at the Kelly's store in Middlesborough. Figure 8.17 illustrates the makeup of the new store's revenue by product line. Different combinations of product mix within the store could be evaluated to generate the highest total revenue or total margin for that store. Figure 8.17 also shows the results of the new store opening across the major superstore brands in this region. Because the Staceys brand had only one existing supermarket within this region, and at some considerable distance from the proposed new store, we can see that the new revenue generated by the store is entirely incremental and results in little cannibalisation for the one existing store already in the region. However, as the network of stores begins to increase the greater the likelihood of cannibalisation from the new store.

Not only does the model predict new and existing store revenue impacts it also allows an evaluation of the impacts at a local geographical level. Figure 8.18 shows the market penetration of available retail expenditure within surrounding postal sectors of the proposed new store. The model has determined precisely where the revenues will be generated from and this allows the supermarket group to undertake effective local marketing to ensure that the predicted revenues are realised.

Overall this type of business model allows for the evaluation of a wide range of potential scenarios simply and quickly. While it does not represent a replacement of traditional local surveying techniques it allows a large number of sites to be evaluated and a short list of sites to be generated that warrant further investigation.

Applied Spatial Interaction Modelling   151

Open a new outlet...

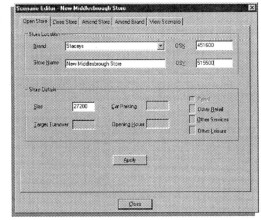

The Model can simulate the opening of a new outlet, of a particular brand.

The user is required to specify the location and attributes of the new outlet.

The Model uses this information to assess how attractive the outlet will be to customers.

Figure 8.11.

Close an existing outlet...

The Model can simulate the effect of closing a store.

For each closure, the user must select a brand and a store.

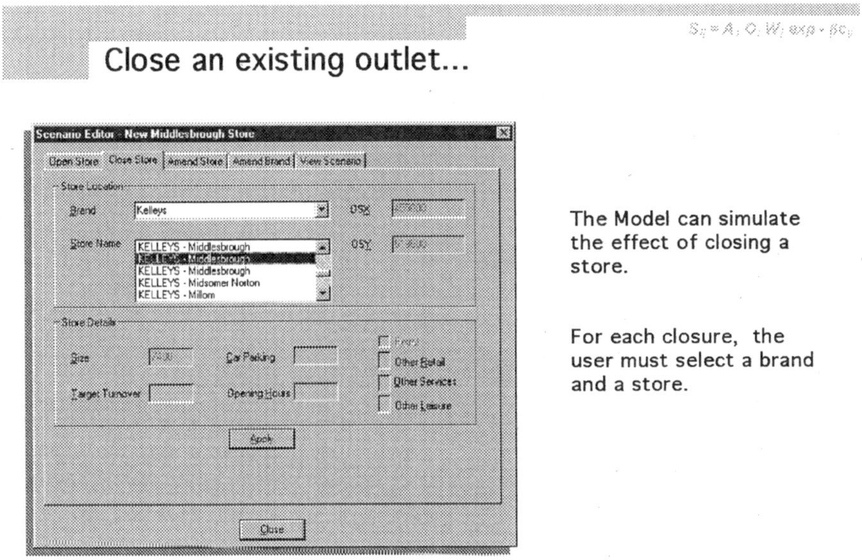

Figure 8.12.

Change the attributes of an outlet...

The effect of changes to outlets may be handled in a number of ways.

The effect of changing physical attributes - like size and opening hours can be assessed.

The effect of re-furbishments can be simulated through alterations in the 'store attractiveness' rating.

The effect of changing product-line mix may also be assessed.

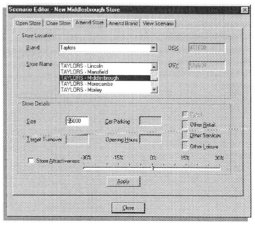

Figure 8.13.

Simulate the impact of re-branding...

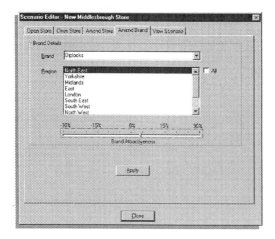

Like outlets, the attractiveness of brands may be altered to evaluate the effects of:

- Local marketing
- Price promotions

This re-branding scenario may be applied 'regionally' or to a complete chain.

Figure 8.14.

Applied Spatial Interaction Modelling 153

An example business scenario...

To demonstrate the operation of the Model a 'new outlet' is opened in a metropolitan market...

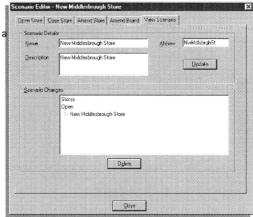

Figure 8.15.

Impact evaluation... store sales

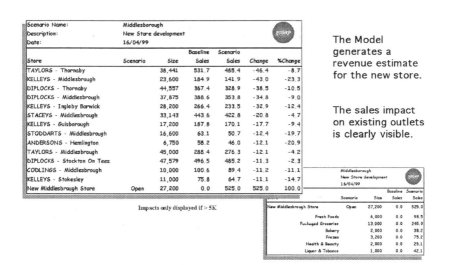

The Model generates a revenue estimate for the new store.

The sales impact on existing outlets is clearly visible.

Figure 8.16.

Impact evaluation ... brand totals

$S_{ij} = A_i O_j W_j \exp - \beta c_{ij}$

Scenario Name:	Middlesborough				
Description:	New Store development				
Date:	16/04/99				
Brand	Size	Baseline Sales	Scenario Sales	Change	%Change
Andersons	47,210	243.6	226.9	-16.7	-6.9
Codlings	30,000	285.8	265.9	-19.9	-7.0
Diplocks	150,011	1252.5	1168.0	-84.6	-6.8
Duleys	72,831	518.3	472.4	-45.9	-8.9
Gillies	24,000	219.4	211.1	-8.3	-3.8
Kelleys	178,800	1538.8	1390.9	-147.8	-9.6
Staceys	60,348	443.6	947.8	504.2	113.7
Stoddarts	30,600	113.5	98.8	-14.7	-13.0
Taylors	83,441	820.1	761.7	-58.4	-7.1
Whittams	39,208	113.6	105.3	-8.3	-7.3

The Model is able to illustrate the impact upon the total network.

What level of cannibalisation is likely?

Figure 8.17.

Impact evaluation..... the detail

$S_{ij} = A_i O_j W_j \exp - \beta c_{ij}$

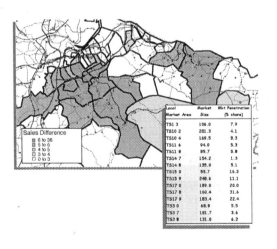

The Model user is able to evaluate a business scenario at the most local level.

What will be the sales and market share impact within individual local markets?

Figure 8.18.

8.7 Conclusions

It is undoubtedly the case that the experience of building models for the private sector has increased our knowledge greatly on the properties of spatial interaction models. Following the diagram bellow, they can also help turn data into intelligent information and action.

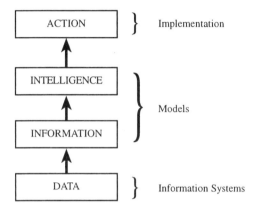

However, we argue that models are best used within a process that involves other components. Using models to reproduce existing data is a start but this alone is not enough. Interpretation of the processes is the key. Spatial interaction modelling should be seen as a set of methods and tools, which are context dependent. Models should be built around client problems not the other way round. We would argue that generalized off-the-shelf modelling packages are unlikely to address the critical issues in each and every unique market. This argument holds for modelling routines which are part of more generic GIS solutions (see Benoit and Clarke 1997 for more detailed arguments here). Every aspect of model design and calibration is context dependent. As well as modelling skills, the analyst must offer the client sector knowledge, ingenuity, imagination and problem-solving skills. Fortunately, there are many people in regional science that possess these strengths!

References

Benoit D. and Clarke G.P. (1997) Assessing GIS for retail location planning, *Journal of Retail and Consumer Services*, 4(4), 239-258

Birkin M. (1995) Customer targeting, geodemographic and lifestyle approaches, in P.A.Longley and G.P.Clarke (eds) *GIS for business and service planning*, Cambridge: Geoinformation, 104-149

Birkin M., Clarke G.P., Clarke M. and Wilson A.G. (1996) *Intelligent GIS: location decisions and strategic planning*, Cambridge: Geoinformation.

Clarke G.P. and Clarke M. (1995) The developments and benefits of customised spatial decision support systems, in P.A.Longley and G.P.Clarke (eds) *GIS for business and service planning*, Cambridge: Geoinformation, 227-245

Clarke G.P. and Clarke M. (1997) Spatial decision support systems for health care planning, in Cropper S. and Forte P. (eds) Enhancing health services management, O.U. Press, Buckingham.

Clarke G.P. and Clarke M. (2000) Trends in UK retailing and the implications for network planning, Working paper, school of Geography, University of Leeds.

Clarke G.P., Langley R. and Cardwell W. *(1998)* Empirical applications of dynamic spatial interaction models, *Computers, Environment and Urban Systems, 22(2),* 157-184

Clarke M. and Wilson A.G. (1983) The dynamics of urban spatial structure: progress and problems, *Journal of Regional Science, 21,* 1-18

Clarke M. and Wilson A.G. (1985) The dynamics of urban spatial structure; the progress of a research programme, *Transactions, Institute of British Geographers, 10,* 427-451

Clarke M. and Wilson A.G. (1987) Towards an applicable human geography: some developments and observations, *Environment and Planning A, 19,* 1525-1541

Crouchley R. (1987) An examination of the equivalence of three alternative mechanisms for establishing the equilibrium solutions of production constrained spatial interaction models, *Environment and Planning A, 21,* 861-874

Diplock G. and Openshaw S. (1996) Using simple genetic algorithms to calibrate spatial interaction models, *Geographical Analysis, 28,* 262-279

Eyre H., Clarke G.P. and Rees P.H. (2000) Retail interaction models in practice: refining the attractiveness term, working paper, School of Geography, University of Leeds.

Fik T.J. and Mulligan G.F. (1990) Spatial flows and competing central places: towards a general theory of hierarchical interaction, *Environment and Planning A, 22,* 527-549

Fischer M. and Getis A. (1999) New advances in spatial interaction theory, *Papers in Regional Science, 78(2),* 117-118

Fischer M., Hlavackova-Schindler K. and Reismann M. (1999) A global search procedure for parameter estimation in neural spatial interaction modelling, *Papers in Regional Science, 78(2),* 119-134

Fotheringham A.S. (1986) Modelling hierarchical destination choice, *Environment and Planning A, 18,* 401-418

Fotheringham A.S. and Knudsen D.C. (1986) Modeling discontinuous change in the spatial pattern of retail outlets: extensions of the Harris-Wilson framework with results from a simulated urban retail system, *Geographical Analysis, 18(4),* 295-312

Fotheringham A.S and O'Kelly M.E.(1989*) Spatial interaction models: formulations and applications*, Dordrecht, Kluwer.

Guldmann J.M (1999) Competing destinations and intervening opportunities interaction models of inter city telecommunication flows, *Papers in Regional Science, 78(2),* 179-194

Harris B. and Wilson A.G. (1978) Equilibrium values and dynamics of attractiveness terms in production constrained spatial interaction models, *Environment and Planning A, 10,* 371-388

Kantorvich Y. (1992) Equilibrium models of spatial interaction with location capacity constraints, *Environment and Planning A, 24,* 1071-1095

Openshaw S. (1993) Modelling spatial interaction using a neural net, in Fischer M. and Nijkamp P. (eds*) Geographical information systems, spatial modelling and policy evaluation,* Springer-Verlag, Berlin.

Openshaw S. (1995) Marketing spatial analysis: a review of prospects and technologies, in Longley P.A. and Clarke G.P. (eds) *GIS for business and service planning,* Cambridge: Geoinformation,

Ottensmann J.R. (1997) Partially constrained gravity models for predicting spatial interactions with elastic demand, *Environment and Planning A, 29*, 975-988

Pacione M. (1974) Measures of the attraction factor, *Area, 6*, 279-282

Phiri P. (1980) Calculation of the equilibrium configuration of shopping facility sizes, *Environment and Planning A, 12*, 983-1000

Pooler J. (1994a) A family of relaxed spatial interaction models, *Professional Geographer, 46*, 210-217

Pooler J. (1994b) An extended family of spatial interaction models, *Progress in Human Geography, 18*, 17-39

Rijk F.J.A and Vorst A.C.F (1983a) Equilibrium points in an urban retail model and their connection with dynamic systems, *Regional Science and Urban Economics, 13*, 383-399

Rijk F.J.A and Vorst A.C.F (1983b) On the uniqueness and existence of equilibrium points in an urban retail model, *Environment and Planning A, 15*, 475-482

Roy J.R. (1999) Areas, nodes and networks: some analytical considerations, *Papers in Regional Science, 78(2)*, 157-178

Sen A. and Smith T.E. (1995) *Gravity models of spatial interaction behavior*, Springer-Verlag, Berlin.

Spencer A.H. (1978) Deriving measures of attractiveness for shopping centres, *Regional Studies, 12*, 713-726

Taylor P. (1975) Distance decay in spatial interaction, CATMOG 2, Norwich.

Timmermans H. (1981) Multi-attribute shopping models and ridge regression models, *Environment & Planning A, 13*, 43-56

Wilson A.G. (1970) *Entropy in urban and regional modelling,* London: Pion.

Wilson A.G. (1974) *Urban and regional models in geography and planning*, Chichester: Wiley.

Wilson A.G. (1983) A generalised and uniform approach to the modelling of service supply structures, working paper 352, School of Geography, University of Leeds.

9 The Objectives and Design of a New Land-use Modelling Package: DELTA

David Simmonds

9.1 Introduction

This chapter considers the objectives, the design and the applications of the DELTA land-use modelling package, which has been developed by the author and his colleagues at David Simmonds Consultancy since the beginning of 1995.

It should be explained at the outset that DELTA is not intended normally to be used as a stand-alone system. It was conceived from the outset as the "land-use" component of a land-use/transport interaction model, with the transport element (and possibly the environmental element) being provided by other software. DELTA needs at least some transport inputs from a transport model to work at all, and to work as intended it needs to interact with a transport model over time. Discussion of the objectives and design of DELTA therefore has to take into account the likely characteristics of the transport models which could be used. The applications of DELTA need to be considered as joint applications of DELTA and the corresponding transport models.

The structure of this Chapter is as follows. Section 9.2 deals with the background to the development of DELTA. This deals with both the policy context – the possible opportunities for the application of a new land-use model – and the historical modelling context – the range of existing modelling approaches and packages against which DELTA has to be positioned. Section 9.3 considers the needs of DELTA's clients. Section 9.4 outlines the contrasting versions of the DELTA package implemented in the two applications examined. Section 9.5 fills in a little of the detail of the sub-models which make up those applications. Section 9.6 summarises some of the results obtained from those implementations. Section 9.7 outlines some research into the responses of practising planners to one of these models, and the way in which these responses and other work have been incorporated into user training. Section 9.8 draws some conclusions.

9.2 Land-use/Transport Models and the Development of DELTA

The development of land-use/transport models in planning, geography and regional science began in the late 1950s and early 1960s and was strongly influenced by 'systems thinking' (later formalised by, for example, McLoughlin, 1969, and Chadwick, 1971). This favoured formal models which could, to some extent, be turned into operational, numerical models using newly available electronic computing power. The most significant of the various early attempts was the 'Lowry model' (Lowry, 1964). This was not by any means the most sophisticated of the early models, but it was probably all the more influential because it took a simple, recognizable economic theory and applied it to the urban situation in a demonstrably practical manner. Something of an academic industry grew up focussed on the improvement and elaboration of the Lowry model – some of which promptly made it much less simple and much less practical! Much of this was subsequently documented by Batty (1976), including applications in planning practice.

Within this background, two streams of development were the 'entropy modelling' approach and the models developed at the Centre of Land Use and Built Form Studies in the University of Cambridge. The 'entropy' approach, associated in particular with Alan Wilson (e.g., Wilson, 1970), was fairly typical of the first generation of models in that it draw upon mechanical analogies to determine the most appropriate functional forms for models. The key analogy related to the most likely pattern of movement of individual particles (persons or gas molecules) when only certain aggregate properties (average travel cost, temperature of the gas) are known. The Cambridge models were (and still are) characterised by particular attention to the stock of buildings as an influence on the location of activities.

The Lowry-type models, and various other parallel activities, can be seen as a 'first generation' of urban models which flourished approximately between 1964 and 1974. A landmark event near the end of this period was the publication of a paper by Lee (1973), the rather premature 'Requiem for large-scale models'. This correctly pointed out that many of the more ambitious modelling projects had failed to deliver all (or sometimes any) of the advantages which had been claimed for the models. Less justifiably, it tended to condemn all models and all modelling as equally guilty of a whole list of deadly sins.

Lee's criticisms were reflected in other papers, for example Sayer (1976), in something of a reaction to models. This reaction was part of a wider trend away from the 'systems view' of planning towards something less rigorous and less demanding. In the following years, it was widely believed that urban land-use modelling had ceased. In fact, research and a limited amount of practice continued, on a relatively modest scale and with more realistic expectations and claims. These activities refined the models of the first generation, and developed a new generation of models based on micro-economic theory rather than on physical analogies. A key theoretical development was the growth of discrete choice theory, which provided a much better basis for the representation and explanation of choices between locations, between modes of transport, and so on. Helpfully, the standard implementation of discrete choice theory, the logit model, used the same mathematical formula as the standard implementation of entropy-maximising. In some cases modellers could adopt a new and more convincing rationale without the inconvenience of having to modify their software (see chapter 6 for more details on discrete choice models).

The most successful of the second generation models in the UK, representing a continuous stream of both research and application, have been the 'Martin Centre models'. This collective term refers to the various spatial-economic models whose origins were in the Martin Centre of the University of Cambridge[1]. These were typified by the integration of three different economic theories in spatial form:

1 extended input-output modelling[2];
2 utility-maximising consumer behaviour, based on Cobb-Douglas and Stone-Geary consumption functions; for non-spatial choices, and
3 discrete choice theory, for locational choice,

and by treatment of interactions between land-uses as the direct determinants of the pattern of transport demand. The history and theoretical basis of these models is documented in more detail in Hunt and Simmonds (1993). Since the mid-1980s, they have been represented by the TRANUS and MEPLAN packages.

Quite separately from all this, the UK Transport and Road Research Laboratory (TRRL) co-ordinated a collaborative study in the late 1970s about the changing demand for public transport. This brought together a wide and international range of experts and knowledge (TRRL, 1980). As a follow-up project, the TRRL initiated a further collaborative exercise on land-use/transport interaction, which developed into the International Study Group on Land-Use/Transport Interaction (ISGLUTI). The initial search for potential participants identified a number of centres of land-use/transport modelling activity, of which nine participated in Phase 1 of the ISGLUTI project. This phase involved discussion and comparison of the models, and a programme of scenario or strategy tests carried out by seven of the teams using their different models of different cities. The resulting discussion of models and their results was published as a substantial book (Webster *et al.*, 1988), commonly known as 'The ISGLUTI Report'.

That book is widely referred to, not least because it remains the only document which brings together theory, model design (in mathematical and non-mathematical terms) and results from a standardised set of tests for a range of different models. However, it should be recognised that the material it contains is now distinctly dated. The models included are those which were already developed when the Group began work in 1981, and most of them therefore date from the 1970s rather than the 1980s. Moreover, the book itself reported only on Phase 1 of the ISGLUTI project. Phase 2 involved some of the models being applied to cities already represented by other models, thus allowing more systematic comparison both of

---

[1]The Martin Centre was formed by renaming the Centre for Land Use and Built Form Studies, mentioned above. [2]Extended in the sense the local consumers' final demand is endogenous rather than exogenous, but not as fully extended as the University of Liverpool models (see for example Madden, 1993 and Chapter 3 in this book).

different models' results for one city, and of one model's results for different cities. The findings of Phase 2 were unfortunately never gathered into book form, but appeared as a series of papers in the journal *Transport Reviews* in 1990 and 1991. Consequently they have received less attention than Phase 1, whilst probably deserving more (Mackett 1990: Webster and Paulley 1990, Paulley and Webster 1991, Echenique *et al.*, 1990, Wegener *et al.*, 1991).

Three models or packages – IRPUD, LILT and MEPLAN – received most examination in Phase 2 of ISGLUTI. This partly reflected the conclusions of Phase 1, though the practical arrangements for Phase 2 and proximity to TRRL may also have had an influence. These three, together with TRANUS, represented much of the state-of-the-art when the development of DELTA was originally considered.

LILT (Mackett, 1983, 1991a,b) can be regarded as a late flowering of the first generation of models, in that it relied mainly on entropy maximising rather than on market – and price-related mechanisms. The lack of economic content made it seem less convincing compared with models which explicitly represented supply and demand in competitive markets, such as housing.

Outside the UK there have been many other important land/use transport model developments (see also chapter 10). IRPUD was by far the most complex of the three models mentioned above. It was developed by Professor Wegener and colleagues at the University of Dortmund strictly as a research tool rather than as a software product. It was considered to require too much implementation effort to be practically capable of application to other cities except through an unaffordably lengthy project. Hence it was not applied to any other city in Phase 2 of ISGLUTI, although the University of Dortmund was able to supply an excellent database for the application of LILT and MEPLAN to Dortmund. However, the IRPUD approach appeared to offer a number of very attractive characteristics:

- a focus entirely on processes of change over time, whereas the Martin Centre models work largely by producing an "instant metropolis" for each point in time;
- explicit treatment of demographic change, rather than "generating" population or households as functions of employment;
- starting from an input database, rather than requiring the initial situation to be reproduced by calibration to base year data;
- adopting different approaches to different aspects of change, rather than attempting to treat all processes of change within a single framework;
- feedback over time between different processes of land-use change, in contrast with an elaborate and extensive equilibrium for each modelled time point;
- relatively short time steps, and greater attention to the rapidity and duration of different responses.

On the other hand, the Martin Centre approach has significant advantages, particularly

- the adjustment of rents as the interaction of supply and demand is carried out so as to find a partial equilibrium condition within one time period, rather than by setting the rents for each period in response to the outcome of the previous period; and
- practical success in applying a model package to a variety of different study areas.

It appeared feasible to develop a model package which would combine the focus on processes and dynamics which characterizes IRPUD, with the practicality and explicit rent-determination of the Martin Centre models. It also seemed desirable for several practical reasons to concentrate on developing a land-use package which could be linked to any appropriate transport model, rather than to invest in providing the new package with a transport model of its own.

One reason for the interest in processes of urban change relates to a curious phenomenon of urban modelling. The developers of land-use models have made surprisingly little reference to the literature of urban geography, urban economics and so on. The large majority of papers and reports on modelling contain many references to other models, and very few to other empirical studies. The design of DELTA was therefore developed with the intention of representing processes of change which at the very least would be recognizable to other researchers. At best, the components of DELTA might represent the state of the art in each particular discipline.

The decision to develop what became DELTA was influenced by perceptions of:

- likely growth in the market for software and studies to examine the interaction between land-use and transport, ie the policy context; and
- the modelling context: the value of developing a new kind of model, distinct from its predecessors.

This section deals rather briefly with the former, much of which will be familiar to most readers, and in rather more detail with the latter.

9.3 Client Needs

The value of land-use/transport modelling in practice, as opposed to research, lies very much in the facility to examine a number of options on a consistent basis. Together with the fairly substantial cost of developing such models, this means that the main market for modelling is among public planning agencies which need to consider a significant number of major options for the development of cities and regions. Modelling work is also undertaken in order to give advice, usually about major investment choices, to private clients. Whilst this is probably more common than the professional and academic literature would suggest – because private clients are less likely to encourage or even permit publication – it was not significant in any of the discussion about the development of DELTA.

The opportunity on the demand side was therefore the likely market for modelling in the government sector. This was seen as being driven by a re-emerging interest in strategic planning and in the interactions between land-use and transport, which in turn reflects several inter-related changes in planning thinking:

- a growing acceptance that trend-based planning for road transport ('predict and provide') was financially unaffordable and publicly unacceptable;
- increasing concern about the decline in many traditional urban centres;
- corresponding concern about pressures for development on the periphery of existing urban areas and in rural locations – including the growth of out-of-town retail centres in direct competition with traditional urban centres; and
- the emergence of 'sustainability' as a key theme in all types of planning, implying a greater need to look at the total impact of the urban system rather than independent initiatives to solve individual problems such as housing needs or traffic congestion.

Evidence that these concerns were being incorporated into analysis, rather than merely being talked about, came from a number of different developments, including:

- the series of strategic transport studies in British and other cities, several of which incorporated some limited analysis of land-use impacts (see Roberts and Simmonds, 1997);
- new planning guidance for England and Wales (PPG13, 1994), which identified land-use as important but only as a component of a strategy, not as a sufficient policy in itself;
- continuing interest in the work of ISGLUTI (eg., CERTU, 1996);
- in the USA, the Clean Air Act Amendments and the Intermodal Surface Transportation Efficiency Act (see for example Transportation Research Board, 1995);

and so on. This conjunction of activities suggested the market for land-use/transport interaction modelling was likely to grow.

The rest of the chapter presents two case studies. The first describes the Greater Manchester Strategy Planning Model (GMSPM). This is the result of a contract let by the Greater Manchester Passenger Transport Executive (on behalf of the Association of Greater Manchester Authorities) and the Highways Agency. Their joint requirement was to obtain a tool for analysis of a wide range of possible impacts of alternative land-use and transport strategies in the Greater Manchester conurbation. GMSPM has been developed by 'MVA', 'DSC' and 'ITS' using a substantially enhanced implementation of DELTA.

The second case study describes the development of the model for strategic environmental assessment in the Trans-Pennine Corridor (SEATPC). This project is being carried out by 'MVA', 'DSC' and Environmental Resources Management for the Regions in Partnership Steering Group, which represents the European Commission, the Department of Environment, Transport and the Regions, and a wide variety of public and private interests in the North-West and the Yorkshire & Humberside regions of Northern England.

A key point in this work is that since the original DELTA design was specifically conceived as an urban model, it requires an addition to represent processes of change at the regional (or in this case, trans-regional) level, and modification of the existing modules so that they apply consistently to the various major conurbations within the study area (Merseyside, Greater Manchester, West Yorkshire and South Yorkshire) as well as to the free-standing cities (eg., York, Hull), smaller settlements and rural areas.

9.4 Model Design

The key point to emerge from consideration of earlier models was that the single, unifying idea of DELTA is that it has no single, unifying idea! The process of design proceeded by trying to identify different processes of change, each of which might be modelled in a different way with its own dynamics. Interaction between different processes or activities would be "instantaneous" only where this involved competition for resources or was required for "physical" consistency (for example, to ensure that a working person who dies in the demographic model is simultaneously removed from the labour supply).

The development of the model design is discussed in more detail in Simmonds (forthcoming). We give here a brief outline of the design that eventually emerged as the initial version of DELTA and of the subsequent variations.

In order to fulfil the objectives discussed above, DELTA is designed to represent processes of change working over relatively short periods. The available transport models to which DELTA is likely to be linked represent the functioning of the transport system at particular points in time. The overall structure within which DELTA works is therefore as shown in figure 9.1. DELTA operates for one period. The transport model is run on the new land-use pattern, and produces (in addition to its standard outputs) accessibility and environmental measures that are added to the database. DELTA then moves forward through another period, and so on. Note that the time-lags in DELTA mean that the initial database must contain data not only for the starting year but also for a sequence of previous years.

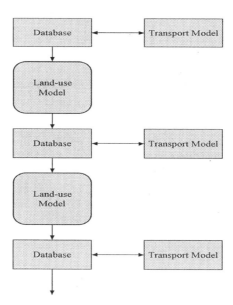

Figure 9.1. Operation of the model over time

9.4.1 Structure of the Land-use Model: Greater Manchester

On the DELTA side, the enhancements focus on the addition of a car ownership model, and the switch to a matrix-based model of employment status and commuting. The structural change is quite slight, as shown in figure 9.2.

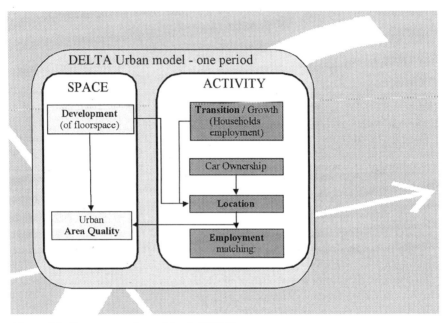

Figure 9.2. Component sub-models of GMSPM

9.4.2 Structure of the Land-use Model: The Trans-Pennine Model

This case study has required:

- adding a process of migration, to represent the longer-distance streams of population movement between areas
- adding a regional economic model, to represent the functional and spatial linkages between different industries (which was assumed to be exogenous in the original urban design, because of the openness of the typical urban economy)
- reconstructing the location and employment status sub-models, in particular, to operate on each of the different labour markets, whilst recognizing that these markets are not closed or self-contained.

The resulting structure is shown in figure 9.3.

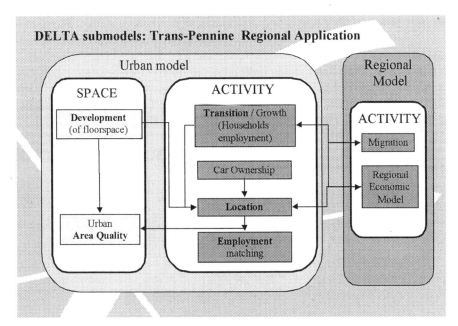

Figure 9.3. Component sub-models: Trans-Pennine application

The need to make these modifications in order to develop a regional form of DELTA contrasts with the Martin Centre tradition. In that approach, one form of spatial-economic model structure is applicable at any spatial scale from the individual city up to a whole continent, with the limited option of treating certain areas in finer detail by means of a hierarchical zoning system of the same choice process (de la Barra, 1989, pp 156-7).

9.5 The DELTA Sub-models

9.5.1 Introduction

The following sections outline the various sub-models. Some of the main equations are included, concentrating on those which describe the behaviour of the various groups of actors represented in the model.

9.5.2 Development Sub-model

The development sub-model seeks to predict the operation of the private sector development process. It takes into account the effect of the planning system, measured through planning permissions and indications of longer-term policy. The model estimates the total amount of development of each kind that will be proposed in each period, constrains it by planning effects, and allocates it to individual zones. Developers are motivated by the expected profitability of development, estimated by comparing current rents with construction costs. Time-lags in the development process mean that developments initiated when rents are high may not become available until rents have fallen again. The model can therefore simulate the 'boom-and-bust' cycle of the development industry. Public sector development and exceptional private sector schemes are exogenously input to the system, together with information about planning decisions and policies.

The sub-model involves two main calculations for each type of floorspace in each period:

- the amount of floorspace which developers will seek to build (see figure 9.4), and
- the allocation of that development to zones (see figure 9.5).

Between the two, the total amount of floorspace may be reduced by the effects of planning policy. This assumes that if the supply of developable land appears to be running out, developers will slow down their rates of development.

The following table summarises which sub-models appear in which of the applications of DELTA to date.

Table 9.1. Sub-models in DELTA applications

Sub-model type	Sub-model	Greater Manchester	Trans-Pennine
Urban model: space sub-models	Development	Yes	Yes
	Quality	Yes	no

Sub-model type	Sub-model	Greater Manchester	Trans-Pennine
Urban model: activity sub-models	Car ownership	Yes	Yes
	Transition	Yes	Yes
	Location	Yes	Yes(a)
	Employment status	Yes(m)	Yes(m)
Regional model: area sub-models	Migration	no	Yes
	Investment	no	Yes
	Production & trade	no	Yes

Key:
a   operates for multiple, overlapping areas
m   multiple labour markets (commuting pattern modelled in DELTA)
s   single labour market (commuting pattern modelled only in transport model)

Figure 9.4. Development sub-model: total floorspace proposed

The initial unconstrained amount of floorspace type u which developers will seek to build in period p is

$$F(U)^u_{p*} = \alpha^u_p \cdot \left(p^u_p\right)^{\beta^u_p} \cdot F^u_{t*}$$

where:

$$p^u_p = \frac{\sum_i \left(r^u_{ti} - c^u_{pi}\right) F^u_{ti}}{\sum_i F^u_{t*}}$$

in which:
$F(U)^u_{p*}$ is total unconstrained floorspace type u proposed to be started in period p
F^u_{t*} is total floorspace type u existing at time t
r^u_{ti} is the rent of floorspace type u in zone i at time t
c^u_{pi} is the cost of building floorspace type u s in i starting in period p;
p^u_p is expected average profitability of starting to build floorspace type u in period p;
β^u_p is a parameter expressing the elasticity with respect to average profitability
α^u_p is a parameter scaling the whole equation.

Figure 9.5. Development sub-model: allocation of development to zones

---

The allocation of development to zones uses a conventional weighted logit formula, ie:

$$F^u_{pi} = F(U)^u_{p*} \frac{F(\max)^u_{pi} \cdot \exp(\gamma^u_p \cdot p^u_{pi})}{\sum_i F(\max)^u_{pi} \cdot \exp(\gamma^u_p \cdot p^u_{pi})}$$

where:

γ^u_p is the sensitivity of development location to expected profitability.

The program then checks that the constraints on development are satisfied, ie that

$$F^u_{pi} \leq F(\max)^u_{pi}$$

Any development in excess of what is permitted in each zone is subtracted, accumulated and reallocated using a revised version of the allocation equation in which the weights are the amount of *additional* development permissible. This is repeated until the total quantity of development has been located.

9.5.3 Transition and Growth Sub-model

This sub-model represents processes of demographic and economic change. Demographic change is expressed in terms of rates of household formations, transitions (from one composition of household to another, e.g., couple-without-children to couple-with-children) and dissolution. More complex changes are implied by combinations of these. Migration to and from the study area is defined by means of a proportion of each household type by socio-economic group that will leave in each period, plus a ratio of arrivals to departures. Economic change is represented as growth or decline by sector. All these changes are (at present) independent of other factors within the model, and therefore represent an economic and demographic scenario rather than a responsive behavioural part of the model.

These changes influence mobility and hence the response to transport changes. Newly-formed and newly-arrived households have to find a home; modified households are more likely to relocate than wholly unchanged households. A proportion of the unchanged households is assumed to be wholly immobile in each period; they and the housing they occupy are excluded from the workings of the location model.

9.5.4 Location Sub-model

The location sub-model is both the 'location and relocation sub-model', and the 'property market sub-model'. Mobile activities respond to changes in four variables:

- accessibility;
- quality of the local environment (as affected by traffic/transport);
- quality of the area (particularly of housing); and
- the cost or utility of consumption, ie of spending income on housing, travel, and other good and services.

The first two of these variables are calculated from outputs of the transport model. Area quality is adjusted by the quality sub-model (see below). Utility of consumption is calculated within this sub-model. Utility of consumption is influenced by the rent of floorspace in each zone, and therefore has to be recalculated within each step of the sub-model as rents are iteratively adjusted. The rent adjustment seeks to equate the total demand for each type of floorspace in each zone (determined by the number of locating activities and the amount of floorspace each occupies, both of which are variable) with the amount of floorspace available (the total stock less that occupied by 'immobile' activities). (Later versions of DELTA allow for a variable proportion of the stock to remain vacant.)

The overall operation of the location model in respect of households and housing is therefore

1) calculate utility of consumption (see figure 9.6) for each household type in each zone, given the current rent per unit housing floorspace (in the first iteration, the previous period's rent); in doing so, find the amount of space that each household will occupy if it locates in each zone;

2) calculate the change in utility of location for each household type in each zone (see figure 9.7);
3) locate all locatable households (see figure 9.8);
4) find the total space used by these households, compare with the available space, and if necessary adjust the rent and repeat.

The market mechanism uses a consumption function and rent-adjustment process similar to those in MEPLAN and TRANUS (see Hunt and Simmonds, 1993). The critical differences are that in DELTA they deal only with the households and housing that are 'in the market' during this period, and they work in an incremental form (with a mixture of timelags) rather than in a cross-sectional form.

The assumptions of the model are such that relocating households will tend to remain in the same location unless there are changes in one of the four influential variables. Newly locating households will similarly tend to locate in proportion to the previous location of similar households. Concentrations of particular kinds of households will therefore tend to persist unless their numbers decline or they are dispersed by other modelled factors.

The location process for employment is similar, but instead of using utility of consumption in the location function it uses a simple measure of cost per employee. The cost is calculated as a function of floorspace rent and floorspace per employee. Floorspace per employee is elastic with respect to rent. As in the residential part of the model, the rents are iteratively adjusted until a combination of density and location changes equilibrates the current demand and supply of floorspace. In GMSPM and TPCM, changes in demand also lead to changing levels of vacancy.

Figure 9.6. Location sub-model: households' utility of consumption

---

The Cobb-Douglas type utility of consumption function in the standard model has just two goods, namely housing space and all other goods and services (commonly known as ogs):

$$U^h_{(t+1)i} = (a^{hH}_{pi})^{\alpha^{hH}_p} \cdot (a^{hO}_{pi})^{\alpha^{hO}_p}$$

where:

a^{hH}_{pi} = space per household type h in zone i (see calculation below);

a^{hO}_{pi} = expenditure on other goods and services per household type h in i (see calculation below);

α^{hH}_p, α^{hO}_p = propensity of households type h to spend available income on space, H, or other goods and services.

Note that the model requires that $\alpha^{hS}_p + \alpha^{hO}_p = \mathbf{1}$ for each household type h, ie that all of the modelled income must be spent either on space or on ogs.

The model finds the maximum utility of consumption for each type of household in each zone, given the income of that household type and the current rent of housing in the zone.

Figure 9.7. Location sub-model: households' change in utility of location

---

The key term in the location sub-model for households is the change in utility of location is calculated as:

$$\Delta V_{pi}^h = \theta_p^{Uh} . \Delta U_{\Delta t,i}^h + \theta_p^{Ah} . \Delta A_{\Delta t,i}^h + \theta_p^{Qh} . \Delta Q_{\Delta t,i}^h + \theta_p^{Rh} . \Delta R_{\Delta t,i}^h$$

where:
$\Delta U_{\Delta t,i}^h$ = change, over a defined past period Dt, in utility of consumption for households type h locating in zone i at time t
$\Delta A_{\Delta t,i}^h$ = similar past change in accessibility of zone i for households type h at time t
$\Delta Q_{\Delta t,i}^h$ = similar past change in quality of housing areas in zone i at time t
$\Delta R_{\Delta t,i}^h$ = similar past change in transport-related environmental quality as perceived by households type h in zone i at time t.

Figure 9.8. Location sub-model: households' location choice

The main location equations are incremental weighted logit functions, which take slightly different forms for 'pool' households (those that have no previous location, eg., newly-arrived immigrants) and for 'mobile' households (those which do have a previous location within the Study Area). These equations are

$$H(LP)_{pi}^h = H(P)_{p*}^h . \frac{H_{ti}^h . (F(V)_{pi}^H / F_{ti}^H) . \exp(\Delta V_{pi}^h)}{\sum_i H_{ti}^h . (F(V)_{pi}^H / F_{ti}^H) . \exp(\Delta V_{pi}^h)}$$

for 'pool' households, and

$$H(LM)_{pi}^h = H(M)_{p*}^h . \frac{H(M)_{pi}^h . (F(V)_{pi}^H / F(M)_{ti}^H) . \exp(\Delta V_{pi}^h)}{\sum_i H(M)_{pi}^h . (F(V)_{pi}^H / F(M)_{ti}^H) . \exp(\Delta V_{pi}^h)}$$

for 'mobile' households, where

$H(LP)_{pi}^h$ = households type h located from the 'pool' of locators to zone i
$H(P)_{p*}^h$ = total 'pool' of households type h to be located;
$H(LM)_{pi}^h$ = mobile households type h located to zone i;
$H(M)_{pi}^h$ = mobile households type h initially located in zone i;
$F(V)_{pi}^H$ = available housing floorspace;
F_{ti}^H = previous occupied housing floorspace; and
$F(M)_{ti}^H$ = initial 'mobile' housing floorspace, ie the space previously occupied by all the households now classified as 'mobile'.

9.5.5 Employment Status Sub-model

The employment status sub-model is the one part of the package which works primarily in terms of persons rather than of households. It first calculates the demand for labour by socio-economic group, given the new number and location of jobs by sector (see figure 9.9). It then adjusts the employment status of individuals and their households until that much labour is supplied. These changes affect the income and hence the locational preferences of households; these take effect in the *next* modelled period. There is thus a short lag, for example, between an increase in the demand for labour and a resulting increase in the demand for housing.

Figure 9.9. Employment status sub-model: demand for workers by socio-economic group

---

The demand for workers by socio-economic group is found for the numbers of workers by sector by

$$E^g_{(t+1)j} = l^{gs}_{(t+1)j} \cdot E^s_{(t+1)j}$$

where:
$l^{gs}_{(t+1)j}$ = employees by socio-economic group g per job in employment sector s in zone j at time (t+1).
$E^g_{(t+1)j}$ = employment (jobs) in group g zone j
$E^s_{(t+1)j}$ = employment (jobs) in sector s zone j

The adjustment of households' employment status assumes that the Study Area represents a single labour market and that additional jobs in any zone may be met by increased supply labour from all zones. Locational effects come about as a result of changing employment status, but have little influence upon such changes. The distribution of travel to work is dealt with entirely in the transport model. The actual process of adjustment in households' status is driven entirely by the demand for labour, ie it does not represent any kind of household behaviour in seeking and taking or rejecting jobs. It has been noted elsewhere that land-use/transport interaction models tend to be weak in their representation of this worker-choice side of the labour market, and to date DELTA is no exception to this.

The employment status model was originally designed to represent a single labour market. GMSPM represents a much larger area, consisting of a number of identifiable though overlapping and interacting labour markets. To take account of this, a more elaborate sub-model was introduced to determine the employment status of residents given the distribution of jobs. This takes into account the likelihood that jobs in one zone will be filled by workers from each residential zone. A consequence is that GMSPM contains matrices of travel-to-work. The person matrix is directly modified in DELTA, and the changes are copied to the trip matrix. The GMSPM version of the employment status sub-model also takes account of the likelihood that a multi-worker household will supply workers of different socio-economic groups (eg., one professional and one 'other' non-manual). This was introduced following analysis of the Census Sample of Anonymised Records (as part of a different project), which suggested that the previous assumption of uniformity of workers within households (both in the original version of DELTA and in MEPLAN and TRANUS) was a potential source of inaccuracy.

9.5.6 Area Quality Sub-model

This area quality sub-model attempts to capture something of the quality of different areas of the city. This is much discussed in planning, but has until now been under-represented in urban modelling. The model hypothesises that the inhabitants of an area themselves influence its characteristics and, over time, affect its desirability as a place to live. Positive influences include maintaining and improving buildings, cultivating gardens, planting trees, etc. Negative effects are neglect and misuse, such as use of residential property for 'nuisance' purposes (such as breaking up cars in the front garden). At present, positive influences are associated with rising average incomes and decreasing vacancy rates, and vice versa.

This sub-model is important to the overall design of the model, because it represents a process of 'positive feedback' to represent the virtuous or vicious circles that tend to maintain or to enhance the differences between prosperous and deprived areas within cities.

9.5.7 Car Ownership Sub-model

The car ownership model is based upon the national car ownership model developed in 1996 by the MVA Consultancy for the then Department of Transport. The model predicts car ownership (in terms of the probability of having no car, one car or 2+ cars) separately for different categories of households, as a function of income, licence holding, area type and socio-economic status. The effect of differences between with-car and without-car accessibility was also calibrated, though they were not included in the recommended national model (partly for reasons of data availability).

In the free-standing form developed for the DoT, this model is applied at a national level, and then a simpler model is used to allocate the results down to a local level, where detailed household forecasts are not available. In the GMSPM context, DELTA provides all the required household detail on a zonal basis. The more sophisticated component of the car ownership model is therefore used in an incremental form, to predict changes in the probability of car ownership for each household type in each zone, as a function of changes in income, accessibility and licence holding.

9.5.8 Migration Sub-model

The migration sub-model has been based upon recent research, particularly the multi-stream models developed by Gordon and Molho (1998). It is based upon 'push' and 'pull' factors which can be calculated from other variables within the model, and on a distance deterrence function. It can represent several different 'streams' of migration influenced by different variables with different distance deterrence effects. It moves households between areas, typically groups of zones approximating to labour and housing markets. Migration is incorporated into the sequence of sub-models between the transition and location sub-models, which continue to deal with demographic change and with local (within labour/housing market areas) location respectively.

9.5.9 Investment and Production/Trade Sub-models

The investment model is intended to represent several streams of investment and one or more processes of disinvestment. The underlying argument is that investment in productive capacity is influenced by a range of factors, differing depending whether the investment is (for example) inward investment, local reinvestment or a small business start-up. Relevant factors may include labour supply, production costs and accessibility to markets and/or suppliers. Disinvestment may come about through the depreciation of equipment or through deliberate closure. The initial Trans-Pennine application uses a very simple incremental treatment whereby investment tends to replace the reductions in capacity due to depreciation but may be relocated by changes in accessibility.

The production/trade sub-model is a spatial input-output model in which the main categories of final demand are exports and consumer demand. Exports are exogenously specified as part of the overall economic scenario. Consumer demand is determined as a function of the total expenditure on other goods and services in the location model (ie other than housing or transport). The pattern of trade is influenced by

- the demand in each area (final demand plus intermediate demand calculated by the input-output process)
- the capacity of each area (resulting from the workings of the investment model)
- the cost of production in each area (costs of inputs plus value added)
- the costs of transport.

Both investment and production/trade work at the same area levels as the migration model. The addition of these sub-models modifies the working of other sub-models at the original urban level. The location sub-model now locates or relocates units of capacity, whilst the numbers of jobs to be filled in the employment sub-model are determined by the combination of capacity and production in each area. The production/trade sub-model has similarities to the Martin Centre regional models. However, an important difference is that changes in capacity are modelled explicitly as changes over time in the investment model, and may be influenced by other factors.

9.6 Data Issues

Urban modellers and analysts can perhaps be divided into two groups: those whose strategy is to work forward from what data is available, having regard to theoretical issues and application requirements, and those who, whilst keeping in mind the likely availability of data, focus mainly on developing a model design which expresses the theory they believe to be appropriate to the kind of application they foresee. The present author is definitely in the latter camp.

The decision that DELTA should represent processes of change starting from a wholly input database, rather than representing an equilibrium situation calibrated in the base year, means that there is a clear distinction between providing DELTA with starting data and calibrating the DELTA parameters.

The DELTA database requires information not only for the base year but also for a series of previous years. The exact requirement depends upon the specification of categories in the model and of the time-lags which influence change. The general expectation is that

- the base year for modelling (at least during model implementation) will be a Census year
- much of the required input will be available from the Census in that year and that a decade earlier
- in principle, all of the data requirement is observable (or, in the case of accessibility measures, a function of observable data)
- some estimation of data will almost inevitably be needed to provide the full pattern of data in the base year, and likewise some interpolation of data to complete the sequence of previous years.

The uncertainties resulting from the need to estimate and interpolate some of the data should if possible be dealt with by carrying out sensitivity tests, and, if necessary, by taking the resulting uncertainty into account in basing decisions upon the interpretation of model results. The principle underlying this is that it is better to have such uncertainties within the model system, where they can be subjected to formal, quantified analysis, than to leave them out and not know what they may imply.

In terms of calibration, the distinction between those who would model only what can be calibrated in a particular study area and those who seek to build more comprehensive models is even more marked. Even within the better-established field of transport modelling, there are very few cities or regions which have extensive databases except of traffic flows on roads; as regards land-uses, systematically available detailed spatial data is largely confined to the decennial Censuses, which are too widely spaced in time to be suitable for the calibration of models describing processes of change affecting households or employment. The alternative approach, in transport modelling as in land-use modelling, is to seek to use relationships and sensitivities from other work and from other study areas. This is sometimes controversial: there is a gulf of opinion between those who regard land-use/transport studies as a cumulative body of knowledge which may be transferable between different situations, and those who regard them as a developing set of techniques which are only valid if their parameters have been calibrated on data for a particular study area. DELTA is strongly associated with the view that there should be, and is, a cumulative body of knowledge. This makes it difficult if not impossible to make the model convincing to those who believe that the only good parameter is a local parameter, but is the only way to satisfy those who want more comprehensive models, given that large-scale surveys and retrospective data assembly are likely to be prohibitively time-consuming and expensive, if not altogether impossible.

9.7 Example Results

9.7.1 Results from Greater Manchester Example Policy Tests

The two tests considered here are a 'highway demonstration' test and a 'public transport demonstration' test.

The 'highway demonstration' test assumed very substantial investment in the highway system between 1996 and 2001:

- motorway capacity increased by 30%
- capacity of their roads increased by 10%
- parking supply (in those central area zones where parking supply is represented as a critical constraint) increased by 30%.

The 'public transport demonstration' test involved

- large-scale extensions to the Metrolink (light rail) network, linking the Regional Centre to most of the conurbation's subcentres and to Manchester Airport
- fuel duty increases continued from 2006 to 2011
- public transport fares held constant at 1996 levels
- bus lanes on major radial roads, reducing highway capacity for other traffic by 15%
- waiting times for bus passengers reduced by 20% (attributable to better reliability, partly due to bus lanes)
- bus running speeds increased by 10% (attributable to improved fare collection systems).

The tests considered here were used only for testing the capabilities of the model, and do not reflect the policy intentions of the model's sponsors. It should also be noted that the results quoted below are based on runs of the transport model with levels of rail and light rail services held constant, whereas the preferred version of the model eventually incorporated a delayed response whereby rail operators would change frequencies in response to demand changes.

The following tables summarise the land-use effects of the two demonstration tests, relative to the Reference Case, in 2011.

It can be seen that the changes in activity location, at this very broad level, are small. What was slightly surprising at first sight was that both Tests had small positive effects on the centre zones – both the Regional Centre and the set of Other Centres; this was contrary to a general expectation that the Highway Demonstration Tests would have a negative effect in those areas. The explanation lies in the fact that the Highway Demonstration Test includes considerable capacity increases on radial routes and, in particular, increases in parking in the Regional Centre and the Other Centres. It is therefore not representative of the pattern of highway investment which is thought to have led to decentralisation in the recent past, which was more focused on capacity increases in outer areas. It can be seen that the Highway Demonstration has a very slight negative effect on the the Inner and Outer Rings, ie the two concentric sets of zones surrounding Manchester City Centre, whilst the Public Transport Demonstration has a slight positive effect on those areas.

Table 9.2. GMSPM: population by area in 2011

Area	Reference Case	Highway Demonstration	Public transport Demonstration
Regional Centre	1.9	2.0(+3.2%)	1.9(+0.4%)
Inner Ring	104.8	103.2(-1.6%)	106.9(+1.9%)
Outer Ring	567.2	560.4(-1.2%)	575.1(+1.4%)
Other centres	23.2	2.6(+1.8%)	23.9(+3.1%)
Rest of study area	1868.6	1876.4(+0.4%)	1857.4(-0.6%)
Study area total	2565.8	2565.6	2565.2

Table 9.3. GMSPM: employment by area in 2011

Area	Reference Case	Highway Demonstration	Public transport Demonstration
Regional Centre	110.9	113.1(+2.7%)	111.4(+1.1%)
Inner Ring	89.5	88.7(-0.9%)	90.5(+1.5%)
Outer Ring	211.0	207.8(-1.5%)	211.0(+0.0%)
Other centres	75.6	79.8(+5.5%)	813.0(+7.5%)
Rest of study area	616.5	613.3(-0.5%)	608.2(-1.3%)
Study area total	1102.8	1102.8	1102.8

Figure 9.10 shows the profile over time of employment in some of the 'other centres'. Three centres are shown – centres 1 and 2 are provided with Metrolink access as a result of the test, whilst centre 3 already has Metrolink access. It can be seen that there is considerable complexity to the results over time. In the case of centre 1, there is a very marked positive effective which reaches a maximum three years after the public transport changes are introduced, and then declines to about half of the maximum effect. In the case of centre 2, the effect takes slightly longer – four or five years – to come about, and the tendency for the gains to disappear is less marked. For centre 3, which was already served by Metrolink, the effects are insignificant.

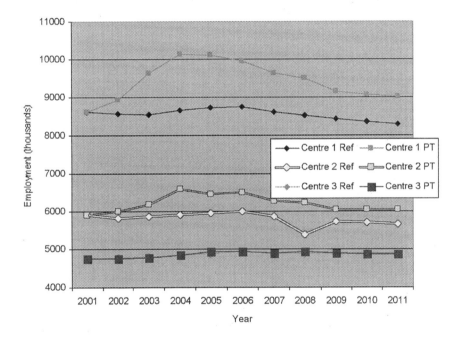

Figure 9.10. GMSPM Results: other centres

The profile of transport changes is shown in figure 9.11 and figure 9.12 which show that the differences between the different tests tend to increase over time. Note that this is due mainly to the operation of the land-use responses. The final preferred version of the model, including additional delayed responses in public transport supply, tends to show more marked divergence of transport test results over time after the initial implementation of an alternative strategy.

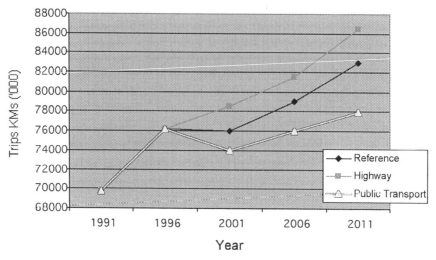

Figure 9.11. GMSPM: transport results (i)

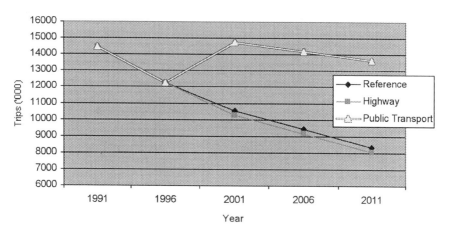

Figure 9.12. GMSPM: transport results (ii)

9.7.2 Results from Trans-Pennine Example Policy Tests

9.7.2.1 Test Definitions

The tests considered here were carried out as explorative exercises during the project for Strategic Environmental Assessment in the Trans-Pennine Corridor, and do not necessarily reflect the policy intentions of the model's sponsors.

The Objectives and Design of a New Land-use Modelling Package: DELTA 181

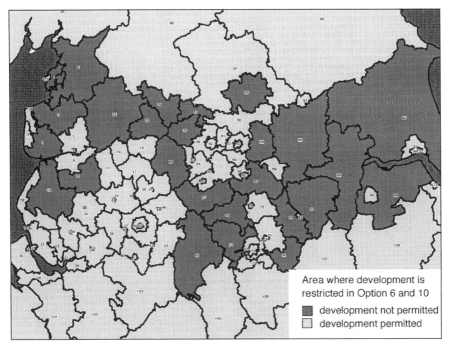

Figure 9.13. TPCM: areas of restricted development

The test considered here (designated Option 6) aimed to concentrate development within the major urban areas, particularly the inner parts of the conurbations. The changes in planning policy relative to the Do-Minimum were as follows:

- all post-2001 allocations of greenfield land were deleted, in all zones;
- all post-2001 allocations of land for development were deleted in zones categorized as rural or of scattered development (these zones are indicated in figure 9.13)
- the allocations of brownfield land for residential in 'inner urban' and 'central' zones were increased by factors of 2 and 5 respectively
- the allocations of brownfield land for retail and office development were halved in 'outer urban' areas, increased by a factor of 1.5 in 'inner urban' areas and doubled in 'central' zones
- the costs of greenfield development (for all types of construction, in all zones) were doubled (making it more expensive than brownfield development) – this was done in order to deter the use of greenfield sites available but not yet built upon in 2001.

Note that all floorspace existing in 2001 is assumed to remain – there is no demolition in the model – so the differences between the tests stem entirely from the differences in how much development of each type is allowed in each zone. Note also that

182 D. Simmonds

- the allocations of land which represent planning policy are expressed as annual allocations which have been set up to be the same, within each test, for each one-year modelled period from 2001-2002 onwards
- because of the time-lags in the model between development 'starting' and becoming 'available' to occupiers, the pattern of activities is not affected until a year or two after 2002
- neither the total Study Area allocations of land, nor the total amounts of development predicted, are necessarily constant between the two tests.

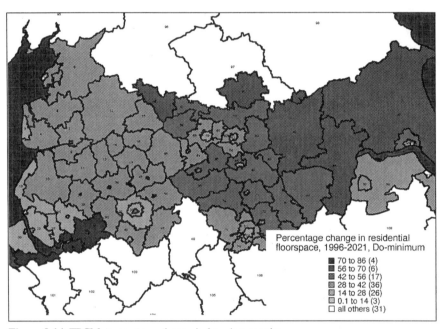

Figure 9.14. TPCM: percentage change in housing supply

9.7.2.2 Test Results

The land-use results from Option 6 start to diverge from those of the Do-Minimum in 2003. As one would expect, given that the Option has direct effects only on the amount and location of development permitted, the differences between Option 6 and the Do-Minimum generally increase over time, though there are some divergences from this pattern.

Figure 9.14 shows the percentage differences in housing supply that result from the operation of the test. Note that these are differences in the total stock compared with that predicted for the Do-Minimum case. It can be seen that the supply of housing is reduced in zones along the Pennine spine, across extensive areas of rural Yorkshire and in more limited areas of Lancashire. The major increases are, as intended, concentrated into much smaller areas, and range up to a doubling in the amount of housing.

Figure 9.15 shows the pattern of percentage changes in 2021 population for Option 6 compared with the Do-Minimum. As one would expect given the importance of the housing market as an influence on household location, there is a clear relationship between these population changes and the preceding housing changes. (Note that within the model, the relationship between housing and population is important but not particularly simple; there is competition for housing between different types of household, of differing size and differing income, and it is possible for more housing to be left vacant if demand decreases.) The most marked changes are in the populations of central areas, which generally increase by +50% to +100%.

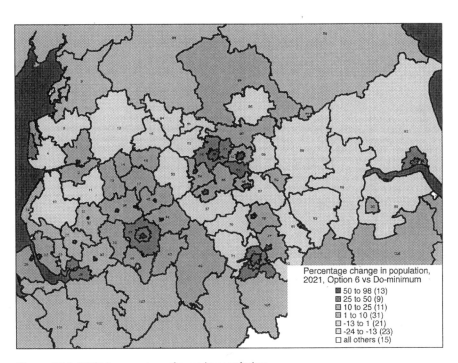

Figure 9.15. TPCM: percentage change in population

One zone, High Peak, appears as an anomaly – it very slightly gains households and population under Option 6. The planning inputs for this zone are exceptional in that no further residential development is allowed in either the Do-Minimum or Option 6. The greater 2021 population in the zone under Option 6 appears to be a spill-over from the growth in Greater Manchester.

The general pattern of change in rents for housing is that rents are higher in zones where the housing supply is more tightly restricted by the Option 6 specification. There are however some zones (mainly on the edges of conurbations) where households decrease but rents increase, eg., 18 (Sefton), and 60 (Leeds North), as result of changes in the household mix. Note that rents in Buffer Zones are assumed to follow the Study Area average in each test. The result is that the Buffer Zones close to Merseyside, Greater Manchester and Sheffield lose population, as a result of the cheaper alternatives available within those conurbations. The Buffer Zones adjoining South Humberside and East Yorkshire, where rents are increased, gain population as a result.

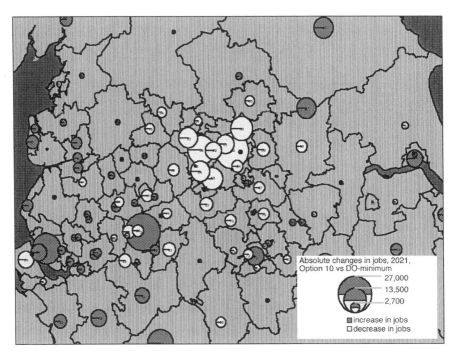

Figure 9.16. TPCM: changes in employment

The pattern of differences in employment is dominated by large increases in all the central areas of the conurbations and decreases in most other areas (see figure 9.16). The general tendency suggests that the Option 6 test has some success in encouraging employment in areas where it is more accessible by public transport. This of course is not the same as encouraging employment to locate so as to minimise the 'need to travel' (the total distance between workers' homes and their places of work). The general scale of the employment changes is distinctly smaller than the scale of the population changes. This is because the supply and price of housing have very direct effects on the location of households, in both their local choices and their longer distance migration, whilst the influence of non-residential floorspace supply and rent on employment location is weaker and part of a more complex chain of responses. Access to labour also plays a part in employment location, and contributes to the centralising effects seen.

One of the consequences of the imbalance between the population changes and the employment changes is that the problem of unemployment in the conurbations is likely to be increased, whilst shortages of labour may occur in other areas. The detailed results for the non-central part of Liverpool (zone 22) show that by 2021 the working-age population in the semi-skilled and unskilled manual group has increased from 75000 persons to 80000 persons, but the number of those persons in work goes up by only a few hundred.

As a final illustration of the workings of the land-use model, the final two figures show changes over time in two variables for one particular zone.

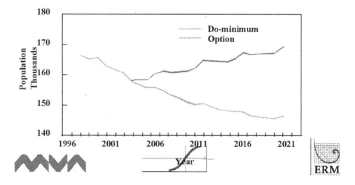

Figure 9.17. TPCM: population of Sheffield inner city

Figure 9.17 shows the changes in population in the Sheffield Inner City area (zone 72) in the Do-Minimum test (light line) and Option 6 (dark line). In the Do-Minimum, a decline in population continues fairly steadily, slowing down towards the end of the forecast and even showing slight increases in one or two years. The range of the vertical scale should be noted – the decrease is from around 165000 to around 145000 over 25 years. Option 6 reverses the decline, bringing about an upward trend which is still continuing at the end of the forecast period.

Figure 9.18 shows the housing rents (per unit of floorspace) associated with the same two tests in the same zone. Again the vertical scale should be noted – the upward trend in the Do-Minimum goes from just under 0.69 (£/m²/week) to just under 0.76 over the whole period, with rather more oscillation than the population changes. The rent in the Option 6 case is initially slightly higher than in the Do-Minimum, indicating that in the short run the effect of Option 6 is to reduce the overall supply in the Sheffield area. After 2011, however, rents under Option 6 are fractionally lower than in the Do-Minimum, suggesting either that supply in the housing market is now greater than in the Do-Minimum, and/or that other factors (eg., changes in employment, migration) have reduced the demand for housing.

Figure 9.18. TPCM: housing rents in Sheffield inner city zone

These last two figures are included as illustrations and as reminders that the 2021 model results are outputs from a process which can show complex behaviour over time. All of the intermediate outputs are stored in the model system and can, given sufficient resources, be analyzed to account for the results in as much or as little detail as appropriate.

9.8 Conclusions

This chapter has presented the perceived requirement for a new land-use/transport modelling package, the model design developed in response to that perception, a selection of results, and some of the practical issues arising in application, in use and in subsequent development. Experience to date suggests that the general approach taken is appropriate to meet a variety of different requirements in planning and research. The modular design of DELTA should make it easier to incorporate future improvements, or indeed wholly different approaches, for particular processes of change. The general tendency of policy debate indicates an increasing need for such tools. A variety of model-oriented projects and programmes around the world show that this need is being recognized and followed up.

Acknowledgements
The author is grateful to all those at ITS, MVA and DSC who have invested time, effort, money and enthusiasm in the development of the DELTA/START system. The support of EPSRC for the CASE studentship and the subsequent ITS project is also gratefully acknowledged. The GMSPM results are quoted by kind permission of the clients for that model, namely the Greater Manchester Passenger Transport

Executive, the Greater Manchester Councils and the Highways Agency. The TPCM results are quoted by kind permission of the Regions in Partnership Steering Group. The selection of tests quoted here does not necessarily represent the policy-making process of any of the client bodies mentioned.

The author alone remains responsible for the design of the DELTA package and for this description of it.

References

Batty M. (1976) *Urban modelling: algorithms, calibrations, predictions.* Cambridge University Press, Cambridge.

Chadwick G. (1971) *A systems view of planning.* Pergamon, Oxford.

de la Barra T. (1989) *Integrated land use and transport modelling.* Cambridge University Press, Cambridge.

Echenique M.H., Flowerdew A.D.J., Hunt J.D., Mayo T.R., Skidmore I.J. and Simmonds D.C. (1990) The MEPLAN Models of Bilbao, Leeds and Dortmund. *Transport Reviews,* vol 10, pp 309-322.

Gordon I.R. and Molho I. (1998) A Multi-stream Analysis of the Changing Pattern of Interregional Migration in Great Britain 1960-1991, *Regional Studies*, vol 32 No 4, pp309-324

Hunt J.D. and Simmonds D.C. (1993) Theory and application of an integrated land-use and transport modelling framework. *Environment and Planning B,* vol 20, pp 221-244.

Lee D.B. (1973) Requiem for large-scale models, *Journal of the American Institute of Planners*, vol 39, pp 163-178.

Lowry I.S. (1964) *A Model of Metropolis*, RM-4035-RC, RAND Corporation, California.

Mackett R.L. (1983) *The Leeds Integrated Land-Use/Transport Model (LILT)*. Supplementary Report 805, Transport and Road Research Laboratory, Crowthorne, Berkshire.

Mackett R.L. (1990) The systematic application of the LILT model to Dortmund, Leeds and Tokyo. *Transport Reviews,* vol 10, pp 323-338.

Mackett R.L. (1991a) A model-based analysis of transport and land-use policies for Tokyo. *Transport Reviews,* vol 11, pp 1-18.

Mackett R. L. (1991b) LILT and MEPLAN: A Comparative Analysis of Land-Use and Transport Policies for Leeds. *Transport Reviews,* vol 11, pp 131-154.

Madden M. (1993) Welfare payments and migration in a non-linear, extended input-output model with an application to Scotland. *Papers in Regional Science*, vol 72, pp 177-199.

McLoughlin J.B. (1969) *Urban and regional planning: a systems approach.* Faber, London.

Paulley N.J. and Webster F.V. (1991) Overview of an international study to compare models and evaluate land-use and transport policies. *Transport Reviews,* vol 11, pp 197-222.

Roberts M. and Simmonds D.C. (1997) A strategic modelling approach for land-use policy development. Paper presented to the World Conference on Transport Research, Sydney, Australia, 1995. Published in *Traffic Engineering & Control*, vol 38, pp 377-384.

Sayer R.A. (1976) A critique of urban modelling. *Progress in Planning*, vol 6, pp 187-254. Pergamon Press, Oxford.

Transportation Research Board (1995) *Expanding metropolitan highways: implications for air quality and energy use.* National Academy Press, Washington DC.

TRRL (1980) *The demand for public transport*. Transport and Road Research Laboratory, Crowthorne, Berkshire.

Webster F.V., Bly P.H. and Paulley N.J. (1988) *Urban land-use and transport interaction: policies and models*. Phase I report of the International Study Group on Land-Use Transport Interaction (ISGLUTI). Avebury, Aldershot

Webster F.V. and Paulley N.J. (1990) An international study on land-use and transport interaction. *Transport Reviews,* vol 10, pp287-308.

Wegener M., Mackett R.L. and Simmonds D.C. (1991) One city, three models: comparison of land-use/transport policy simulation models for Dortmund. *Transport Reviews,* vol 11, pp 107-129.

Wilson A.G. (1970) *Entropy in urban and regional modelling*. Pion, London.

10 Transport and Urban Development

Gunter Haag and Kathrin Gruetzmann

10.1 Introduction

Transport systems can only be seen in their regional context. Therefore the interactions between the transport system consisting of different transport modes and the regional structure (especially the settlement pattern) have to be investigated. On the one hand for example, changes in the population distribution influences the volume and the distribution of traffic flows. On the other hand the volume of transport within a region and the mix between public and private provision has an impact on the migration behaviour of households and firms.

The number of real-world applications of integrated transport and land use models has increased steadily over the last decade. There has been a continuous reflection of purposes, directions and theoretical frameworks of land use and transport modelling as witnessed by Hutchinson *et al.*, (1985), Hutchinson and Batty (1986), Webster *et al.*, (1988), by reviews of Harris (1985), Wegener (1986, 1987), Kain (1987), Boyce (1988), Berechman and Small (1988), Aoyama (1989), Batty (1994), Harris (1994), Wilson (1997) and Wegener (1994, 1998). Using a framework for the classification and evaluation of the current state-of-the-art of urban and transport modelling developed by Wegener, 15 contemporary operational models are compared within the Eurosil Project (Eurosil, 1998) using criteria such as comprehensiveness, overall structure, theoretical foundations, modelling techniques, dynamics, data requirements, calibration and validation, operations and applicability. Among the many candidates TRANUS, MEPLAN and the following described STASA model (STASA[1]) have shown their leading competence in the summary of model capabilities. For a full review of land-use/transportation models see Chapter 9.

In this chapter the interactions between transport and urban/regional development are investigated for a peri-urban area, where the effects are expected to be very strong. The study area lies within the region of Stuttgart, along the Stuttgart – Sindelfingen/ Böblingen – Herrenberg corridor. In this corridor we find the A81 motorway which runs from Stuttgart to Singenvia Herrenberg. The A81 is part of the TEN network

---

[1]STASA (**S**teinbeis **T**ransfer Centre **A**pplied **S**ystems **A**nalysis)

(Transeuropean Network) and links Germany with Switzerland and Italy. It has been in use since 1978. Moreover, in 1992 the rapid transit rail system S1 was built. The aim of this chapter is to use the transport model to measure the effects of the A81 and the S1 on the population distribution and the changes in the traffic load on the A81 in this corridor building in changes in modal split (the specific impacts of S1) (see map in Apendix A10 at the end of this chapter).

The region of Stuttgart is one with the highest population densities in Germany and a high traffic density as well. This has implications for the capacity of the transport network in this region, especially the A81 during peak hours. The improvement of the public transport system has some impacts on the traffic volume in the considered region by changing the modal split. Nevertheless, the greater efficiency of the whole transport system can lead to higher capacities on the different modes. In turn, this may result in an increase in traffic performance caused by the improvement of accessibilities (direct induced traffic) on the one hand, and a change in population's behaviour concerning the location choice (indirect induced traffic) on the other hand (SACTRA 1994, Der Bundesminister für Verkehr 1996).

For this case study the STASA transport model is used to investigate both public and private transport developments. This transport model is able to describe non-linear multi-modal transport systems and their impact on urban/regional development. The results of the STASA model are examined in the light of a panel study based on appropriate questionnaires concerning the location choice behaviour of households and firms within this region. Furthermore, the effects of the A81 and the S1 on the location choice behaviour and the mobility behaviour of the population are determined. For this purpose, the population distribution is analysed using a multinomial logit model, since the database consists only of zone specific population numbers. In the case of available migration between zones or cells, a more sophisticated modelling of the population redistribution is possible and recommended. Changes in the accessibility of the traffic cells will lead to a population redistribution resulting in long-term changes in the traffic flows. These applications have been carried out on a commercial basis, in order to assist the clients with fundamental information about the mechanisms behind urban and transport development.

10.2 Client's Needs

As with many academic programmes, the development of the STASA models has been funded from a variety of sources. Most recently, the research programme has been funded by the EU. This European project is called EUROSIL (EUROpean Strategic Intermodal Links).[2] Although EU monies have funded the developemnt of the projects there are now other agencies keen to explore the potential outputs of the model. The

---

[2]This case study is the result of an co-operation of STASA with Steierwald, Schönharting und Partner, Heßbrühlstr. 21, 70565 Stuttgart, Germany, Dr. L. Neumann, Th. Pischner and A. Gresser.

case study described in the rest of the chapter was funded by the Ministerium für Umwelt und Verkehr Baden-Württemberg. The presented results are based on a simulation case study **SILUS** (Strategic Intermodal Link Urban area of Stuttgart) including empirical investigations and surveys such as before-afterwards-examination.

The client is broadly interested in the following subjects:

- development of socio-economic structures (population, employees, jobs)
- development of settlement structures (residential areas, industrial and business areas, nature/recreational areas)
- interoperational diversions from the parallel road network in private vehicle traffic
- intermodal diversion between private vehicle traffic and public transport (including eventual modal shifts since the completion of the rail rapid transit)
- significant impacts on traffic (generated traffic, vehicle mileage, modal-split travel times, journey distances)
- secondary impacts relevant to the assessment (time costs, operational costs, emission of pollutants, traffic accidents)
- connection of TEN with the regional and urban network.

More specifically the Ministry of Environment and Transport for Baden-Württemberg was interested in:

- regional impacts and economic implications due to the expansion of the A81
- impact of the expansion of the A81 on other transport modes (e.g., the rail system S1)
- integration aspects of TEN and the urban transport network
- test of the STASA urban and regional model with respect to other planned investment projects
- recommendations

From this SILUS case study, experiences concerning intermodal and interoperational strategies for the elaboration of the TEN can be expected. The intermodal impacts affect the interdependencies between private vehicle traffic and public transport in the TEN, as well as those between regional and urban traffic. In the future, the expansion of the rail rapid transit system (installation of new lines, increase of headway frequencies) can be expected, which is of European importance: on the one hand it provides relief for the A81, relevant to the TEN, on the other hand, it means the connection of regional and urban traffic with the high-speed railway network (ICE) and Stuttgart Airport.

10.3 The Study Region

The Stuttgart Region is situated in the south-west of Germany and covers five state districts with 179 communities of an overall population of 2.6 million inhabitants. Together with its state capital, it represents the economic and cultural centre of the state of Baden-Württemberg. With an area of 3,600 km² this region is one of the most densely populated regions of Germany. The transport system is made up of private vehicle traffic routes ranging from city streets to highways, and public transport systems ranging from city buses to high-speed trains (ICE).

Due to the volume of urban and regional traffic, the network (especially the A81) is highly congested on the outskirts of Stuttgart. The consequences are serious delays in long-distance traffic, as well as high environmental pollution levels in the Stuttgart area.

In 1992 the rail rapid transit line Herrenberg – Stuttgart was built on the conventional railway tracks between Stuttgart-Schaffhausen, parallel to the A81. This rail rapid transit line serves mainly commuting traffic directed towards Stuttgart – and in addition to this, is connected to the ICE-link Mannheim-München, as well as to Stuttgart Airport through the city's central railway station. The existing rail rapid transit line is extremely congested in peak hours, – as is the A81 – although it still seems to have potential capacity reserves (see map in apendix A10).

Population has grown rapidly since 1978 in the Stuttgart area, especially in the periphery of the city. In addition, an enormous growth in industrial and business zones took place in the area of Böblingen/Sindelfingen, which, besides its traffic-generating impact on passenger transport also affects freight transport.

The framework of the STASA model is given in section 10.4. The development of the population distribution is modelled by a multinomial logit model. The calibration process is described in section 10.5, whilst section 10.6 describes the simulations. The results are described in sections 10.7 and 10.8.

The data base necessary to estimate the parameters of the STASA model requires traffic flows (a basic trip matrix) and travel times between the 58 traffic zones or cells of the area under investigation as well as some socio-economic variables like population distribution and workplaces.

10.4 The Framework of the STASA Model

For the modelling of traffic flows, the region under investigation is subdivided into L non-overlapping zones or cells (Bundesminister Verkehr 1996). The number of trips for the trip purpose α from traffic cell i to traffic cell j at time t (hour of the day) with mode r is denoted by $F_{ij}^{\alpha r}(t)$. That traffic volume $O_i^{\alpha r}(t)$ for each traffic cell i is gained by summation of the traffic flows:

$$O_i^{\alpha r}(t) = \sum_{j=1}^{L} F_{ij}^{\alpha r}(t) \qquad (10.1)$$

The number of trips $D_j^{\alpha r}(t)$ into the traffic cell j at time t reads correspondingly:

$$D_j^{\alpha r}(t) = \sum_{i=1}^{L} F_{ij}^{\alpha r}(t) \tag{10.2}$$

The population distribution is denoted by

$$\vec{E} = \{E_1, ..., E_i, ..., E_L\}, \tag{10.3}$$

where E_i is the number of agents of traffic cell i. E_i will be modified by the decisions of the individual agents to carry out a trip between the cell i and any one of the other cells. Therefore, the population distribution $\vec{E}$ is connected via traffic related activities of the agents with a great number of individual decision processes.

The micro level (individuals, households, companies) appears initially to determine the dynamics of the macro level, while there is no feedback of the macro level to the micro level. However, this is by no means the case. Rather, a mutual influence of the two levels occurs. Among other things, the actions of the individual agents of the urban system (activities) express themselves in the dynamics of the traffic flows and therefore in the time-dependent population redistribution of the system. Therefore, micro behaviour and macro dynamics are strongly coupled.

Since $E_i(t)$ agents are at time t in the traffic cell i, the probability for a trip to another traffic cell is proportional to $E_i(t)$. Let $p_{ij}^{\alpha r}(\vec{E}, \vec{x})$ be the transition rate from i to j for trip purpose α with mode r. Of course, this transition rate depends among other things on the explicit distribution of the agents $\vec{E}$ and cell specific characteristics $\vec{x}$ of the infrastructure. The infrastructure includes job supply, the housing market, services available for companies and households as well as leisure facilities (Haag, 1989). In this way, the number of the trips between i and j is given by:

$$F_{ij}^{\alpha r}(t) = E_i(t) p_{ij}^{\alpha r}(\vec{E}, \vec{x}, t), \tag{10.4}$$

where particular transition rates represent trip probabilities from one traffic cell i into j.

Three factor sets are essentially important:

- attractivities $u_i^\alpha(\vec{E}, \vec{x})$ of the particular traffic cells which depend across-the-board on variables such as the population distribution and the distribution of the work places. The significance of the different socio-economic variables $\vec{x}$ is determined by means of a multivariate regression procedure. Obviously the composition of the particular set of (significant) key variables depends strongly on the trip purpose α.
- resistance's $g^{\alpha r}(w_{ij}^{\alpha r})$ depending on trip purpose α and the transport mode r, where $w_{ij}^{\alpha r}$ represents a generalised resistance parameter $w_{ij}^{\alpha r} = t_{ij}^r + b_1^\alpha c_{ij}^r + b_2^\alpha K_{ij}^{\alpha r}$ for a trip from i to j, and t_{ij}^r is the traffic density dependent variable for mode r c_{ij}^r, are the travel costs, $K_{ij}^{\alpha r}$ are appropriate comfort parameters, and b_1^α, b_2^α are corresponding weight factors. The following resistance function seems to be

adequate in order to describe modal choice behaviour (second term) as well as decisions concerning the beginning of the trip (third term):

$$g^\alpha(w_{ij}^{\alpha r}(\tau)) = c_0^{\alpha r}(w_{ij}^{\alpha r}(\tau))c_1^{\alpha r} \frac{\exp(-c_2^{\alpha r}w_{ij}^{\alpha r}(\tau))}{\sum_{s=1}^{R}\exp(-c_2^{\alpha s}w_{ij}^{\alpha s}(\tau))} \frac{\exp(-c_3^{\alpha r}w_{ij}^{\alpha r}(\tau))}{\sum_{\tau''=\tau-1}^{\tau+1}\exp(-c_3^{\alpha r}w_{ij}^{\alpha r}(\tau'))} \quad (10.5)$$

- and a time-dependent scaling parameter $\varepsilon^{\alpha r}(t)$ that correlates with mobility behaviour.

For the traffic flows within the STASA model, the following functional dependence has proved to be effective:

$$F_{ij}^{\alpha r}(t) = E_i(t)\varepsilon^{\alpha r}(t)g^{\alpha r}(w_{ij}^{\alpha r})\exp(u_j^\alpha(\vec{E},\vec{x}) - u_i^\alpha(\vec{E},\vec{x})) \quad (10.6)$$

The general observation that agents compare the attractivities of the traffic cells with respect to their travel purpose and that the probability for a trip $i \to j$ increases with increasing differences $(u_j^\alpha(\vec{E},\vec{x}) - u_i^\alpha(\vec{E},\vec{x})) > 0$ of attractivities per unit of time are reflected in the above formula. In addition to such rational behaviour, uncertainties have to also be considered. Therefore, the addition of a stochastic consideration of these processes seems reasonable. For this reason, the probability $P(\vec{E},t)$ is introduced to find a certain distribution of the agents at time t, taking into account the complicated interactions of these agents. Therefore, the agents are not considered to be independent from each other. Knowing the transition rates for the population distribution (transition probabilities per unit of time) an equation of motion for the temporal evolution of the probability distribution can be built. This is also known as the master equation approach (Weidlich and Haag, 1983; Haag, 1989). The dynamics of the (mean) population redistribution at the macro level can be derived directly by an averaging procedure from the master equation:

$$\frac{dE_i(t)}{dt} = \sum_{\alpha,r}\sum_{j=1}^{L}E_j(t)p_{ji}^{\alpha r}(\vec{E},\vec{x}) - \sum_{\alpha,r}\sum_{j=1}^{L}E_i(t)p_{ij}^{\alpha r}(\vec{E},\vec{x}) \quad (10.7)$$

The change of the population number $E_i(t)$ of a traffic cell i from one hour to the next hour can be calculated on the basis of the traffic flows $F_{ij}^{\alpha r}$ and $F_{ji}^{\alpha r}$ between the different traffic cells i and j. The flows themselves in turn depend on the respective attractivities $u_i^\alpha(\vec{E},\vec{x})$ of the traffic cells, the resistance functions $g^{\alpha r}(w_{ij}^{\alpha r})$ and the scaling parameter (mobility parameter) $\varepsilon^{\alpha r}(t)$. In this way, the population distribution and therefore the car (mode) distribution are represented in a time-dependent manner. In particular the determination and estimation of daily traffic peaks require a time-dependent estimation of the flows.

However, the above equation describes only short-term redistribution effects of the population. Long-term effects such as migration of the population between cells as well as structural development effects have to be considered too. The dependence of the attractivities $u_i^\alpha(\vec{E}, \vec{x})$ of the traffic cells on the population distribution $\vec{E}$ and further socio-economic variables $\vec{x}$ has the consequence that the evolution equation now becomes a non-linear differential equation or difference equation. In this case, non-linearities reflect the complexity of the decision making processes. Depending on the initial conditions i.e. the agents and/or vehicle distribution at a given time and the further system parameters of the urban/regional areas, the non-linear dynamics can lead to a complex variety of self-organising traffic flows. Consequently, the dynamics of the population distribution is described by traffic flows and feedback effects resulting from the land-use pattern.

For the EUROSIL case study the STASA transport model has been modified appropriately. The trips could not be disaggregated according to different trip purposes. The traffic flows from cell i to cell j by mode r are given by:

$$F_{ij}^1(t) = E_i(t) X_{ms}^{IT}(t_{ij}^1 - t_{ij}^2)\varepsilon^1(t)(t_{ij}^1)^{b^1} \exp(-c^1 t_{ij}^1) \exp(u_j(\vec{E}, \vec{x}) - u_i(\vec{E}, \vec{x})) \quad (10.8)$$

for the individual transport ($r=1$) and

$$F_{ij}^2(t) = E_i(t)(1 - X_{ms}^{IT}(t_{ij}^1 - t_{ij}^2))\varepsilon^2(t)(t_{ij}^2)^{b^2} \exp(-c^2 t_{ij}^2) \exp(u_j(\vec{E}, \vec{x}) - u_i(\vec{E}, \vec{x}))(10.9)$$

for the public transport ($r=2$), with the population number E_i of cell i and the travel time t_{ij}^r from cell i to j by mode r. The factor $X_{ms}^{IT}(t_{ij}^1 - t_{ij}^2)$ gives the share of individual transport, in this case given by a linear function of the difference of the travel times of both modes (figure 10.1).

The model parameters are:

- attractivities u_i of the cells i, which depend on characteristics of the cells like population numbers or numbers of employees,
- scaling parameter (mobility parameter) ε_r for the modes r and
- the parameters of the resistance function b^r and c^r.

Given the transport flows there is a redistribution of the population over the day (short-time process), described by $E_i(t)$. In the long-term (driven by events such population ageing and migration), the 'housing population', denoted by E_{i0}, is changing as well. This slowly changing 'housing population' E_{i0} provides the starting conditions for the short-time redistribution:

$$E_i(t = 6am) = E_{i0} \quad (10.10)$$

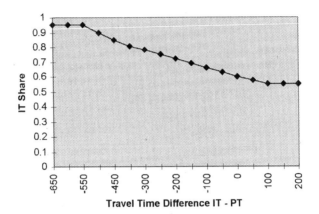

Figure 10.1. Modal split function

The long-term population redistribution is analysed in this case study by a multinomial logit model, since migration data were not available. Thereby the population number E_{i0} of cell i (housing population distribution) is given by

$$E_{i0} = E_{ges} \frac{\exp(V_i(\vec{x}(t)))}{\sum_i \exp(V_i(\vec{x}(t)))} \tag{10.11}$$

with the total population number E_{ges} and the 'population attractivities' $V_i(\vec{x}(t))$ of the cells. The population attractivities can be connected with characteristics (appropriate indicators) of the spatial areas by a multiple regression model. Possible explanatory indicators (key indicators) include the number of work places, the number of dwellings, but also accessibility indicators of the traffic cells. Therefore a link to the traffic system can be identified.

10.5 Parameter Estimation

The parameters of the STASA model, such as the mobilities, attractivities and the parameters of the resistance function occurring in the approach for the trip frequency can be directly linked to the trip decision processes by different optimisation procedures. Therefore, the model has to be matched to the empirical traffic flow matrices, the population and travel time matrices, which in turn are dependent on the traffic volume of the hour groups, respectively, for the hour groups under consideration $\tau = 1,...,4$. The aim is to minimise the function

$$H_t\left[\varepsilon\alpha, c_1^\alpha, c_2^\alpha, \vec{u}\alpha\right] = \sum_{i,j=1}^{L}\left[F_{ij}^{\varepsilon\alpha}(\tau) - \varepsilon\alpha(\tau)b(\tau)E_i^\varepsilon(\tau)g\alpha(w_{ij}(\tau))\exp(u_j^\alpha(\tau) - u_i^\alpha(\tau))\right]^2 = Min. \quad (10.12)$$

This enables us to calculate the optimal attractivities $u_i^\alpha(\vec{E}, \vec{x})$ and further system parameters for each trip purpose and each time of day. Depending on the time of day, the trip purpose distribution changes within the (macroscopic) traffic flows. For example, trips between home-work and work-home may predominate in the morning hours and the late afternoon respectively, while leisure time trips or shopping trips show another temporal distribution over the day. Therefore, the estimated parameters are dependent on the trip purpose and time of day.

The searched site factors $x_i^n, n = 1,...$ of the traffic cells are obtained from two different areas: on the one hand from the class of the so-called synergy variables, describing general group effects such as pigeons effects and/or band-wagon effects and on the other hand from a sequence of potential attractivity factors e.g., number of jobs available, the number of vacant dwellings, regional income per capita, and other local infrastructure depending factors. The set of key-factors (selected socio-economic variables) is in a second step determined via a multiple regression analysis:

$$u_i^\alpha(\vec{E}, \vec{x}) = \sum_n b_n^\alpha(\tau) x_i^n(\tau) \quad (10.13)$$

The elasticity's $b_n^\alpha(\tau)$ assigned to the socio-economic variables $x_i^n(\tau)$ are dimensionless numbers and indicate the influence of the independent variables on the dependent variable. The selection of relevant site factors occurs by means of the corresponding statistical characteristics (T-values and other significance tests).

10.6 Model Simulations and a Stepwise Procedure

The model can be used to evaluate many different planning scenarios. The base or starting point is the situation as it was in 1990, when the motorway A81 was already open but the rail rapid transit was not yet in use. For this case there exists a data base consisting of traffic flows and travel time matrices for private as well as public transport, population numbers, numbers of employees and other characteristics of the traffic cells. This data provides the starting point for the analysis of the transport system and population distribution.

As planning scenarios we consider the following cases:

- Scenario A: without A81/without S1
- Scenario B: with A81/with S1
- Scenario C: without A81/with S1

For these scenarios changes in the transport system and population distribution have to be calculated. The procedure is summarised in the following and illustrated in figures 10.2, 10.3 and 10.4.

Step 1
In the first step the traffic flows F_{ij}^r from cell i to cell j for the individual and the public transport are assigned on the network with the A81 and without S1 (the base scenario). From this the travel times t_{ij}^r, as well as the accessibility measures for both modes between the single cells, are determined. In the same way the travel time matrices t_{ij}^r and the corresponding accessibility measures are determined both from the traffic flows F_{ij}^r and the corresponding networks for the different planning scenarios A – C.

Step 2
Based on the traffic flows F_{ij}^r for both private and public transport, the corresponding travel times t_{ij}^r and the population numbers E_i (all for the base scenario), the parameters of the STASA transport model are estimated by a non-linear estimation procedure:

- attractivities u_i of the cells i,
- scaling parameter (mobility parameter) ε^r for the modes m and
- the parameters of the resistance function b^r and c^r.

Statistical tests (such as correlation, Fisher's F-value) are used to evaluate the quality of the parameter estimation.

Step 3
The dependency of the attractivities of the cells on the characteristics of the cells (population and employee numbers, accessibilities) are determined by means of multiple regression models. The corresponding parameters b_i^n and β_i^n describe the influence of the independent variables. Statistical tests values are also calculated.

Step 4
To analyse the population distribution and its dependency on the accessibilities of the cells, related 'population attractivities' V_i (MNL model) are calculated based on the population distribution in 1990. For the considered scenarios, the influence of the changed accessibility of the cells on the "population attractivities" are determined by a multiple regression procedure using accessibility measures for the base scenario.

From step 5 an iteration procedure (step 5 to 8) is applied:

Step 5
In the next step for the planning scenarios A - C the change of population attractivities V_i and of the population distribution E_i *(scenario X, iteration 1)* due to different accessibilities are calculated.

Step 6
In parallel to step 5 the (transport) attractivities u_i *(scenario X, iteration 1)* of the cells are calculated for the different considered scenarios A - C by taking into account the corresponding accessibilities and population pattern.

Step 7
Based on the attractivities calculated in step 6, the travel times t_{ij}^r *(scenario X, iteration 1)* and the population numbers E_i *(scenario X, iteration 1)* of step 5, the traffic flows F_{ij}^r *(scenario X, iteration 1)* are determined for both modes r. The mobility parameter and the parameters of the resistance function are assumed to be constant.

Step 8
The traffic flows F_{ij}^r *(scenario X, iteration 1)* are now assigned for each scenario A - C on the corresponding network to obtain new travel times t_{ij}^r *(scenario X, iteration 2)* and accessibility measures. With these travel times and accessibility measures for each mode, the population configuration (dependent on the accesibilities) and the traffic flows change again, so that an iterative procedure has to be applied. Steps 5 to 8 are repeated until the change of the flows and travel times have become negligible. As a result, a consistent set of traffic flows F_{ij}^r (sc. X) for the single modes m and the different scenarios A - C are obtained.
End of the iteration procedure

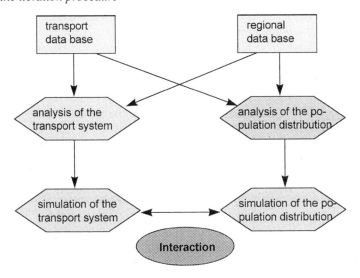

Figure 10.2. Procedure of the STASA model in general

Step 9

A comparison of the traffic flows F_{ij}^r (sc. X) for the scenarios A - C with the base scenario shows the effects on modal split and its interaction with population development. Even direct and indirect induced traffic shares are considered by the model. From the traffic flows F_{ij}^r (*case X*) in cases 1 - 4 and their assignment on the corresponding network, traffic performances and relevant emissions, travel costs, accident costs, etc. are determined for the single schemes.

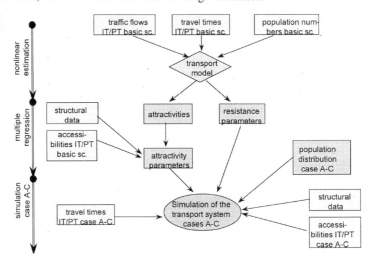

Figure 10.3. Flow chart of the transport model

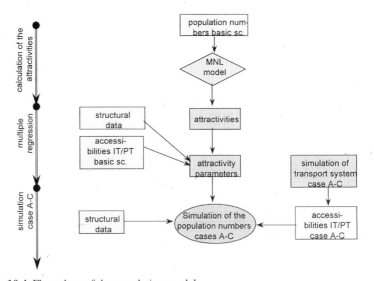

Figure 10.4. Flow chart of the population model

10.7 Results of the SILUS Case Study

10.7.1 Estimation of the Parameters

As described in section 10.5, the parameters of the STASA transport model are estimated from the traffic flows of the base scenario via a non-linear regression method. The results are:

- attractivities u_i of the cells i,
- scaling parameter (mobility parameter) ε^r for the modes m and
- the parameters of the resistance function b^r and c^r.

In the following step the attractivities of the cells are investigated by a multiple regression procedure. The results of this analysis are presented below and form the basis for the simulation of the different scenarios.

Table 10.1 presents the correlation and the Fisher's F-value for the comparison of the given ('empirical') flow matrix with the 'modelled' matrix, calculated by the STASA transport model with the estimated parameters. These values can be seen as a measure that indicates the correspondence between the original data base and the estimated values and therefore the quality of the results of the estimation. The result for the estimation in this case study show very high significance with a correlation coefficient of 0.87.

Table 10.1. Statistical tests for the estimation of the STASA transport model

CASE 1	with A81 / without S1
Correlation (R^2)	0.87
Fisher's F value	703.44

The dependence of the resistance functions on travel time for both modes is shown in the figures 10.5 and 10.6. For public transport we find much higher travel times, coupled with a smaller value of the resistance function than for the private transport. This leads to less trips being made by public transport than by individual transport.

For the transport attractivities of the cells we find (for the base scenario) high attractivities in the region of Sindelfingen/Böblingen, cells with many work places. If the attractivity of a cell is high, the probability for a trip into the cell increases. However the southern part of the study area (the cells near Herrenberg) show negative attractivities. This is a plausible result because the data base concerns the time interval from 6a.m. to 10a.m., when the trip purpose home-work dominates and a lot of people drive from the south of the studied region to work in the north.

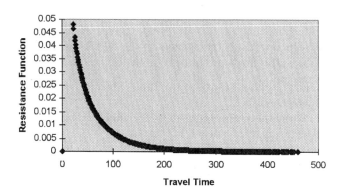

Figure 10.5. Resistance function: individual transport

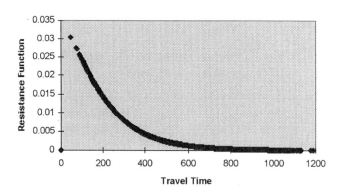

Figure 10.6. Resistance function: public transport

The attractivities are related to characteristics of the traffic cells (such as population numbers or accessibility indicators) by a multiple regression model. On the base of the 58 traffic cells the following (independent) variables are used:
- population numbers
- numbers of employees
- number of persons accessible within the time of 30 minutes by IT/PT
- number of employees accessible within the time of 30 minutes by IT/PT

- 'Luftliniengeschwindigkeit' (travel times per distance, weighted by the number of trips) for private and public transport

These variables were normalised according to the attractivities. The variable with the highest significance is the variable 'number of employees'. That means that if the number of employees within a certain traffic cell is high, then the attractivity to make a trip into this cell is also high (which is plausible for the considered time interval from 6 a.m. to 10 a.m). The population numbers show only a very low positive significance and only in connection with the variable 'number of employees'. Very similar and positive influence show the variables 'accessible population numbers' and 'accessible employee numbers' by private and public transport. The influence of the accessibility measures of private transport is a little bit higher than for public transport. The regression result with the best statistical tests is again the one with the variables

- number of employees
- population size
- number of persons accessible by private transport

These variables are used for the simulation of the different planning scenarios.

10.7.2 Analysis and Simulation of the Population Distribution

The 'population attractivities' u_i of the cells i are determined from the population distribution of the traffic cells for the base scenario via a multinimial logit model (see section 10.4). Thereby the population attractivities are calculated from the empirical population numbers of each cell. In the second step, the population attractivities can again be connected with the characteristics of the cells by a multiple regression.

The statistical tests show that the significance of the variables used is not very high. Other variables such as number of shopping facilities and so on could be used. However, for the application in this case study it seems to be not necessary to investigate the dependency of the population attractivities on other characteristics of the cells because they do not change for the planning scenarios A - C. The 'number of employees' has negative influence (separation home-working), the 'number of accessible employees' and the 'accessible population numbers' (correlated variables) have positive influence. For the simulation the following variables will be used:

- number of employees
- number of persons accessible by public transport

With this set of variables only the influence of public transport on the population distribution can be simulated. The increase of population from 1978 to 1990 in the study area is very different from the population development in other but comparable areas of the region of Stuttgart. Here, we find a strong sub-urbanisation process which is partially caused by the greater accessibility of the region of Herrenberg to the A81 (Der Bundesminister für Verkehr 1996, Göbel 1997).

10.8 Results of the Simulations

Here, we present the results of the simulation of the two-mode transport system and the population development for the three planning scenarios.

10.8.1 Population Distribution

For the base scenario we find a high population density in the areas of Stuttgart and Böblingen/Sindelfingen. If we look at the changes of the population distribution for the different planning scenarios we find some differences. For the cases without the A81, the population is more concentrated in the area Böblingen/Sindelfingen (where the A81 ended before 1978). For the area of Herrenberg the population density without the A81 is not as high as in the case with the A81. The effect of the motorway on the population distribution is obvious. We find a strong shift of population in the direction of Herrenberg caused by the extension of the A81.

For the scenario with the A81 and with S1, the population density increases further compared to the base scenario mainly along the rail rapid transit line, i.e. in the vicinity of stops of the S1. The effects caused by the S1 are not as strong as those caused by the motorway.

10.8.2 Transport System

For the base scenario the attractivities of the cells in the areas of Stuttgart and Böblingen/Sindelfingen are high. For the planning scenarios without the A81, the attractivities of Böblingen/Sindelfingen increase by at least 5 to 10 %, whereas the attractivity of the area of Herrenberg decreases compared to the basic scenario. This leads to corresponding shifts in the traffic flows. The transport attractivities of the cells do not change for the cases with or without S1.

10.8.3 Assignment Results

The assignement results show that the construction of the A81 leads to huge additional traffic loads in the Stuttgart-Herrenberg corridor augment. Table 10.2 presents the assignment results of the various cases. With the increase in traffic volume and the overall mileage, average travel distance and travel time increase as well in the study area. Private vehicle traffic increases more than public transport usage due to the construction of the A81.

The construction of the rail rapid transit line S1 increases public transport use and thus the modal split. Through diverting public transport trips from previously mainly bus transport in the Herrenberg area to rail, the extension of the rail rapid transit line to

Herrenberg leads to a clear decrease in public transport mileage as well as of average travel distances in public transport. A slight decrease in the mileage of private vehicle traffic can also be registered, which is caused by the construction of the rail rapid transit.

Table 10.2. Traffic effects in the study area

		Basic scenario: (with A81 without S1)	Planning scenario A: (without A81 without S1)	Planning scenario B: (with A81 with S1)	Planning scenario C: (without A81 with S1)
			Change in relation to the basic scenario		
Passenger trips [trips / 24h]	priv. veh. traffic	276.000	- 14,8 %	+ 0,2 %	- 14,6 %
	public transport	85.000	- 4,3 %	+ 1,9 %	+ 3,5 %
Modal-split [%]		23,5 %	+ 9,2 %	+ 1,3 %	+ 15,4 %
Mileage [km / 24h]	priv. veh. traffic	2.906.000	- 23,0 %	- 0,4 %	- 21,7 %
	public transport	775.000	- 7,6 %	- 27,1 %	- 24,5 %
Average travel distance [km]	priv. veh. traffic	11	- 9,6 %	- 0,7 %	- 8,3 %
	public transport	9	- 3,5 %	- 28,5 %	- 27,1 %
Average travel time [min.]	priv. veh. traffic	10	- 1,9 %	- 0,4 %	- 2,0 %
	public transport	26	- 3,0 %	- 3,8 %	- 2,7 %

10.9 Conclusions

The STASA transport model is based on the framework of a dynamic decision approach which is behavioural oriented and also highly non-linear. The STASA model contains several links between the traffic subsystem and the urban/regional subsystem and is able to describe a multi-modal transport system.

Contrary to the classical gravitational approach and/or four-stage model (generation, distribution, mode-choice, assignment) the STASA model proved to be able to estimate not only direct induced traffic components but also indirect effects. This is based on the fact that the trip frequency and the destination choice are related to a great number of individual decision processes, and therefore determined by the individual activities of the different agents of the system. For each trip purpose attractivities to each traffic cell can be assigned, representing the socio-economic structure (inhabitants, number of work places, purchase possibilities, leisure time possibilities and others) of the appropriate traffic cells. In addition to these cell related socio-economic factors, traffic demand and traffic supply dependent variables (e.g., accessible work places within a certain time interval) may be considered. This leads to a strong coupling between traffic levels and the urban/regional system. In this way, the time-of-the-day-dependent redistribution of the population between the traffic cells can be calculated as well as the long-term evolution of the urban/regional subsystem. Moreover, trip generation, distribution and mode-choice can be modelled in one step. Therefore, the STASA transport model corresponds to an aggregated trip-chain model.

The following conclusions can be drawn in relation to the specific case study:

- The motorway A81 has strong effects on the long-term population distribution in the considered corridor. A slight sub-urbanisation process can also be identified.
- The changed population distribution leads to a higher traffic load in the area.
- The rail rapid transit S1 has effects on the population distribution only in close vicinity to the train stops.
- The rail rapid transit S1 leads to a relief of the A81 and therefore of the TEN road network.

Therefore the simulations clearly demonstrate that an intensive interaction between the urban transport system, the different modes of transport system, and the long-term development of the region has to be addressed.

With regard to population development, during the first years after the introduction of the A81, only the study area faced additional increase in population, whereas the population numbers in the surrounding area and the region of Middle Neckar decreased slightly.

Possible effects of the trunk-road are even more obvious in the development of settlement structures in the study area. During the first years after the opening of the A81, the areas for housing and traffic use expanded considerably in the study area, whereas the surrounding areas in both of the comparison regions expanded only slightly during the same period of time. This development was mostly caused by increased activities in house-building. However, increased designation of building areas for trade use was also present in the study area.

In relation to the cost of land-use (that are one of the main determining factors in the development of settlement structures), it is observed that the most extreme rise in prices occurred especially in those parts of the region having a good access to the A 81. That period was also marked by the biggest difference in growth between the planned and the comparison areas.

Similarly, strong impacts of the trunk-road are obvious in the development of workplaces and employment structure. The manufacturing trade, the sector of economy that is most dependent on convenient traffic connections, is the most affected in that region. After the opening of the A81, this sector showed significantly higher growth dynamics in the study area than in the surrounding areas. The number of companies and the number of compulsorily insured employees active in the manufacturing trade, increased not only relatively but also absolutely. In light of this, both of the comparison areas complained about losses in the numbers of the employed in the manufacturing trade, whereas the planned region experienced significant increase rates in this field.

References

Aoyama Y. (1989), A historical review of transport and land-use models in Japan, *Transportation Research* 23 A, 53-61

Batty M. (1994), A chronicle of scientific planning: the Anglo-American modelling experience, *Journal of the American Planning Association* 60, 7-16

Berechman J. and Small K.A. (1988) Research policy and review 25: modelling land use and transportation: an interpretative review for growth areas, *Environment and Planning A*, 20, 1283-422

Boyce D.E. (1988) Renaissance of large-scale models, *Papers of the Regional Science Association* 65, 1-10

Der Bundesminister für Verkehr (Hrsg.) (1996): Qualifizierung, Quantifizierung und Evaluierung wegebauinduzierter Wegebauprozesse, Studie im Auftrag des Bundesministeriums für Verkehr, FE-Nr. 90436/95, Stuttgart

Eurosil, WP7 (1998): Modelling Multimodality, Intermodality, Interoperability, Accessibility and Area Impacts of Transport Infrastructure: State of the Art, Steinbeis Transferzentrum Angewandte Systemanalyse Stuttgart (STASA), Institute of Transport Science University of Cologne, Insitute of Spatial Planning (IRPUD), EU, to be published

Eurosil, WP6 (1998): SILUS Case Study, Steierwald Schönharting und Partner GmbH (SSP), Steinbeis Transferzentrum Angewandte Systemanalyse Stuttgart (STASA), EU, to be published

Göbel G. (1997) Untersuchungen über die Wechselwirkungen zwischen Siedlungsstruktur und Verkehrssystem am Beispiel des Ausbaus der A81 im Korridor Stuttgart - Herrenberg, Diplomarbeit, Trier

Haag G. (1989) Dynamic Decision Theory. Application to Urban and Regional Topics, Dordrecht

Harris B. (1985) Urban simulation models in regional science, *Journal of Regional Science* 25, 545-67

Harris B. (1994), Science in planning: past, present, future, *Journal of the American Planning Association* 60, 31-4

Hutchinson B., Nijkamp P. and Batty M. (eds.) (1985) *Optimization and Discrete Choice in Urban Systems*, Springer Verlag, Berlin/Heidelberg/New York

Hutchinson B. and Batty M. (eds.) (1986) *Advances in Urban Systems Modeling*, North Holland Publishing Company, Amsterdam/Oxford

Kain J.F. (1987) Computer simulation models of urban location, in Mills, E.S. (ed.), *Handbook of Regional and Urban Economics*, vol. II, Elsevier Publishers, Amsterdam, 847-75

Ministerium für Umwelt und Verkehr Baden-Württemberg (Hrsg.) (1996) Telematik im Verkehr, Regionales Verkehrsmanagement Stuttgart, Das STORM-Projekt, Verkehrswissenschaftliche Bewertung, Schriftenreihe der Straßenbauverwaltung, Heft 7, Stuttgart

SACTRA (1994) The Standing Advisory Committee on Trunk Road Assessment: Trunk Roads and the Generation of Traffic, Department of Traffic, London

Studiengesellschaft Nahverkehr mbH (SNV) (1990): Vereinfachte Nachfrageermittlung und Kosten-Nutzen-Abschätzung für Park-and-Ride-Anlagen, Teil 1: Nachfrageermittlung, in: Verkehr und Technik, Heft 3

Webster F.V., Bly P.H. and Paulley N.J. (eds.) (1988) *Urban Land-Use and Transport Interaction. Policies and Models, Report of the International Study Group on Land-Use/ Transport Interaction (ISGLUTI)*, Avebury, Aldershot

Wegener M. (1986) Integrated forecasting models of urban and regional systems, in *Integrated Analysis of Regional Systems*, London Papers in Regional Science 15, 9-24

Wegener M. (1987) Transport and location in integrated spatial models, in Nijkamp, P. and Reichman S. (eds.), *Transportation Planning in a Changing World*, Gower, Aldershot, 208-25

Wegener M. (1994) Operational urban models: state of the art, *Journal of the American Planning Association* 60, 17-29

Wegener M. (1998) Applied models of urban land use, transport and environment: state of the art and future developments', in Batten D., Kim T.J., Lundqvist L. and Mattsson L.-G. (eds.) *Network Infrastructure and the Urban Environment: Recent Advances in Land-Use/Transportation Modelling,* Springer Verlag, Berlin/Heidelberg/New York (forthcoming)

Weidlich W. and Haag G. (1983) *Concepts and Models of a Quantitative Sociology,* Berlin, Heidelberg, New York

Weidlich W. and Haag G. (1988) *Interregional Migration*, Berlin, Heidelberg, New York, 1988

Wilson A.G. (1997) Land use/transport interaction models - past and future, *Journal of Transport Economics and Policy* 32,1

Appendix A.10. The study area

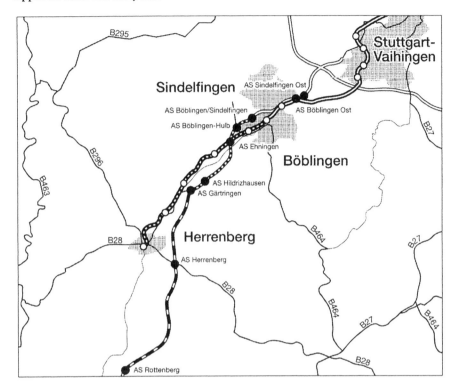

Regional Railways
≡≡≡ Opened 1985
▰▰▰ Opened 1992
○ Station

Road Network
Motorway:
≡≡≡ Already open
▰▰▰ Opened 1975-1977
▰▬▰ Opened 1978
● Junction
── Federal road
──── Minor road

11 Boundary-Swapping Optimisation and Region Design

Bill Macmillan

11.1 Introduction

The project described in this chapter, which was conducted in the mid 1990s, had some unusual characteristics. From a regional science perspective, it lay outside the core area of spatial economics. Analytically, it had as much to do with algorithms as models. Procedurally, it was more concerned with education than model implementation. And for the client, its value was as much institutional as technical. The project was conducted for a national Electoral Agency. It was part of a redistricting exercise on the boundaries for national elections. The country, which for reasons of client confidentiality must remain nameless, was regarded as having suffered in the past from widespread gerrymandering and the Electoral Agency was responsible for preparing a new, defensible, electoral map. The larger part of this exercise, in terms of labour and computing time, involved the production of a spatial database containing digitised boundaries and statistical information for census tracts. The particular element with which we[1] were concerned was the assembly of census tracts into constituencies. We were not asked to set up and run a procedure for generating constituencies but to advise on the scope for, and difficulties inherent in, the use of so-called active redistricting methods. In particular, we focussed on the use of boundary-swapping optimisation methods in a geographical information system (GIS) environment (for more details on GIS see chapters 15-17).

Despite the peculiarities of the project, it had at its core one of the classical problems of regional science: the production of a set of regions from a larger set of zones in order to satisfy certain criteria. This problem has numerous applications across the private and public sectors. In its basic form, it is easy to state but hard to solve because of its combinatorial nature. The next section sets out the bare bones of the problem, explains why it has been resistant to approaches involving optimisation, and gives a brief outline of a computational strategy that circumvents the principal impediment.

The difficulties over the use of optimisation are not merely technical. It is not at all clear at first sight what it might mean to say that a particular set of regions is optimal

---

[1] The consultancy exercise involved the author and Dr. Todd Pierce, a former graduate student of the School of Geography, University of Oxford.

in a socio-political context. Moreover, the nature of the constraints and their technical representation are equally problematic. The third section explores these issues in the context of a discussion of the client's requirements.

Section 11.4 contains a brief review of data issues, part of which follows directly from the observations on possible objectives and constraints. The review also includes a description of the type of locational data that is needed if the GIS-based strategy described in section 11.2 is to be employed.

The development of methods tailored to meet the complex requirements of redistricting is discussed in section 11.5. The methods include different strategies for dealing with multiple, often ill-defined objectives.

The final section is concerned with the outcome of the project, which had an important institutional dimension. The consultancy exercise appeared to reinforce the position of the technical team within the Electoral Agency by providing external validation for their work. To use an academic analogy, the role we were asked to assume became more like that of external examiners than contract teachers.

11.2 Designing Regions

The problem of region design may be conceptualised in a number of different ways. The territory to be divided into regions can be treated as a continuous plain, as a collection of unconnected points, as a set of nodes on a network, or as a plain partitioned into a finite number of zones. In the latter two cases, the spatial relations represented by the map need be no more than connectivities or contiguities. In political districting, this is what matters. Distance is not a major factor (although it can feature in measures of compactness and other design criteria) but contiguity is. Typically, there is a legal requirement for constituencies to be contiguous. In addition, constituency boundaries need to be well defined and are often required to follow the boundaries of smaller statistical or administrative areas and it is these areas for which basic redistricting data – such as voter populations – are available. It is sensible therefore to conceptualise redistricting problems in terms of a map partitioned into a finite number of well defined zones.

11.2.1 The Design Problem

It is the contiguity requirement that makes the region-design problem challenging. It functions as a constraint on the design process which is easy to understand but awkward to specify mathematically and difficult to handle algorithmically (see Macmillan 1999). The primary objective in districting problems tends to be the achievement of fair representation. This can be operationalised in a variety of ways around the theme of equalising voter populations. There may well be other criteria, in the form of legislative requirements or custom and practice, that have to be allowed for as additional constraints or subsidiary objectives, but it is convenient to restrict the discussion for the time being to what can be regarded as the basic redistricting problem: given a map consisting of N zones, the voter populations of which are known, find M regions, where M<N, such that each zone is allocated to one region, each region is contiguous, and the deviation of the regional voter populations from the average is minimised.

11.2.2 Solution Strategies

Having defined the problem, the next task is to outline a solution strategy. One of the simplest ways to circumvent the difficulties associated with the contiguity restriction is to grow regions from initial 'seeds', where the seeds are M randomly selected zones. Contiguity is guaranteed, at relatively little computational cost, by ensuring that a zone is added to a growing region only if it is a neighbour of a zone already in the region. The disadvantage of this approach is that each run generates a single trial solution so a very large number of runs is required in a well-structured trial framework to make any headway.

Another strategy is to begin with a contiguous plan and to modify it step by step through the movement of a single zone from one region to another. This boundary – swapping approach has both technical and practical attractions. Technically, it limits the contiguity problem to one of checking that a prospective move does not breach the contiguity restriction. Practically, it is often the case that there is an existing plan that requires modification and it is helpful to be able to demonstrate that the new plan is an improvement on the old. Moreover, given that the structure of the problem tends to generate multiple, local solutions, it may be advantageous to search in the vicinity of the existing plan so that the improvement can be regarded as incremental. Radically changed constituency plans can fuel suspicions of gerrymandering.

Notwithstanding these observations, it may be possible to achieve large improvements in the equality of voter distribution by making radical changes. The existing plan may be discredited in any case. Thus, a good solution strategy must be capable of escaping local optima and must generate some confidence that it can find solutions that are at least close to being globally optimal. Computationally, this is a hard task because it requires a search through a very large number of solutions. Infeasible solutions are eliminated automatically by both of the strategies described above but, for many practical problems, the number of feasible solutions is still daunting.

11.2.3 Solution Spaces

It is possible to think of the problem's solution space as having a tree-like structure, where the tip of each twig represents a single plan. Coming down the twig to its base and going back up to the tip of a neighbouring twig represents a change in the plan through the movement of a single zone from one region to another. The junction between the twigs can be thought of as the point at which the decision over the allocation of the single zone is pending. The next junction down the branch represents the point at which the allocation of that zone and one other both need to be made. The set of twig ends that can be reached from this new junction represents the set of all possible ways of making the two allocations. The argument repeats as the tree is descended, with the number of unallocated zones increasing each time.

The analogy is strained by the fact that the tree has the strange mutation that the tips of different twigs are fused; thus, the same tip can be arrived at going up the tree in different directions. Similarly, the points where branching occurs can have 'arriving' limbs as well as 'departing' limbs. With this exception, the analogy works reasonably well. The globally optimal solution can be thought of as the highest twig tip but there are local optima all over the tree.

Growth algorithms consist of the repeated selection of a low branching point and an ascent of the tree to the tip of a twig. They avoid the problem of becoming trapped at a local optimum by repeating the search process from scratch for each trial solution. Boundary swapping consists of a movement around the crown from one twig tip to another. For this procedure to avoid local traps, it must have the capacity to backtrack to solutions with inferior objective-function values (lower twig tips). Bowdry (1990) was the first to suggest the use of simulated annealing for this purpose in a redistricting context.

Simulated annealing gets its name from an analogy with a physical process involving the slow cooling of a metal. Slow cooling translates into a large number of trial solutions, a significant proportion of which will involve a deterioration in the objective function value. Thus, the problem of avoiding local traps is circumvented at a cost and that cost is considerable if the contiguity-checking procedure is slow.

11.2.4 The Switching Point Method

The standard method for checking contiguity has been that of Openshaw and Rao (1995). It is effective but time consuming. When a zone is identified as a candidate for shifting between regions, the procedure is to check that every remaining zone in the donor region can still be reached from a given zone without crossing the regional boundary. This is a non-trivial task and it increases in magnitude as region size increases.

The so-called switching point method relies on a different strategy based on a topological observation. When a zone is removed from a region, there is only one new section of regional boundary if contiguity is not lost but there are two sections if it is – one each in the two parts of the newly-divided region. By inspecting the boundary of a zone flagged for removal, the number of points at which the boundary changes from being internal to the region to external (i.e., to being part of the boundary of the region itself) can be counted. If there are no such switching points, the zone is in the interior of the region and can be removed without loss of contiguity but cannot be attached to another region without making the latter discontiguous. Similarly, if there are four such points, then there are two sections of boundary and contiguity would be lost. Two switching points, on the other hand, indicates that the zone could be removed and attached to a neighbouring region without loss of contiguity. The switching point method, coupled with a simulated annealing search procedure, provides the basis for an efficient algorithm in which a fair degree of confidence can be placed. What is more, it is a procedure which can work efficiently in a GIS environment, provided the boundary data are in an appropriate form (see below).

11.3 Clients Needs

The work with the Electoral Agency arose from the publication of a paper on the above issues in the proceedings of the I14 Meeting of the NCGIA in San Diego (Macmillan and Pierce, 1994). That paper alerted the Agency to the possibility that so-called active redistricting procedures might be practicable in a GIS environment. Passive redistricting involves the manual re-assignment of zones to regions and a number of proprietary GIS have a utility for this purpose. Active redistricting can be thought of as an automation of this point-click-move procedure, using an algorithm of the kind described above.

Prior to our involvement, the Electoral Agency had made a substantial investment in developing an electoral GIS using Intergraph technology. Working to a very tight deadline, the technical group decided to explore the feasibility of providing some automation of the next stage of the process, that of constituency design. Our brief was to provide advice, primarily through a series of presentations and consultations on the socio-political and technical aspects of redistricting. We were not asked to supply a turn-key redistricting package or to build a new utility to graft on to the client's system.

At the time of the consultancy exercise, Macmillan and Pierce (1994) appears to have been the only paper published on the problem of active redistricting in a GIS environment. This is significant because it indicates that the approach was by no means well established. In addition to its technical novelty, it challenged the conventional wisdom of the political science community. Indeed, at a later conference on the problem of boundary determination, the proceedings of which were published by McLean and Butler (1996), the whole notion of seeking an optimal solution to what is, after all, a socio-political problem, was called into question.

Part of the difficulty is that redistricting problems often involve multiple criteria. Moreover, the criteria can be difficult to articulate as measurable objectives or constraints. Amongst the criteria that have been used in various contexts are maximisation of compactness, promotion of minority representation, protection of incumbents, and preservation of natural communities. The first two pull in opposite directions. The pursuit of compactness is supposed to guard against the emergence of strung out constituencies, like Governor Gerry's salamander (see Butler and McLean, 1996). On the other hand, an attempt to secure minority representation (which is usually taken to require the construction of one or more constituencies in which a minority group nationally forms a majority locally) can generate strung-out constituencies like Congressional District number 12 in North Carolina (see The Economist 1992). Incumbent protection is rather like minority representation in that it is unambiguous and has a clear intended beneficiary. This contrasts with the preservation of natural communities, which is widely regarded as being important but is highly ambiguous. Any number of ways can be found to measure the naturalness of a region and some measures run directly counter to others. Rivers and watersheds are capable of being treated as both natural boundaries and natural centre lines for communities. Journey-to-work patterns can be used to define a natural community associated with a workplace but there are plenty of places for which such patterns would be unhelpful or downright misleading. The same can be said of other candidate measures. Even if these difficulties could be overcome, legislative frameworks tend to give a fair amount of scope for choosing between subsidiary criteria and it is often necessary to make such a choice because 'fair distribution' is insufficiently discriminating (it can generate a large number of solutions that are more or less equally good).

The general thrust of this argument is that it can make sense for an interested political party, say, to talk about an arrangement which is optimal or near optimal from its perspective but that it is difficult to take an Olympian view of what is optimal with respect to the public interest. The line we took in discussions with the Electoral Agency was that one aspect of serving the public interest is transparency. If an active redistricting system is used to generate an optimal plan, the criterion with respect to which the plan is optimal is clear. Moreover, the performance of any plan can be assessed with respect to all proposed criteria, provided they are measurable, and plans that are optimal with respect to any one criterion, or any particular combination of criteria, can be generated. Thus, if claims are made by an interested party that a particular plan best satisfies some public interest criterion, those claims can be tested. In addition, if an interested party argues that a certain objective is desirable in one place (related, say, to what constitutes a natural community) then the consistency of their attachment to that objective can be tested in other places.

Transparency and consistency are undoubtedly important but they constitute rules for judging plans and plan-making rather than design criteria. Thus, there remains some force in the argument that it is not possible to find an optimal redistricting plan. However, it is important to appreciate the nature of the impediment. There may be no optimal choice of criteria or optimal choice of weights to be attached to them but, for any given set of criteria and weights, an optimal plan can be generated. It is necessary, therefore, for there to be a social process by which criteria are chosen within which a technical process of plan generation is embedded. The social process might lead, for example, to clear legislation on what is required of a plan. On the other hand, the process might be partly legislative and partly consultative. The U.K. system, for example, has provision for public enquiries, which allow arguments over such matters as the nature of local communities. In some contexts, the legislative framework is extremely fluid. The position of the Electoral Agency on this spectrum is noted in section 11.4.

The choice of design criteria is only part of the problem. Reference was made above to the possibility of attaching weights to criteria but, as readers familiar with multiple-objective programming will know, there is rather more at issue than this. In optimisation problems, design criteria can appear in both the objective function and the constraints. If there are several criteria in the objective function, they tend to be combined linearly with the weights summing to one. The sizes of the weights indicate the relative importance attached to each objective. This notion carries over in a modified form to criteria that appear in constraints. In the redistricting case, the contiguity criterion is a Boolean criterion because it is either satisfied or not satisfied but other criteria do not have this property. For example, compactness tends to be measured by a real number so its incorporation as a design constraint requires a threshold level. This level might be set as a percentage of an optimal value obtained by solving the problem of finding the most compact plan in the absence of all other non-Boolean objectives and constraints. The relative sizes of the percentage figures used in a set of constraints of this kind are indicative of the relative importance attached to the criteria embedded in the constraints.

Unless legislation is very restrictive, as in some U.S. contexts where the only admissible criterion is equalisation of voter populations, there will be a need for interactions between the social and technical processes of redistricting. Active redistricting technology can contribute to, rather than substitute, for these interactions.

11.4 Data Requirements

The data issues facing the Electoral Agency with respect to the redistricting exercise were part of the more general problem of assembling an electoral database. This was related, in turn, to the formidable task of registering electors. For the redistricting exercise, the basic requirements were to obtain spatial data on a set zones and demographic and social data on the electors living within them. To use our active redistricting techniques, including the switching point method, it was necessary for the spatial data to have specific properties.

11.4.1 Locational Data and the Switching-point Method

The switching point method depends on the availability of data in an arc-node form. This type of spatial data is standard in ARC/INFO but it is not the preferred format in all GIS. An arc is a line feature defined by the co-ordinates of a 'from' node, a 'to' node, and a set of points along the line between them. A polygon consists of a set of arcs joined at their nodes to form a closed loop. A set of polygons consists of a system of arcs such that each arc defines the whole of the boundary between two polygons, where the 'outside world' is regarded as a polygon that surrounds the remainder of the system. In such a system, each arc can be characterised by the labels of the polygons or zones that lie either side of it. All the switching point method requires is two tables based on these data. The first table has one row for each zone in which the columns are filled by the numbers of the zones around its boundary, written in order from an arbitrary starting node. The second table has an identical structure except that the column entries are the numbers of the regions to which the corresponding zone currently belongs. Thus, the first table contains locational data and the second contains the problem's variables. Details of the method are given in outline in Macmillan and Pierce (1996) and in full in Macmillan (1999).

In ARC/INFO, the data table can be generated using the macro language AML but it is necessary to begin with a system of polygons as described above. The Electoral Agency's database contained quite a number of zones that were not closed polygons because the boundaries of census tracts were not all defined; this applied particularly in mountainous areas where there was no population to allocate to one zone or another. In addition, there were a number of islands for which rules were needed to establish the mainland zone(s) with which they were notionally contiguous. There was a further technical twist in that some areas currently had one constituency wholly surrounded by another and there was no legislative reason why such an arrangement should not feature in the new plan; our algorithms had not been designed with this possibility in mind. These challenges were surmounted relatively quickly.

11.4.2 Measuring Socio-political Objectives

The problems associated with the use of socio-political measures in the redistricting process were partly due to the availability of data but also sprang from the relative openness of the legislative framework. Data on the distribution of population by ethnic group were available but there were few other series that could have been used in connection with the idea of natural communities. Measures of compactness could have been derived from the locational data or, with more subtlety, the locational data combined with zonal populations. However, it was unclear from the legislative framework what kinds of measures could or should be used, other than total voter populations.

11.5 Development Strategies

The relative openness of the problem could have been handled in a number of different ways. The objective of minimising the deviation in voter populations from the norm could have been used on its own. The advantage of this approach is its simplicity and transparency. However, it has two potential disadvantages, as suggested earlier. First, it may throw up constituencies that fall foul of informal but generally held notions of natural community. For example, it may produce a solution in which a suburb of one city forms a constituency with another town and the connecting rural area. Second, it can generate a large number of solutions with quite different appearances in map form that do not differ greatly in terms of their optimality. Thus, any attempt to propose a solution that is only marginally better than an alternative is open to question if there is any allowance for the use of subsidiary criteria. The first disadvantage may compound the second if the optimal solution contains 'unnatural' constituencies.

11.5.1 Multiple Objective Methods: Ex Post Weights

This suggests that an approach involving multiple objectives would be advantageous but there are difficulties here as well. In addition to the problem of articulating criteria as measurable objectives or constraints, there is the problem of arbitrariness in the choice of objective weights. If a single objective approach can yield a number of candidate plans then so can one with multiple criteria if it proceeds by combining the criteria into a single objective function. Given that there is arbitrariness in the choice of weights, preference for one member of the set of candidate plans over another is also, to some degree, arbitrary. The same argument applies when some of the criteria are included in the form of constraints.

An alternative strategy, which reduces the problem of arbitrariness, is to proceed without weights. The strategy relies on the production of a large number of alternative plans and their evaluation with respect to each member of a set of selected criteria. Generating many plans is straightforward when using simulated annealing. In a single-objective problem, the solution process is to store the best plan found so far whilst amending a working plan (making it better or worse at each step as the algorithm dictates). With multiple criteria, there are a number of possible variations on this theme.

One such variation is to continue to use the single 'fair representation' criterion to drive the search procedure but with the storage rule modified to keep track of every plan that is no worse than the stored plans in a well-defined sense. To talk about a plan being no worse than others when there are multiple criteria implies that some assumptions are being made about a so-called optimality frontier. No problem arises in classifying a new plan if it performs less well than some existing plan on all criteria – it is clearly dominated by the existing plan and can be rejected. However, if it performs better on at least one criterion, it may be worth retaining.

To decide whether it is or not, depends on the optimality frontier. To picture the frontier, assume that there are K criteria and that each plan may be represented by a point in a K-dimensional space, such that the co-ordinates of the point represent the scores of the plan on the K criteria. Assume also that, on all criteria, high scores are better than low (it is easy to transform minimisation criteria into equivalent maximisation criteria). Now imagine putting an envelope around the cloud of points such that all points lie on or beneath it (i.e., on the envelope or between it and the origin). The envelope is the optimality frontier. For simplicity, it may be assumed to be defined as a piecewise linear surface in which each facet is a linear combination of a subset of the points. If a new plan lies beneath the frontier, it may be rejected as being worse than a linear combination of certain existing plans. For *any* system of weights attached to the criteria, there would be an existing plan that was better than the new plan.

There is a difficulty that ought to be mentioned straightaway. It is a nontrivial task computationally to keep revising the frontier as it is pushed out by the discovery of improved plans. It is likely to be more efficient to keep all plans that are not dominated and to generate the frontier at the end of the search procedure using Data Envelopment Analysis (DEA). The net result will be the plan that performs best (of those enumerated) on the fair representation criterion plus a surface in criterion space across which trade-offs can be made. Note that the surface defines the range of weights under which each of the plans that mark its vertices would be optimal. Thus, starting from the single-criterion best plan, it is possible to move to a neighbouring plan if the sacrifice in terms of fair representation is within an acceptable range, given the improvement that can be made in other criteria.

11.5.2 Practical Exigencies

Other variations on this algorithmic theme are possible but none had been developed at the time of the project with the Electoral Agency. Indeed, the above structure was little more than a thought experiment. It represented a development path for redistricting problems in which the technical process is firmly embedded in a social process for determining acceptable trade-offs between clearly defined criteria. The problems faced by the Electoral Agency were somewhat different. Their general objective was to produce a modern electoral system with clearly defined constituencies and properly registered electors. Their immediate priority was to take – and to be seen to be taking – all reasonable steps to ensure that the redistricting process was fair. Moreover, they were facing a very tight timetable and the scope for experiment was extremely limited.

11.6 Conclusions

By utilising a GIS, the Electoral Agency had already placed itself at the sharp end of the redistricting world. To have decided to explore the limits of the technology by examining the possibility of employing active redistricting algorithms demonstrated the Agency's pioneering spirit. The technical team was dedicated to achieving a good outcome through the employment of the best technical means available.

11.6.1 The Technical Dimension

As noted above, the core of the consultancy exercise was a series of presentations and discussions on both the technical and socio-political aspects of redistricting. To this extent, the product of the exercise was the process itself and the knowledge exchanged therein. However, the core sessions were interspersed with two other types of occasion.

On the technical front, there were meetings with the Agency's programmers to look at the feasibility of constructing a contiguity checker compatible with their system and the possibility of building a simulated annealing structure for it to plug into. Somewhat to our surprise, the contiguity problem was tackled relatively quickly. The simulated annealing structure presented a more difficult challenge because of system incompatibilities. As it was not possible to import a completed package, it proved necessary to provide a specification for an algorithm for later encoding.

11.6.2 The Institutional Dimension

The other type of occasion was bureaucratic and political. Through a series of meetings with increasingly senior members of the Agency, it became clear that the institutional dimension of the exercise was every bit as important as the technical.

There was a shared understanding from the outset that redistricting processes evidently involve a balance between the technical and the socio-political. One of the distinctive characteristics of redistricting is that it is easy to make judgements about plans without any knowledge of the technical process used to generate them. At one level, therefore, the nature of the technical activity is unimportant politically. What is important, however, is to establish that the technical process is one in which confidence can be placed. Within the Electoral Agency, part of our function as consultants, in practice, was to provide external validation for the work that the technical team was doing. Happily, that work was impressive. They were well versed in the use of GIS in electoral planning and, through the consultancy, were in a position to develop and implement an active redistricting utility. Moreover, the team acquired a good understanding of the limitations of the technology and seemed unlikely to fall prey to false expectations.

The use of GIS in redistricting is still in its infancy and active redistricting utilities are best thought of, at this stage, as research tools. The most exciting aspect of the consultancy exercise with the Electoral Agency was their keenness to explore the capabilities of such tools. Adventurous clients with well-informed, technically competent teams are vital to the process of transferring research ideas into practical propositions.

References

Bowdry M. (1990) Simulated annealing – an improved computer model for political redistricting, *Yale Law and Policy Review*, 8, 163-179.

Butler D. and McLean I. (1996) The redrawing of parliamentary boundaries in Britain. In I. McLean and D. Butler (Eds.) *Fixing the Boundaries: Defining and Redefining Single-Member Electoral districts*, Dartmouth, Aldershot, pp 1-38.

Economist, The (1992) Congressional redistricting: no grey areas. *8 February 1992*, pp. 48-49.

Macmillan W. (1999), Optimal zone design: SARA 1 (forthcoming).

Macmillan W. and Pierce T. (1996) Active computer-assisted redistricting. In McLean I. and Butler D. (Eds.) *Fixing the Boundaries: Defining and Redefining Single-Member Electoral districts*, Dartmouth, Aldershot, pp 219-234.

Macmillan W. and Pierce T. (1994) Optmization modelling in a GIS framework: the problem of political redistricting. In Fotheringham S. and Rogerson P. (Eds.) *Spatial Analysis and GIS*, Taylor and Francis, London, pp 221-46.

McLean I. and Butler D. (Eds.) (1996) *Fixing the Boundaries: Defining and Redefining Single-Member Electoral districts*, Dartmouth, Aldershot, pp 1-38.

Openshaw S. and Rao L. (1995) Algorithms for reengineering 1991 Census geography. *Environment and Planning A 27,* pp 425-46.

12 Optimal Distribution Strategies

Mark Birkin and Richard Culf

12.1 Introduction

This chapter examines the relationship between methods developed in the regional science community, and the application of those methods within practical business planning contexts. It is based on the experience of the authors over the last decade within a commercial organisation, GMAP Limited, which has grown out of the School of Geography at the University of Leeds. The linkages between GMAP and the University of Leeds continue to be strong ones. GMAP's historical 'core competence' has been in the field of spatial interaction modelling, and this in itself builds directly on a strong research tradition within the School of Geography at Leeds (e.g., Wilson, 1970, 1974; Harris and Wilson, 1978; Clarke and Wilson, 1983). Without doubt the greatest achievement of our business has been to find practical, robust and effective ways to implement these largely theoretical urban planning models within large commercial organisations, including banks, supermarkets, motor vehicles distributors, telecommunications and oil companies. GMAP has generated a turnover in excess of £20,000,000 over the last ten years through location modelling (see also Clarke and Clarke, chapter 8).

Consider a retail or service activity which is expanding and therefore needs to identify new sites for development. To fix ideas, it may be helpful to think of a specific organisation, such as the McDonalds Quick Service Restaurant (QSR) chain which currently faces this issue in most if not all European countries. The impact of a new site may be evaluated through the construction of a market model which can assess the relationship between the local population and its propensity to consume hamburgers, the provision of similar products by competing retailers, and the activity (spatial interaction) patterns which connect retailers to their customers. The model may be 'calibrated' against the experience of existing customers and outlets to provide estimates of the potential turnover of a new outlet and perhaps its impact on the trading patterns of other outlets in the network.

In addition to modelling outlet performance, there are many complementary and related questions which demand resolution through other methods. For example, in planning the development of a network, where will the population be in ten or twenty years time? How is it possible to reconfigure a whole network of outlets to improve profitability or service levels? How do local market areas interact with one another and what effect does this have on advertising or sales territory planning?

In this chapter, we will look specifically at how methods of spatial optimisation can add value within an applied business context. There are two aspects to spatial optimisation. First, we will look at applications of regionalisation techniques to commercial problems in terms of client needs in Section 12.2. This builds on the material introduced and discussed more broadly by Macmillan in chapter 11. Second, in Section 12.3, we look at approaches to the problem of branch network optimisation. In the latter case particularly, we believe that the methods developed here are of genuine scientific interest and bear comparison with any of the techniques reported elsewhere within the academic literature. In Section 12.4, we describe a variety of examples that will show how both of these technologies have been used in practice. Concluding remarks will be offered in Section 12.5.

12.2 Client Needs

12.2.1 Territory Optimisation

Consider a situation in which a pharmaceuticals company is about to launch a new drug. A sales force must be recruited and each salesperson assigned responsibility for key accounts within hospitals or health centres. In order to set targets, monitor performance and to balance workloads, it would be reasonable to create territories which are equal in 'size', where size is likely to be determined with reference to the number of accounts in each territory. In order to maximise the time spent with clients, it makes sense that responsibility for an account falls to the nearest sales representative.

A second example can be discussed in relation to the motor industry. Most of Europe's car market is heavily over-dealered (see table 12.1) and dangerously unprofitable. This leads to a situation of 'intra-brand' competition in which dealers for an individual manufacturer are competing for trade against one another, as well as against dealers of competing brands. This is clearly unhealthy from a manufacturer's point of view. Whilst at first sight it may appear that this kind of competition is healthy from a customer choice point of view there are also drawbacks for customers:

- often a customer may be tied in to an aftersales warranty with an inappropriate dealer – for example, if a purchase is made on the basis of a dealer near to one's place of residence, but the vehicle need to be serviced from the place of work
- excess network capacity means wasted cost in facilities which ultimately the customer will bear.

Table 12.1. Comparative statistics for European dealer networks

Country	Population	New Vehicle Sales	Dealers	Dealers/head	Cars/Dealer
Italy	57,062,670	1,718,800	5,726	10	300
Portugal	9,868,037	219,518	1,094	11	201
UK	54,853,618	2,025,420	7,712	14	263
Sweden	8,816,381	185,084	1,719	19	108
Finland	5,098,754	94,785	1,031	20	92
Ireland	3,525,719	86,232	749	21	115
Netherlands	15,344,668	472,704	3,614	24	131
Spain	39,433,942	973,670	10,486	27	93
Denmark	5,213,472	140,656	1,722	33	82
Germany	81,262,432	3,315,681	28,911	36	115
France	56,738,473	2,133,958	21,865	39	98
Norway	4,344,482	131,916	1,862	43	71
Austria	7,795,786	275,230	3,511	45	78
Belgium	9,941,896	420,810	5,050	51	83
Switzerland	6,888,037	269,245	3,855	56	70

For these reasons, there is now a trend towards the creation of substantially larger dealer areas of responsibility (DARs) in which an individual dealer will offer a combination of facilities i.e. new car showroom, vehicle servicing, parts distribution and storage etc.

In order to address this problem, GMAP has created the concept of a Customer Marketing Area (CMA). A CMA is an area that is highly self-contained in terms of the activity patterns of the people who live in them, whether those activities are trips to work, shops, leisure facilities and so on. CMAs need to be comparable in terms of size, so that similar opportunities are available to neighbouring dealer franchisees. This constraint may need to be relaxed in rural areas if an area becomes too large to be physically manageable.

CMAs are equally applicable for analysis and planning purposes in each of the different markets outlined in Section 12.1, as well as motor vehicles distribution.

Our usual procedure is to define containment in terms of journey-to-work flows:

T_{ij} – the number of people living in zone i and working in zone j;

CMAs can now be defined in two ways – the proportion of people who are resident in a macro-area (I) who also work in that area:

$$C_I^1 = \sum_{i \in I, j \in I} T_{ij} \bigg/ \sum_i T_{ij} \qquad (12.1)$$

or the proportion of people who work in an area who are also resident in that area:

$$C_J^2 = \sum_{i \in J, j \in J} T_{ij} \Big/ \sum_j T_{ij} \qquad (12.2)$$

Given that the purpose of CMAs is to identify areas of customer activity, a focus on C_I^1 is more logical and therefore generally adopted.

The number of CMAs to be created may be determined in relation to the minimum or average level of containment which is sought; or by the average market size within each area; or by any other appropriate criterion. An iterative process may be adopted in which the problem is solved for a given number of CMAs, and the number of areas is then reviewed in the light of the quality of the solution produced. The optimisation problem may be written as follows:

$$\max_{\{I\}} \sum_I C_I^1 P_I \Big/ \sum_I P_I \qquad (12.3)$$

s.t.

$$C_I^1 = \sum_{i \in I, j \in I} T_{ij} \Big/ \sum_i T_{ij} \qquad (12.4)$$

$$\sum_{i \in I, j \in I} d_{ij} \geq 1 \qquad (12.5)$$

$d_{ij} = 1$ if i is contiguous with j

The problem can be solved using an algorithm that has been created and refined by GMAP over the last three to four years. A partition of the UK into 120 CMAs is shown in table 12.2.

Table 12.2. CMA partition of UK

Region	Area Code	Area	JTW Containtment	DM Containment
East	006	Cambridge	82.5	65.8
East	010	Colchester Area	82.3	78.9
East	022	Lincoln	88.2	75.9
East	025	Milton Keynes	77.6	71.8
East	035	Norwich	94.7	84.8
East	041	S Essex	67.1	74.8
East	056	Watford	64.4	70.9
East	057	Welwyn Garden City	69.2	67.0
Gtr London	009	City	50.5	52.3
Gtr London	012	E London	53.3	66.3
Gtr London	026	N E London	46.6	54.8
Gtr London	029	N London	37.3	51.2

Cont...

Region	Area Code	Area	JTW Containtment	DM Containment
Gtr London	032	N W London	39.9	49.2
Gtr London	039	S E London	45.3	55.5
Gtr London	042	S London	41.4	48.9
Gtr London	044	S Thames	46.3	50.3
Gtr London	045	S W London	62.0	61.8
Gtr London	052	W London	53.6	56.0
Gtr London	058	West End	61.6	55.8
Midlands	002	Birmingham	87.2	88.9
Midlands	004	C Midlands	86.9	74.6
Midlands	013	E Midlands	86.9	77.8
Midlands	027	N E Midlands	91.7	87.4
Midlands	031	N Midlands	83.1	73.9
Midlands	043	S Midlands	81.7	76.0
Midlands	053	W Midlands	73.6	78.1
North	005	C Yorkshire	72.2	82.4
North	014	E Yorkshire	91.9	87.5
North	019	Halifax	83.4	86.0
North	021	Leeds	85.3	82.3
North	034	Northumbria	95.4	93.9
North	047	S Yorkshire	89.5	91.2
North	050	Teesside	89.8	85.6
North West	007	Cheshire	71.3	73.5
North West	008	Chester, Wirral &	87.9	85.8
North West	011	Cumbria/Fylde	94.3	86.5
North West	023	Manchester	78.6	78.2
North West	024	Merseyside	85.9	85.2
North West	030	N Manchester	71.0	76.3
North West	037	Preston	84.3	81.2
S Wales & W	001	Avon	94.4	84.6
S Wales & W	016	Exeter/Somerset	94.0	78.8
S Wales & W	017	Glouc/ N Wiltshire	89.7	80.6
S Wales & W	036	Plymouth/Cornwall	96.8	87.0
S Wales & W	040	S E Wales	94.2	88.6
S Wales & W	046	S Wilts/Poole Ba	88.8	72.4
S Wales & W	055	W Wales	93.0	87.6
Scotland &	003	C & E Scotland	80.9	81.9
Scotland &	015	Edin, Loth & Bord	94.4	91.6
Scotland &	018	Grtr Glasgow & A	92.8	88.2
Scotland &	028	N Ireland		98.0
Scotland &	033	NE Scotland	98.1	92.9
Scotland &	054	W Scotland & Lan	80.0	90.3
South East	020	Kent	81.9	79.3
South East	038	Redhill	70.8	62.0
South East	048	Solent	90.1	83.1
South East	049	Sussex Coast	84.7	73.3
South East	051	Thames Valley	79.7	76.4

The structure of this problem is rather complex and it cannot be guaranteed that the solutions produced are optimal or near optimal. In practice, benchmarking is achieved against containment levels within existing networks, or against randomly generated configurations. The problem and its solution are clearly similar to a number of attempts within the regional science literature to create regional partitions from small area data, and in particular to the Travel to work areas (TTWAs) which were created for the Department of Employment in the UK at CURDS in the 1980s and still widely used as a basis for government statistics.

12.2.2 Branch Network Optimisation

The modelling approach becomes an order of magnitude more complex when the interactions – previously taken as given – are themselves dependent on the underlying network structures. The nature of the problem is to find a set of outlets (or network) which is optimal but in which the quality of the solution can only be evaluated with reference to a complex consumer behaviour model.

Suppose, for example, that the local market for motor vehicles can be described accurately using a spatial interaction model (see section 12.3 and chapter 8). A manufacturer or franchised motor vehicles distributor wishes to identify a network of locations which generates maximum sales through a profitable suite of locations. This is precisely the problem faced by manufacturers throughout Europe who are struggling with distribution networks which are typically too large and often unprofitable. Similar issues are faced in the retail financial services industry under pressure from new forms of distribution such as call centres and internet banking; and in the automotive fuels market where the number of outlets in the UK has fallen from 26,000 to 14,000 over the last ten years (Guardian, July 16, 1998).

12.3 The Optimisation Models and Results

12.3.1 Territory Optimisation

Let us recap on the problem of the pharmecutical firm outlined in section 12.2.1. This problem can be described algebraically as follows:

$$\underset{\{\delta_i\}}{Min} = \sum_i \left\{ (\sum_j p_{ij} y_j) - \frac{\sum_j y_j}{N} \right\} \qquad (12.6)$$

s.t.

$$\sum_i \delta_i = N \tag{12.7}$$

$$p_{ij} = 1 \text{ if } \delta_i = 1 \text{ and } d_{ij} < d_{ik} \quad \forall k \neq j \tag{12.8}$$

where:
N is the number of representatives
y_j is the workload associated with area j
δ_i is the probability that there is a representative at location I (either zero or one)
d_{ij} is the distance from i to j.

For a given number of territories, this problem can be solved using an algorithm of the following type:

1) Locate N representatives at random
2) Set weights for each representative to be 1
3) Assign each zone to the nearest representative (where 'nearest' is evaluated after applying weights to each distance)
4) Locate each representative at the centre of his/her assigned region. Iterate through Steps 3 and 4 until there is no change in the representative locations
5) Calculate workloads for each representative
6) Adjust weights upwards for representatives with above target workloads. Adjust penalties downwards for representatives with below target workloads.
7) Iterate through Steps 3 to 6 until there is no change in workloads.

A simple real life application of this problem is considered now. Here a new drug has been launched and initially it is targeted at users within 42 Key Cancer Centres (KCCs) within the UK. The workload associated with each KCC is assumed to be proportional to the number of hospital consultants employed. Eight sales representatives are to be recruited to meet the requirement. Each representative should have an equal workload, except for one who will also have managerial responsibilities, and will therefore work part-time as a salesman. The ideal solution to this problem is shown in figure 12.1. The part-time representative is located within West Yorkshire, and the others as shown. The resulting workloads are as shown in table 12.3.

In passing, it is worth noting that even such a straightforward problem is not without ambiguities. Compare Scotland with South East England in figure 12.1. It is clear that the Scottish representative must resign himself to spending considerably more time in his motor car than the representative for the South East. At the same time, any attempt to try and quantify the trade off between access time and 'sales time' must be heroic in nature. For this reason it has not been attempted in the study.

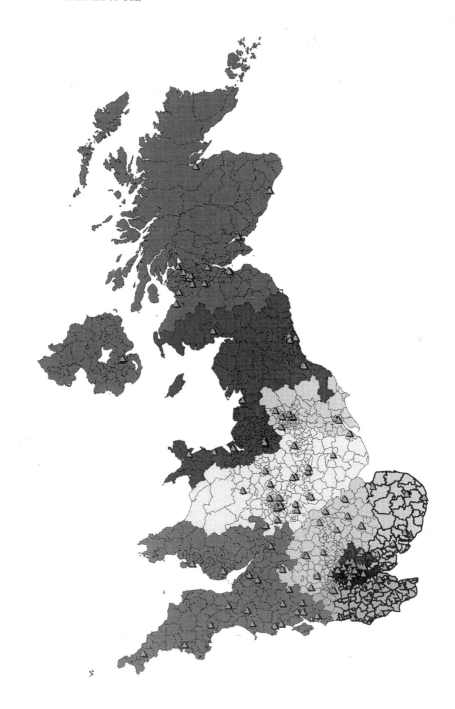

Figure 12.1. Ideal solution to territory allocation problem (different shadings represent assigned territories)

Table 12.3.

Oncology Fieldforce Territory summary statistics							
Territory	Consultant Numbers	Optimisation Score	% Consultants	KCC Numbers	% KCCs	Total Population	% Population
SOUTH WEST	42	0	13.7	7	12.5	5873480	10.4
SCOTLAND & NORTH	42	0	13.7	7	12.5	7247683	12.8
MIDLANDS & SOUTH	42	0	13.7	9	16.1	9212629	16.3
N. LONDON & E. ANGLIA	43	1	14.1	10	17.9	7263457	12.9
S. LONDON, KENT & SUSSEX	42	0	13.7	8	14.3	6375530	11.3
NORTH WEST & N. IRELAND	43	1	14.1	3	5.4	8065910	14.3
N. MIDLANDS, MID-WALES & E/N YORKS	41	1	13.4	11	19.6	9861806	17.5
COOKRIDGE KCC	11	0	3.6	1	1.8	2531734	4.5
TOTAL	306	3	100.0	56	100.0	56432229	100.0

Note: Optimisation score = Target territory consultant total (42) compared to the computed consultant total

12.3.2 Spatial Optimisation

The problem described in section 12.2.2 in relation to the car industry will again be familiar to many regional scientists (and indeed management scientists) being in essence a straightforward location-allocation model (see Ghosh and Harche, 1993, for a fairly recent review). A more complex version of the problem arises when the properties of the zones which need to be aggregated into territories (e.g., the KCCs in the example above) involve some kind of interaction component – such as a count of shopping trips or journey to work flows.

An algorithm to address a problem of this type has been provided all of thirty years ago by Teitz and Bart (1968), albeit with a more straightforward interaction component. There is also a strong linkage to the frequently cited theoretical work of Harris and Wilson (1978) and the associated numerical experiments of Clarke and Wilson (1983) and others. A large scale variant of the Teitz and Bart algorithm with a complex behavioural model has been developed and tested at GMAP. This model exploits parallel computer architectures to find network solutions over many thousands of geographical zones. This work is reported elsewhere (Birkin, Clarke and George, 1995; George et al., 1997). In this section of the paper, we revisit this model and show how it may be solved using a genetic algorithm, which is flexible, efficient and reasonably fast. We show how this algorithm may be applied to generate valuable business scenarios for car distributors in a particular European country.

Suppose that the relationship between a facility and its customers can be represented using a model of the following type:

$$S_{ij} = A_i O_i W_j \exp^{-\beta c_{ij}} \qquad (12.9)$$

$$A_i = \frac{1}{\sum_j W_j \exp^{-\beta c_{ij}}} \qquad (12.10)$$

where:

S_{ij} is spending by residents of i at location j

O_i is expenditure of residents of location I

W_j is the attractiveness of destination j

c_{ij} is the cost of getting from i to j

β is a parameter

A model of this type may be used to gauge the impact of a new facility (or closure of an existing one); and also as a basis for benchmarking the performance of each outlet against its potential (e.g., Birkin *et al.*, 1996, and Clarke and Clarke this volume for examples).

A more radical question to ask is what an ideal distribution of facilities might look like, given that their performance is governed by the interaction equations (12.9) and (12.10). This is a relevant question to ask when major network restructuring is planned, or when new markets are under development (e.g., many western European service providers now entering eastern Europe for the first time). The optimisation might be applied to a single CMA (see discussion of Section 12.2.1 above) or a to a complete network.

A solution to this problem using a hill-climbing heuristic in the style of Teitz and Bart was presented by Birkin, Clarke and George (1995). The authors demonstrated the quality of the solutions which were generated by comparing the results for an abstract and simplified problem with a 'brute force' approach which evaluated all available solutions before selecting the best. The key results are summarised as table 12.4.

An alternative procedure has now been developed using the concept of a 'genetic algorithm'. Suppose that the problem is to select the best 50 locations from a universe of 8000 possible locations. This solution can be represented as a binary string of 8000 digits representing the possible locations. 50 of the digits are set to 1, representing the selected locations; the remainder are set to zero – these locations have not been selected. The 'fitness' of each solution is the number of cars which can be sold when dealers are placed at the specified locations (which are in turn calculated by summing the interactions for each dealership).

The first step in the procedure is to create a population of random solutions, in which the solution takes the form of an 8000 bit string, as described above. Suppose that 1000 solutions are created. At each step of the algorithm, new solutions are created by selecting a pair of solutions from the initial population and combining them to form two new solutions. At regular intervals within the process, a 'mutation' may be introduced whereby any dealer is moved at random to a new location.

The process by which solution pairs are selected from the initial population is determined by a Monte Carlo selection process according to the fitness of the parent solution. At first there is little differentiation between solutions, but the selection process is made more aggressive as the solution develops, so that the characteristics of the fitter parents proliferate with time. The algorithm can be terminated after a fixed length of time (number of iterations) or when the change between iterations falls below a certain level. The structure of the algorithm is summarised as a flow diagram in figure 12.2.

Table 12.4.

Comparison of outcomes under IRP and MBF								
No. of dealers	Run code	MBF				IRP		
		optimal sales	mean	SD	no. of solutions	sales	rank	centile
ß = 0.05								
3	TRIANG31	4765.3	4294.5	487.41	30	4689.4	8	937
4	TRIANG32	4775.2	4511.8	251.38	30	4773.5	3	997
6	TRIANG33	4791.4	4416.6	297.19	435	4791.1	3	1000
7	TRIANG34	4794.9	4518.2	204.78	435	4794.9	1	1000
9	TRIANG51	4789.7	4351.8	359.59	4060	4788.4	4	999
10	TRIANG52	4790.8	4433.9	295.68	4060	4784.9	5	995
ß = 0.10								
3	TRIANG21	4319.4	3567.6	555.31	30	4319.4	1	1000
4	TRIANG22	4437.5	3864.4	311.93	30	4059.4	8	751
6	TRIANG23	4537.5	3953.9	368.17	435	4464.0	12	970
7	TRIANG24	4517.8	4081.9	258.64	435	4491.5	9	988
9	TRIANG50	4596.0	4027.0	399.19	4060	4591.3	3	998
10	TRIANG53	4605.6	4123.1	325.06	4060	4594.1	6	995

Figure 12.2. Description of the genetic dealer location algorithm

0. Assume a geography in which the demand for a product (cars) is distributed evenly across a grid of 91 cells, located on a triangular lattice. Competing outlets are also evenly distributed across the grid.
1. A candidate solution comprises a selection of N unique locations from the 91 available (where N=3,4,6,7,9 or 10).
2. An initial population of 50 candidate solutions is created by selecting locations at random.
3. For each candidate solution, total dealer sales are evaluated for that pattern of dealerships. This evaluation is monotonically transformed to provide the basis of the fitness function. This transformation becomes more extreme as the algorithm progresses.
4. Parents are selected from the existing candidates using the Monte Carlo sampling in proportion to parental fitness. Parental gene material is exchanged in strings (i.e. two parents are selected: a certain number of dealer locations are exchanged from one parent to the other to create two offspring which are different to both parents).

5. There is a probability of 'mutation' (i.e. one dealer location within the offspring is relocated at random). The probability of mutation is relatively low but increases as the algorithm progresses.
6. The fittest candidate in each population automatically reproduces itself to the next generation.
7. Five hundred successive generations are created by the algorithm unless the procedure is arbitrarily terminated.

The performance of the genetic algorithm in the idealised problem of table 12.4 is extremely impressive – see table 12.5. In all cases, the best solution is found. In one case, the GA is able to find a solution which is better than any previously thought to exist. (Note: this is possible because the 'brute force' method adopts a short-cut, which assumes that any optimal solution for 7 or 10 locations must have a dealer at the centre. The GA shows this assumption to be inaccurate).

Table 12.5. (BFM = brute force methods; IRP = iterative proportional fitting; GA = genetic algorithms)

COMPARISON OF GA, IRP AND BRUTE FORCE ALGORITHMS							
BETA	NO OF DEALERS	OPTIMAL VALUES			NO OF SOLUTIONS		
		BFM	IRP	GA	TOTAL	GA	IRP
0.05	3	4765.3	4689.4		121E + 3		50?
0.05	4	4775.2	4773.5		2.7E + 6		100?
0.05	6	4791.4	4791.1	4791.4	667E + 6	2750	200?
0.05	7	4794.9	4794.9	4794.9	8.1E + 9	4250	300?
0.05	9	4789.7	4788.4	4789.3	784E + 9	25000	500?
0.05	10	4790.8	4784.9	4790.8	6.4E + 12	22500	600?
0.10	3	4319.4	4319.4	4319.4	121E + 3	1000	50?
0.10	4	4437.5	4059.4	4437.5	2.7E + 6	2500	100?
0.10	6	4537.5	4464.0	4537.5	667E + 6	3500	200?
0.10	7	4517.8	4491.5	4552.4	8.1E + 9	6000	300?
0.10	9	4596.0	4591.4	4596.0	784E + 9	22000	500?
0.10	10	4605.6	4594.1	4606.5	6.4E + 12	3000	600?

It is with a certain amount of confidence, therefore, that the GA technique can be applied to 'real world' optimisation problems. In this chapter, the outputs from the procedure are demonstrated for a variety of car dealer networks in Denmark. Table 12.6 shows the existing dealer structure and network performance for six different manufacturers – Ford, VW/Audi, GM/Opel, Mercedes, Rover and Nissan. The first challenge for the algorithm is to relocate the existing dealers to their ideal locations. In the case of Ford, 105 dealers can be relocated in such a way that average sales are increased from 158 to 202 vehicles, an increase of 27.7%. It would appear that the most efficiently configured is that of Nissan (and it is probably no coincidence that this is the most sparse of the four networks). Even in this case, however, an average sales increase of more than 18% can be achieved through idealised relocation.

Table 12.6.

Full Network IRP Run												
Manufacturer	Existing Dealers			Full network			Change			% Change		
	Sales	Dealers	Average	Sales	Dealers	Average	Sales	Dealers	Average	Sales	Dealers	Average
Ford (01)	16,615	105	158	21,217	105	202	4,601	0	44	27.7	0.0	27.7
Audi-VW (03)	19,745	101	195	26,024	101	258	6,279	0	62	31.8	0.0	31.8
GM/Opel (09)	17,590	100	176	23,371	100	234	5,782	0	58	32.9	0.0	32.9
Mercedes (17)	697	45	15	1,543	45	34	846	0	19	121.3	0.0	121.3
Nissan (19)	6,152	84	73	7,278	84	87	1,126	0	13	18.3	0.0	18.3
Rover (23)	1,275	23	55	3,736	23	162	2,461	0	107	192.9	0.0	192.9

The second challenge for the algorithm is to strip each network of dealers, so that the same number of cars can by sold through a restricted network of ideally located outlets. The results from these experiments are shown in table 12.7. This can be achieved by a reduction in dealer numbers of between one-third and two-fifths the original quota for the various manufacturers. A comparison of the existing Ford network with the rationalised network is shown as figure 12.3. The same comparison is shown for Nissan in figure 12.4.

Table 12.7.

Rationalised network IRP run												
Manufacturer	Existing Dealers			Rationalised network			Change			% Change		
	Sales	Dealers	Average	Sales	Dealers	Average	Sales	Dealers	Average	Sales	Dealers	Average
Ford (01)	16,615	105	158	16,893	70	241	277	-35	83	1.7	-33.3	52.5
Audi-VW (03)	19,745	101	195	19,478	59	330	-267	-42	135	-1.4	-41.6	68.9
GM/Opel (09)	17,590	100	176	17,676	60	295	86	-40	119	0.5	-40.0	67.5
Mercedes (17)	697	45	15	916	19	48	218	-26	33	31.3	-57.8	211.0
Nissan (19)	6,152	84	73	6,943	50	139	792	-34	66	12.9	-40.5	89.6
Rover (23)	1,275	23	55	2,078	9	231	803	-14	175	63.0	-60.9	316.5

Figure 12.3a. Ford Network - existing dealers

Figure 12.3b. Ford Network - optimal network

Optimal Distribution Strategies 237

Figure 12.4a. Nissan Network - existing dealers

Figure 12.4b. Nissan Network - optimal network

12.4 Business Benefits

So far, the chapter has tended to focus on the technical aspects of spatial optimisation. This section offers a complementary perspective on the applied value of the technology.

Before proceeding, two caveats should be expressed. First, it is fair to say that when spatial modelling technologies are applied within large organisations, it is typically as part of a broad process of business analysis with many components. Secondly, most business decisions would be far easier were it possible to reduce problems to clearly quantifiable measures of cost, benefit, or other metrics. In practice, the world tends to be much messier than this, and it is unrealistic to assert (for instance) that x thousand pounds of investment in modelling technology generates twice x in cost savings for the business.

Nevertheless, it is possible to indicate by example the importance of the technologies outlined in this chapter, and to demonstrate the nature and range of their application. Examples are classed in the paragraphs below according to the way in which the methods have been used in practice.

12.4.1 Benefits of Territory Optimisation

12.4.1.1 Benchmarking and Performance Analysis

In the pharmaceuticals case study discussed above the most important use of ideal territory structures is as a benchmarking tool. Since each area has an equal workload, then the representatives who operate within the territories have balanced incentives in the generation of new business. The performance of each representative can be evaluated against one another and against a common set of business potentials. This is very much the exception rather than the rule – in many businesses, poor performance is masked by the underlying fertility of a territory; while conversely, effective performance is thwarted by the poverty of conditions in a market.

12.4.1.2 Investment Appraisal

This follows naturally from benchmarking, especially when sales are mediated through a physical distribution network. In this case, a CMA approach may be used to create an appropriate business geography. Zones of poor market share may be divided into two types – those in which there is adequate representation but the performance of the individual outlets is unsatisfactory; and others in which representation is just not adequate. Investment may be prioritised towards areas where the network is simply too sparse in relation to the market opportunity. This approach is being used by a multinational oil company across a variety of European countries to target investment;

and also by a top-ten UK financial services provider which is historically weak as both a brand and a network across large parts of the UK.

12.4.1.3 Feasibility Assessment

A manufacturer of automotive components has decided to create a mobile relay service, allowing those components to be distributed via franchises to customers within a fixed access time from each service centre. A process of territory optimisation was undertaken to construct a network of franchise locations. The economics of the proposition dictated that there are certain areas – such as parts of Wales, Scotland and the South-West – which will never be viable. Nevertheless, by optimising the territories, it is possible to ensure that the greatest possible customer expenditure may be accessed from a given number of franchise locations. A similar approach was adopted by GMAP to help a pharmaceuticals manufacturer to devise a network of locations for travel vaccination clinics.

12.4.1.4 Efficient Distribution

The most common reason for wanting to define a set of homogeneous or self-contained territories is to increase the efficiency of distribution within each area. It is fair to say that the majority of service networks in most western European countries are inefficient through being too extensive. This is the reason for the rapid consolidation within the UK petrol industry, as noted in Section 12.2. Although the Spanish market is following a similar process of consolidation, other countries such as Germany and Italy remain cluttered with over-representation. The same is equally true in markets such as financial services and motor vehicles.

The CMA approach has been used in each of these industry sectors to identify a more efficient number of network points, or to focus and specialise the network. For example, for motor vehicles it is possible to distinguish new car sales, pre-owned vehicle sales, and parts and servicing; in financial services, between account management, ATMs, estate agencies, mortgages and investment products; petrol station forecourts as CTNs, mini-markets, car wash or QSR.

12.4.2 Benefits of Branch Network Optimisation

12.4.2.1 Network Reconfiguration

To produce a framework for analysis and planning of distribution, as implied by Section 12.4.1.4, can be tremendously helpful. What is even more powerful is a capability to reconfigure the network in each territory to its ideal shape. GMAP has worked with a major automotive manufacturer to undertake this process throughout Europe. National sales companies have been guided not only in the creation of a 'Network Representation Plan' but also offered support in the implementation of the plans. Implementation is now at an advanced stage in most countries.

12.4.2.2. Network Development

Network analysis need not always be a prelude to branch rationalisation! Another financial services client of GMAP has used this technology to build up its market through 'old-fashioned' branch representation on the UK High Street. Unusually, this approach has also focused on building extra representation within existing core regions, leaving other parts of the country untouched. In this way, the organisation has begun to challenge major players in certain parts of the country, while having no physical presence whatsoever in other places.

12.4.2.3 Depot Locations

A retail client has been using network optimisation technology to appraise the location of the warehouses which supply goods to each of its High Street outlets. The organisation has over 800 shops, which are stocked through 20 warehouses. GMAP has demonstrated that the pattern of distribution would be no less effective with only 13 warehouses provided they are optimally located.

In a similar vein, an automotive client has historically operated with a single parts depot for the whole country, leading to poor customer delays and many delays in obtaining replacements for all but the most routine parts. GMAP has redesigned the distribution network to allow replacement parts to be obtained from any UK dealer with a maximum delay of 24 hours.

12.5 Concluding Remarks

It is the view of the authors that a wide variety of Regional Science methods are applicable to business problems. While the complexity of methods employed by regional scientists may not always be appropriate in the commercial sector, there are some business problems which demand the application of leading edge methods. In this chapter we have shown two particular problems for which this holds true, but it is our belief that the potential for more work of this kind is very great indeed.

References

Birkin M., Clarke G., Clarke M. and Wilson A.G. (1996) *Intelligent GIS: Location decisions and strategic planning*, Geoinformation, Cambridge.

Birkin M., Clarke M. and George F. (1995) The use of parallel computers to solve non-linear spatial optimisation problems: an application to network planning, *Environment and Planning A*, 27, 1049-1068.

Clarke M. and Wilson A.G. (1983) Dynamics of urban spatial structure: progress and problems, *Journal of Regional Science*, 23, 1-18.

George F., Radcliffe N., Smith M., Birkin M. and Clarke M. (1997) Algorithms for solving a spatial optimisation problem on a parallel computer, *Concurrency: Practice and Experience*, 9(8), 753-780.

Ghosh A. and Harche F. (1993) Location-allocation models in the private sector: progress, problems and prospects, *Location Science* 1, 81-106.

The Guardian (1998) Safeway links with BP to pump up sales, July 16.

Harris B. and Wilson A.G. (1978) Equilibrium values and dynamics of attractiveness terms in the production-constrained spatial interaction model, *Environment and Planning A*, 10, 371-388.

Teitz M. and Bart P. (1968) Heuristic methods for estimating the generalised vertex median of a weighted graph, *Operational Research*, 16, 901-1092.

Wilson A.G. (1970) *Entropy in Urban and Regional Modelling*, Pion, London.

Wilson A.G. (1974) *Urban and regional models in geography and planning*, John Wiley, London.

13 An Applied Microsimulation Model: Exploring Alternative Domestic Water Consumption Scenarios

Paul Williamson

13.1 Microsimulation in a Regional Science Setting

Microsimulation is now a long-established tool for the evaluation, projection and retrospective analysis of a diverse range of subjects including kinship networks (Zhao, 1996), income redistribution (Nelissen, 1998), firm evolution (Tongeren, 1995) and mammography screening programs (Szeto and Devlin, 1996). This chapter describes the application of a microsimulation approach to the estimation of household water demand for small areas. Section 13.2 reviews the nature of this problem and outlines the structure of the remainder of the chapter in more detail. Section 13.3 presents a justification for the preference of a microsimulation approach over other demand forecasting techniques in this context. But first it is appropriate to consider the role of microsimulation within the wider regional science context.

The essence of microsimulation (also known as micro-analytic simulation) is a focus upon the rôle of individual decision-making entities. Each entity (e.g., individual, household, firm) is represented by a list of attributes (e.g., age, sex, marital status or number of employees, turnover, equity) and exhibits a range of behaviour(s), such as birth, death and migration. These behaviours can be treated as either deterministic (i.e. a fixed response to a given stimulus/input) or stochastic (i.e. a probabilistic response to a given stimulus/input). As originally conceived by Orcutt (1957) (although see also Hägerstrand, 1952), microsimulation also includes a dynamic element, with the attributes of each entity being updated between model time-steps.

Harding (1993) has identified three key variants of the microsimulation approach. Dynamic population microsimulation, in-line with Orcutt's original vision, 'ages' entities. Between consecutive model time-steps behaviours such as birth, death, migration, marriage and divorce are modelled for each entity in turn. Over time this can lead to the emergence of a range of higher level phenomena including, for example, household formation and dissolution. Examples of this approach include Duley and Rees (1991) and Falkingham and Lessof (1992). Dynamic cohort microsimulation differs only in its consideration of a limited section (cohort) of the population. Such an approach is useful when focussing, for example, solely upon changes in pensioner assets over time (e.g., Hancock *et al.*, 1992), which does not require the consideration of non-pensioner households. In contrast, static microsimulation, instead of ageing individuals forward one step at a time, ages an initial population of individuals forward in one step to a fixed point in the future, using a mixture of reweighting and uprating (revising monetary values to take account of inflation). The static approach, although clearly failing to capture many important processes, such as capital accumulation or firm evolution, is the one most widely adopted world-wide due to its ease of implementation.

Many western governments use static microsimulation to investigate the likely impact of proposed tax or benefit changes. In consequence such models are often referred to as tax-benefit models. Despite falling short of Orcutt's original conception, these tax-benefit models are widely regarded as coming under the microsimulation umbrella, principally on account of their focus on individuals rather than aggregates. More generally, the microsimulation approach lies somewhere between the two extremes of data imputation (adding extra variables to survey microdata), which lacks any temporal aspect or interaction between individuals (e.g., Dale *et al.*, 1995), and actor or agent-based modelling, in which individual entities are modelled as agents with memory, goals and rules which can evolve over time (e.g., Gilbert and Doran, 1994).

In a regional science context, it is interesting to note that microsimulation models in general are often highly aspatial. This is in part a reflection of intended purpose. An investigation into the impact of a proposed national tax change often requires only a summary national answer. It is also partly to do with the size of the population samples used to represent the wider population at-large. For example, in the UK the Inland Revenue bases their tax-benefit model on an annual government survey of approximately 10,000 households (Eason, 1996), a sample felt to be too small to yield statistically robust regional or sub-regional estimates. But this aspatial approach should sit uncomfortably with regional scientists. In almost any tax-benefit system there are local as well as national tax elements. Other problems requiring increased spatial detail include the location of public and commercial facilities such as schools, doctors surgeries and retail outlets (Birkin *et al.*, 1996) and service delivery planning, which ideally needs to take into account not only the number but also the location of potential service recipients. As Caldwell (1986, p59) noted with respect to microsimulation in the United States "... the gains accumulate exponentially when the level of spatial disaggregation becomes even finer than state level'.

One solution to the problem can be simply to exploit the underlying regional spatial basis of many microsimulation models. For example, the dynamic population microsimulation model CORSIM utilises 1960 U.S. Public Use Microdata spatially coded by state, and already incorporates an inter-regional migration model (Caldwell and Keister, 1996). Yet to date only one paper has been published presenting results at a sub-national level (Caldwell *et al.*, 1998). An alternative approach is to create spatially detailed synthetic microdata using one of a range of emerging techniques (Beckman *et al.*, 1996; Mitchell *et al.*, 1998; Simpson and Middleton, 1998; Wanders, 1998; Williamson *et al.*, 1998). The remainder of this chapter illustrates the application of a static microsimulation approach to the estimation of household water demand, incorporating one of the latest techniques for estimating spatially detailed synthetic microdata. For a broader review of applied microsimulation models within a regional science context, see Clarke (1996).

13.2 Client Needs

In 1989 the UK government privatised the water industry, creating 29 privatised water supply companies. After a period of consolidation, 22 companies are now responsible for supplying the industrial and domestic demand for water. Although privatised, the 'new' water companies are legally obligated to meet consumer demand and face restrictions on maximum permissible rates of growth in charges for supply. Within this heavily regulated environment, a crucial element in maximising company profit is the accurate forecasting of future water demand, so that unnecessary investment in additional water treatment plant, reservoirs and water-mains can be avoided, whilst ensuring that legal obligations concerning consumer supply are still met.

However, profit maximisation is not the only reason for seeking to minimise investment in new water supply infrastructure. Policy makers and planners are increasingly becoming concerned about notions of sustainability. Unchecked, growth in industrial and domestic demand for water could lead to the need for unsustainable increases in water abstraction from rivers and water sources, as well as the environmentally damaging creation of new reservoirs. Consequently, not only the privatised water companies, but also water regulators and other planning bodies are becoming increasingly interested in accurately forecasting future water demand. In particular they are realising the need for an enhanced detailed spatial dimension to such forecasts, in order that supply and demand can be more closely matched. This chapter reports on the development of a new domestic water consumption forecasting tool, designed to explore the potential impact upon future water demand of a range of water consuming behaviours. The development of this tool was co-funded by the Engineering and Physical Research Council and a potential end-user, the privatised water utility Yorkshire Water Services Ltd. A companion model of future industrial water demand is reported elsewhere (Mitchell and Wattage, 1998).

The resulting model can loosely be described as a static microsimulation, at the heart of which lies a set of spatially detailed synthetic population microdata. These synthetic microdata are estimated in such a way that they agree with published small area Census tabulations (Section 13.4.1), but an adjustment is then made for Census under-enumeration, by reweighting the microdata to agree to revised small area age/sex counts (Section 13.4.2). The synthetic microdata are also augmented to include two non-Census variables known to affect overall household water consumption; washing machine and dishwasher ownership (Section 13.5). Total annual water consumption is imputed for each household in the synthetic population, drawing upon data supplied by the industrial collaborator, Yorkshire Water Services Ltd.

(Section 13.6). The result is a spatially detailed estimate of domestic water demand for the Census year, 1991. From this base, forecasts of domestic water consumption can be made for alternative demand scenarios (Section 13.8). To do so, it is necessary only to update the imputation functions used, in order to reflect forecast trends in the various components of household water demand (Section 13.7). Note that a more conventional static microsimulation model would also 'age' the base population microdata by reweighting them to agree with projected population counts.

Each of these steps in model development will be considered briefly in the sections that follow, before a range of summary outputs from the model are introduced and explained. But first, it is necessary to briefly outline the industry standard 'micro-components' approach to water demand forecasting, and to justify the adoption of a static microsimulation approach.

13.3 Modelling Spatially Detailed Water Consumption Behaviour

13.3.1 The Micro-components Approach

Herrington (1987) reviewed the range of approaches currently adopted when forecasting future water consumption levels. He concluded that the two most common approaches were based upon an extrapolation of current trends or some kind of subjective analysis. Subsequently, Herrington (1996) developed a specific form of subjective analysis for use in the forecasting of domestic demand, known as the 'micro-components' approach, which has been widely adopted within the UK water industry. In this approach, forecast trends in water consumption are made separately for a range of water consuming activities. Activities range from personal washing and toilet flushing through to car washing and garden watering. A number of water using activities are further broken-down to permit even more detailed forecasts of future trends. For example, dishwashing is subdivided into dishwashing by hand (manual) and dishwashing by machine (automatic) (see table 13.1 for a full list). Each of the trends are themselves the product of forecast trends in activity prevalence, frequency and efficiency (volume of water consumed by activity). By combining the forecast trends in these micro-components of demand with any chosen population projection, forecasts of total future domestic water consumption may be produced.

The merits of the micro-components approach cannot be disputed. The range of assumptions involved in forecasting total water consumption are made explicit, whilst the impact of alternative component trends upon the forecast total can be readily evaluated. However, whilst the base-line national and even regional prevalence, frequency and efficiency of the various water consuming activities might be known, information on sub-regional prevalence, frequency and efficiency are typically unavailable. This is a crucial weakness if spatially detailed forecasts are required.

Further base-line information could be collected but, if the scale of spatial resolution required is very high, the additional information requirement might, in fact, amount almost to a mini-census. As a result, alternative means must be considered.

Table 13.1. The micro-components of domestic water consumption

Component	Litres per capita per day[1]	%
Toilet use	**35.5**	*24.1*
Full-flush	27.8	*18.9*
Dual-flush	7.7	*5.2*
Low-flush	—	
Personal washing	**46.5**	*31.6*
Bath only	18.2	*12.4*
Bath (+ shower)	8.0	*5.4*
Ordinary shower	3.7	*2.5*
Power shower	1.6	*1.1*
Hand	15.0	*10.2*
Clothes washing	**21.7**	*14.8*
Automatic	19.8	*13.5*
Non-automatic	0.5	*0.3*
Washer-drier	0.7	*0.5*
Hand	0.7	*0.5*
Dish washing	**11.8**	*8.0*
Automatic	3.0	*2.0*
Manual	8.8	*6.0*
Waste disposal unit	**0.4**	*0.3*
Car washing	**0.9**	*0.6*
Hose	0.5	*0.3*
Bucket	0.4	
Lawn sprinkling	**2.5**	*1.7*
Other garden use	**3.8**	*2.6*
Miscellaneous	**23.9**	*16.3*
TOTAL	**147.0**	*100.0*

[1] Estimated figures for the population of South and East England, 1991
Source: after Herrington (1996) *Climate change and the demand for water*, HMSO, London, Table A.1

13.3.2 Adding Spatial Detail – The Need for a Microsimulation Approach

A common solution to the problem of estimating spatially detailed water consuming behaviour is to make use of geodemographic profiling. Each small area within a region is classified into one of a number of possible geodemographic types, on the basis of the demographic and socio-economic characteristics of its constituent population. The water consuming behaviour of households from areas of each geodemographic type is then ascertained by means of a small sample survey. In doing so, it is assumed that all households within an area are of a similar type and that the observed behaviour for households of each geodemographic 'type' does not vary spatially. The latter assumption is only plausible in so far as the variables determining the geodemographic type of an area also fully determine variations in water consumption. The former assumption is even less tenable as the great majority of spatial areas, whatever their size, display a considerable degree of internal heterogeneity. Most problematically, however, by using a summary area classification to characterise each household, geodemographic profiling offers no direct link to the micro-components of water demand (see chapter 14 for more on geodemographics).

An alternative solution is to divide the households in an area into discrete water consuming 'types', on the basis of characteristics associated with the micro-components of demand. Conventionally, counts would be made of the number of households falling within each detailed water-consuming category. However, such an approach rapidly reveals its limitations as the number of determinants of water demand considered is increased. For example Pearson *et al.,* (1993), after an analysis of data collected by Severn Trent Water, report an estimated household water demand function based upon twenty-four variables. Assuming that each continuous variable in the Pearson *et al.,* model was aggregated into only five discrete classes, this would be equivalent to specifying a household classification with well over 10 million distinct categories. When it is further borne in mind that domestic water consumption needs to be estimated separately for each small geographical unit, this intuitive approach begins to lose its appeal, even though counts for many of the categories are likely to be zero or near zero.

Instead, a static microsimulation approach is proposed. For every small geographic area, synthetic population microdata are generated, constrained to agree with the known characteristics of the population in that area. As part of this process, functions estimated from a variety of datasets are used to impute household ownership for a range of water consuming goods, and to impute overall household water demand given these and other characteristics. The synthetic microdata are held as a list of individuals (plus attributes) within households (plus attributes), in much the same way that household survey results are normally stored. The list representation of household and individual characteristics means that, instead of the 10 million or so cells per area that would be required to represent the same data in tabular format, for the average enumeration district of 240 households only 24 (variables) x 240 (households) = 6000 'cells' would be needed. In addition, list representation avoids the need to discretise continuous data.

It is true that, just as for geodemographic profiling, the static microsimulation approach outlined here assumes that all households of a similar type exhibit similar water consuming behaviour, irrespective of spatial location. However, the microdata-based approach at least allows different 'types' of households to exist within the same spatial unit. In addition, whilst most geodemographic profiles classify areas (and, therefore, households) into between 10 and 250 categories, microsimulation allows the use of a number of categories orders of magnitude greater. As a result the potential for spatial bias within microsimulation is somewhat reduced. Even more importantly, it can be ensured that the synthetic microdata used include variables that relate directly to the micro-components of water demand. The following section considers further how such spatially detailed synthetic population microdata may be estimated.

13.4 Small-area Population Estimation

In the UK protection of respondent confidentiality means that spatially detailed microdata are not released for any area containing less than 120,000 residents. As a result, therefore, spatially detailed microdata have to be synthetically generated. Over time a wide variety of alternative approaches to creating spatially detailed microdata have been adopted, but they are all reducible to one of two main approaches; imputation and reweighting. Both rely upon the availability of cross-tabulations of local population characteristics for areas smaller than those for which survey microdata are released. For example, in the UK selected cross-tabulations of Census data are available for areas with as few as fifty residents. An earlier version of the domestic water consumption model reported here drew upon synthetic microdata created using an imputation-based approach (Clarke *et al.*, 1997). The current model draws instead upon synthetic microdata created using a variant of the reweighting approach. The change in method arose from a need to include a greater range of variables in the microdata, whilst at the same time reducing data storage requirements.

13.4.1 Combinatorial Optimisation

Williamson *et al.*, (1998) outline a new approach to the creation of spatially detailed population microdata, known as 'combinatorial optimisation'. This involves the selection of a combination of households from a sample survey which, when aggregated, best satisfy a range of known small area constraints (published cross-tabulations of population characteristics). Combinatorial optimisation is, in effect, an integer reweighting approach in which a large sample of survey households are 'weighted down' to represent the population within a small geographic area. Each small area is then represented by a list of household identifiers, acting as pointers to the appropriate positively weighted household records in the sample survey. In this way data storage requirements are minimised.

Using combinatorial optimisation, synthetic population microdata were generated for the entire region served by Yorkshire Water Services Ltd.; 1.83 million households sub-divided into over 9500 different enumeration districts (average size 188 households).

Possible combinations of households were drawn from the 1% Sample of Anonymised Records from the 1991 UK Census, whilst eleven small area Census tabulations were used as constraints in the optimisation process (see table 13.2). The constraining tabulations all involve Census variables thought to be important, directly or indirectly, for the imputation of total household water consumption. Interactions between these variables not captured in the constraining tables are, of necessity, assumed to be spatially invariant. Household and individual attributes from the survey microdata not constrained in any way during the optimisation process were summarily discarded.

Table 13.2. Small-area census tabulations used as constraints in microdata estimation

Variable	SAS Table										
	08	22	31	35	42	57a	57b	58	59	74	91
Individual level											
Age	•		•							•	
Economic activity	•									•	•
Marital status				•							
Occupation										•	
Sex	•		•							•	•
Social Class											•
Household level											
Age of children			•								
Property type (detached, semi, terraced or flat)					•	•	•	•			
Residents (number of)		•				•					
Rooms (number of)		•				•					
Tenure		•			•			•			
Household composition			•		•				•		

One overall measure of goodness-of-fit for the synthetic microdata is the observed total absolute error between the constraining census tabulations and their aggregated synthetic equivalents. Ideally, the sum of the absolute error across all tables should equal zero. In practice this is not achievable, as all of the census counts used as constraints have been subject to data blurring (the quasi-random addition or subtraction of 0 or 1), in order to protect respondent confidentiality. In fact the average total absolute error for each of the 9746 enumeration districts in the region was 261, equivalent to an average error of 0.3 per cell. Table 13.3 illustrates in more detail the goodness-of-fit observed for two of the eleven tables used as constraints, for an enumeration district with a total absolute error equal to the regional average. Those interested in a greater in-depth consideration of the strengths and weaknesses of this approach are referred to Williamson et al., (1998).

Table 13.3. Evaluation of estimated population microdata for an 'average fitting' enumeration district

a) Residents' age, sex and marital status

i) 1991 Census counts *ii) Estimated counts* *iii) Difference between estimated and census counts*

	Male		Female			Male		Female			Male		Female	
Age	SWD	Mrrd	SWD	Mrrd	Age	SWD	Mrrd	SWD	Mrrd	Age	SWD	Mrrd	SWD	Mrrd
0-4	25	0	13	0	0-4	25	0	13	0	0-4	0	0	0	0
5-9	26	0	9	0	5-9	26	0	9	0	5-9	0	0	0	0
10-14	13	0	15	0	10-14	13	0	13	0	10-14	0	0	-2	0
15	3	0	7	0	15	4	0	6	0	15	1	0	-1	0
16-17	6	0	9	0	16-17	6	0	8	0	16-17	0	0	-1	0
18-19	12	0	11	0	18-19	13	0	10	0	18-19	1	0	-1	0
20-24	22	1	16	3	20-24	22	1	16	3	20-24	0	0	0	0
25-29	9	16	5	16	25-29	10	16	4	13	25-29	1	0	-1	-3
30-34	6	10	4	21	30-34	5	9	4	21	30-34	-1	-1	0	0
35-39	1	12	0	12	35-39	1	12	0	12	35-39	0	2	0	0
40-44	1	20	2	16	40-44	1	20	1	17	40-44	0	0	-1	1
45-49	0	18	0	20	45-49	0	18	1	20	45-49	0	0	1	0
50-54	0	19	4	18	50-54	1	18	4	18	50-54	1	-1	0	0
55-59	1	21	4	17	55-59	1	20	4	17	55-59	0	-1	0	0
60-64	0	11	1	18	60-64	0	9	1	18	60-64	0	-2	0	0
65-69	1	22	6	8	65-69	1	22	6	8	65-69	0	0	0	0
70-74	0	7	3	7	70-74	1	6	3	7	70-74	1	-1	0	0
75-79	0	2	3	0	75-79	0	2	3	0	75-79	0	0	0	0
80-84	1	0	2	0	80-84	0	1	4	0	80-84	-1	1	2	0
85-89	0	0	0	0	85-89	0	0	1	0	85-89	0	0	1	0
90+	0	0	2	0	90+	0	0	2	0	90+	0	0	0	0

Total residents (census): 558 Total residents (estimated): 551 Total absolute error: 29

b) Household composition

i) 1991 Census counts

	No. of dependent children in household					
		one		two or more		
Household composition	0	0-4	5+	all 0-4	all 5+	mixed
1 male	6	0	0	0	1	0
1 female	19	0	1	1	0	1
2 (1 male and 1 female)	69	5	1	8	17	12
2 (same sex)	2	0	0	0	0	0
3 or more (male(s) and female(s))	31	0	16	0	4	0
3 or more (same sex)	0	0	0	0	0	0

252 P. Williamson

ii) Estimated counts

Household composition	No. of dependent children in household					
	one			two or more		
	0	0-4	5+	all 0-4	all 5+	mixed
1 male	7	0	0	0	1	0
1 female	19	0	1	1	0	1
2 (1 male and 1 female)	69	5	2	8	17	11
2 (same sex)	2	0	0	0	0	0
3 or more (male(s) and female(s))	31	0	17	0	4	0
3 or more (same sex)	0	0	0	0	0	0

iii) Difference between estimated and census counts

Household composition	No. of dependent children in household					
	one			two or more		
	0	0-4	5+	all 0-4	all 5+	mixed
1 male	1	0	0	0	0	0
1 female	0	0	0	0	0	0
2 (1 male and 1 female)	0	0	1	0	0	-1
2 (same sex)	0	0	0	0	0	0
3 or more (male(s) and female(s))	0	0	1	0	0	0
3 or more (same sex)	0	0	0	0	0	0

Total households (census): 194 Total households (estimated): 196 Total absolute error: 4

13.4.2 Adjustment for Census Under-enumeration

The 1991 UK Census suffers, as do most surveys, from the problem of under-enumeration. Importantly, this under-enumeration is thought to have been spatially concentrated predominantly in poorer urban areas. Clearly, synthetic microdata based upon the Census will reflect this bias. Previous spatially detailed synthetic microdatasets (e.g., Birkin and Clarke, 1988; Duley, 1989), including the one created for an earlier version of the domestic water consumption model (Williamson *et al.*, 1996; Clarke *et al.*, 1997), have simply ignored the problem altogether. An alternative solution is adopted here, made possible by the availability of a widely accepted set of revised census counts. These revised counts, disaggregated by age and sex, correct for the suspected pattern of under-enumeration within wards (average size 3300 households), and sum to agree with officially revised district level counts (average size 83000 households), similarly adjusted to compensate for under-enumeration (Simpson *et al.*, 1994). Unfortunately, no revised household counts have been released. Instead the 'true' number of households in a ward in 1991 is calculated assuming that the observed household formation rate for that ward (by age and sex) applies to the 'missing' people being added back to the area. Official household projections for

England published by the Department of the Environment (DoE, 1995) adopt the same approach. Although this assumption is unlikely to hold true in most areas, insufficient information exists to justify adopting any alternative strategy.

After aggregation to ward level, the originally estimated synthetic population microdata were reweighted to agree with the revised ward level household totals and age-sex population counts, using an algorithm initially proposed by Merz (best described in Merz 1994) and subsequently developed by Gomulka (1992), based upon the Minimum Information Loss principle. An alternative approach would have been to adopt a regression based approach (e.g., Chambers 1996). In reweighting the microdata it is assumed that, having taken account of age and sex, all households within an area are under-enumerated equally. In fact, even within areas as small as wards, poorer households are still more likely to be under-enumerated than affluent ones. Thus this assumption might be expected to lead to a slight homogenising effect. In any case, at least the missing population is reintroduced, and the general pattern of under-enumeration is accounted for.

13.5 Imputation of Non-census Household Attributes

The spatially detailed synthetic population microdata estimated using the approach outlined above contain, by definition, only household and individual attributes collected in the 1991 UK Census. Whilst this covers a wide range of demographic and socio-economic attributes, key determinants of a number of the micro-components of domestic water consumption are excluded. These include the ownership of water-using appliances such as dishwashers, washing machines, washer-driers, showers and power showers. Washing dishes and clothes by machine uses on average two and half times as much water as washing dishes and clothes by hand. Conversely, a shower consumes only one-third as much water on average as a bath (Herrington, 1996).

For washing machines and dishwashers, logistic regression analysis of data from the General Household Survey (OPCS, 1993) permits the estimation of the probability of appliance ownership, given a range of household characteristics. For washing machines, ownership was determined to be a function of age of household representative, property type, tenure, region, number of rooms and family type (including number of adults and number of children). For dishwashers, ownership was found to be dependent upon not only on these six factors, but also upon household representative's employment status and occupation (if any). The probabilities of ownership derived from these functions were used to impute washing machine and dishwasher ownership for households in the estimated population microdata, yielding an estimate of the spatial variation in ownership rates over space. The estimate has, of course, to be viewed with caution, as the implicit assumption is made that regional ownership rates (differentiated only on a metropolitan / non-metropolitan basis) apply locally. However, in so far as the imputation functions include locally varying variables (age, household composition, proxies for affluence/poverty), such an imputation process is perhaps not unreasonable. Certainly it is this very assumption that is used in a wide range of small-area estimation procedures (Ghosh and Rao, 1994), even including the imputation of characteristics for 'missing' households in the Census (Mills and Teague, 1991).

Unfortunately, no currently available government survey data permit the derivation of prevalence rates for the other micro-components of water demand listed in table 13.1. Consequently, household water consumption due to these additional items cannot be directly imputed. Instead, the remaining micro-components of demand are necessarily subsumed within an overall household water consumption function, the estimation of which is the topic of the next section.

13.6 Imputation of Household Water Consumption

13.6.1 Estimating a Consumption Function

Knowledge about patterns of domestic water consumption in the UK is, perhaps surprisingly, rather limited. Until recently, all households were supplied with water for a flat rate annual fee, which varied only as a function of rateable value (a measure of house price at some fixed point in the past). In consequence, water utilities in the UK had no incentive to meter water supplies to individual households. Since 1990 the situation has changed slightly, with legislation increasingly encouraging companies to supply households on a metered basis. Currently, 14 % of households in the UK are metered. Yet even for these households most water companies hold little or no information that might help to throw light upon the actual determinants of water consumption. Instead individual water companies have been forced to launch specially designed mini-surveys, often known as 'Monitors'. Crucially, in addition to monitoring household water consumption, basic information, including number of residents and number and type of water-using appliances, is also collected for each household in the survey.

The industrial partner in the research reported here, Yorkshire Water Services Ltd. made available data from their own 'Daily Consumption Monitor'. These data comprised records of monthly total water consumption for 1017 households, collected over an 18-month period. From this information it was possible to estimate, via regression analysis, a detailed household water consumption function based upon number of residents, number of bedrooms, washing machine and dishwasher ownership, property type (flat or other) and household tenure. This function captures approximately 44% of the observed variation in household water demand. The remaining variation is attributable to known determinants of water demand not recorded by the monitor, including physical household attributes (e.g., garden size), behavioural attributes (e.g., frequency of garden watering) and contextual attributes (e.g., climatic variability).

A calibrated version of the function has been used to impute household water demand for all households in the synthetic population microdata. An additional stochastic element is included to model the 'unexplained' variation in household water consumption. Without this, whilst aggregate totals would still be correctly imputed, overall distributions of household water demand would shrink towards the mean. The imputation process assumes, of course, that household water demand varies spatially solely as a result of those household attributes captured in the water demand function. However, it is felt that in this case the assumption is perhaps not too unrealistic, as the only two obvious place-specific determinants of water demand unaccounted for are climatic variability and garden size. There is a relatively small range of climatic variation across the Yorkshire region, and it can be argued that number of bedrooms and property type (flat) act as a crude proxy for garden size. Other unaccounted for determinants of water demand, such as frequency of garden watering, are thought to vary at random across space.

13.6.2 Data Issues

The use of Daily Consumption Monitor data raises a number of important issues. First, there are considerable problems in obtaining a 'representative' sample of households. High demand households tend to be reluctant to have their water consumption metered, despite assurances that their supply will continue to be charged on a flat-rate basis. In addition, it is not possible to construct a conventional sampling frame, as the underlying distribution being sampled from is not known: no-one knows precisely how many households there are in the Yorkshire area headed by a professional, with four residents, three bedrooms, one washing machine and no dishwasher.

The DCM also suffers from sample attrition as time goes by. On average one household in seven in the UK moves every year. This means that DCM-type surveys can expect to suffer an attrition rate of at least 14% per annum. Attrition bias arises because the more mobile households are typically those with younger residents, in professional careers or full-time education. Further attrition arises as a small proportion of households, having been metered for a short while, realise that they would be better-off being charged on a per unit than flat rate basis and ask to change their charging arrangements. As the survey is specifically targeted at households charged on a flat rate basis only, these households have to be dropped from the survey. This leads to a disproportionate loss of small and low water using households. Sample attrition can also arise due to the failure of the automated data loggers being used to record household water consumption. For affected households this leads to gaps in the series of recorded monthly consumption figures.

Fortunately, one of the key strengths of the microsimulation approach is that it does not matter whether or not the DCM dataset is representative of households across Yorkshire. Synthetic population microdata already provide a representative estimate of the number and nature of these households. All that is needed, therefore, is a DCM sample size large enough to ensure an unbiased estimate of the water consumption for different types of household. In this sense, the more detailed (multivariate) the water consumption function, the less biased the overall estimate of household water consumption is likely to be.

The problem of missing data is more difficult to overcome. Seasonal variations in water demand mean that, as the target is the estimation of average annual household water consumption, even one single missing month can invalidate a household's entire record. Yet, given the relatively small size of Yorkshire Water's DCM, no data can afford to be thrown away if a reliable household water demand function is to be estimated. Consequently replacement values for missing consumption figures have been imputed, drawing upon both household specific consumption figures for nearby months and month specific variations in average water consumption per household.

A third set of problems relates to issues of data reliability. In particular, it turns out to be very difficult to establish whether observed variations in household water consumption are real or artefactual. Legitimate sources of variability in household demand include seasonal variations (e.g., due to lawn watering in summer months), periods of absence (ranging from weekend breaks through holidays to prolonged sojourns elsewhere for work reasons), and changes in water consuming behaviour (e.g., starting to use the dishwasher more). Artefactual sources of variability include data logger malfunction, pipe-bursts and permanent changes in household circumstances. The latter covers the arrival/departure of household members (through births, deaths and partnership formation/dissolution) and changes in water appliance ownership (e.g., the purchase of a dishwasher or lawn sprinkler). Such changes in household circumstances give rise to 'artefactual' variation in water demand as household characteristics, currently surveyed only at the start of the monitor period, are necessarily assumed to remain constant throughout the survey period.

The solution to data reliability problems centres on very careful data checking and validation. All extreme variations in household water demand were investigated, as were significant changes in consumption behaviour (e.g., doubling or halving of demand) which persisted for some time. Pipe-bursts and data-logger malfunctions were relatively easy to confirm, but changes in household circumstances were more difficult. Moving households were identified via changes in billing details, and were re-surveyed. However, the desire to avoid burdening survey participants meant that other households could not be contacted to investigate possible changes in circumstances. Instead, records for such households had to be subjectively deleted, retained or amended, depending upon the size and/or persistence of the observed variation in consumption behaviour.

13.6.3 The Spatial Distribution of Water Demand

Taking the synthetic population microdata estimated in Section 13.3 and combining them with the appliance ownership and household water consumption functions outlined in Sections 13.5 and 13.6.1 permits the estimation of overall domestic water consumption across Yorkshire in 1991. The estimated pattern of domestic consumption is not only spatially detailed (552 wards or ward clusters) but, importantly, gives a direct estimate of total domestic consumption. Previous estimates of this total have necessarily been based upon the assumption that average household water demand as observed in the DCM dataset applies equally to all households in Yorkshire, despite the knowledge that these data are likely to be substantially biased. An alternative solution has been to deduct estimates of leakage from known supply to the domestic supply network. However, estimates of leakage are thought to be even less reliable than the average household water consumption figures derived from the DCM. By overcoming these problems, the new estimate is potentially of great value.

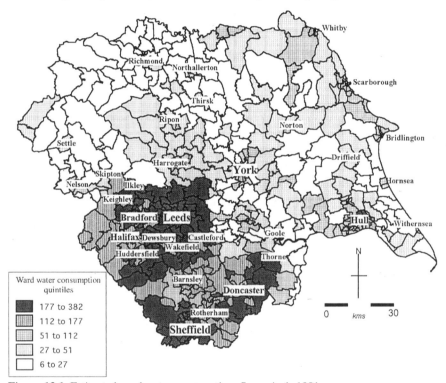

Figure 13.1. Estimated ward water consumption: Scenario 1, 1991

Even more importantly, the geographical detail inherent in the estimation process also allows a detailed map of domestic water consumption to be constructed. Figures 13.1 and 13.2 map the estimated spatial distribution of domestic water demand across the Yorkshire Water Services Ltd. region in 1991 (including the City of York, which in reality is served by the separate York Water Company). As might be expected, figure 13.1 closely resembles a map of the distribution of population across the region. The picture is one of greatest total domestic water consumption in the highly urbanised south of the region. However, the underlying picture is not perhaps as simple as it might first appear. Indeed, a map of average water consumption per household (figure 13.2) reveals a totally different situation: the central urban wards turn out to be those with the lowest per household consumption, as do a few of the remoter rural wards in north and north-east, whilst two bands of high per household water consumption radiate out north and east from the centre of the region. This pattern of consumption is effectively a combination of a map of wealth (richer households own more water consuming appliances and live in houses with larger gardens) and a map of residents per household. A map of per capita water consumption would reveal a different pattern again.

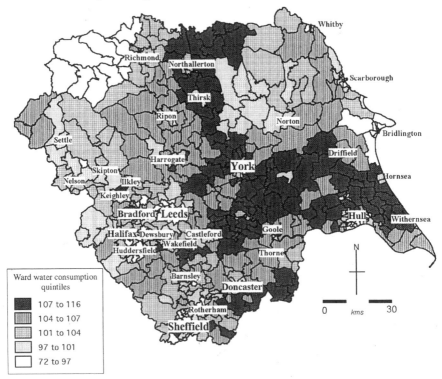

Figure 13.2. Estimated water consumption per household: Scenario 1, 1991

The estimation of the volume and detailed spatial pattern of domestic water consumption is a considerable achievement. However, whilst interesting, estimates for 1991 are clearly of limited benefit. They are neither current, nor do they shed any light on possible future demand. Yet it is precisely the likely patterns of future demand that are of most interest to those engaged in planning and regulating water supplies. The means by which alternative scenarios of future water demand can be modelled are explored next.

13.7 Modelling of Alternative Water Use Futures

Washing machine and dishwasher ownership is on the increase. At the same time, the proportion of washing machines that are actually washer-driers is forecast to rise. This is important because washer-driers consume over 50% more water per wash than a conventional washing machine. Herrington (1996) also forecasts increases in other areas of household water consumption, including personal washing (due to the increased use of power showers), garden watering (due to increased ownership and use of garden sprinklers) and miscellaneous usage. Appliance efficiency, on the other hand, looks set to improve over time. A benefit of the microsimulation approach is that these changes in the prevalence, frequency, and efficiency of the various micro-components of demand can be readily incorporated, simply by rescaling output from the relevant imputation functions. For most micro-components, this is a relatively trivial task. The modelling of changes in appliance ownership rates, in contrast, is a little more complex.

13.7.1 Washing Machine and Dishwasher Market Penetration

Section 13.5 set out the basic algorithms used to impute whether or not a household owns a washing machine or dishwasher in 1991. For imputation of ownership in years subsequent to 1991 it is necessary to select a model of how ownership patterns (market penetration) are thought likely to change over time. Four main alternatives exist. First, a detailed consumer behaviour model could be constructed, in which market penetration levels over time are predicted for a wide range of market sectors. These market sectors would be differentiated on the basis of household characteristics contained in the synthetic population including, for example, age and occupation of household representative. Although perhaps the most desirable approach in the long-term, the construction of such a model is in itself a major research project, which lies beyond the scope of the funded research reported in this chapter. A second and much simpler approach considered, therefore, was simply to iteratively identify an increment that could be added to every household's probability of ownership in order to raise overall ownership rates to projection year targets. The effect of this would be to force ownership rates for those sectors with the highest prevalence in 1991 to rise rapidly to complete market saturation (probability >= 1.0), whilst market penetration in the less saturated end of the market would increase relatively slowly. A third alternative entails increasing the probability of owning an appliance in direct proportion to the size of the original probability, with the proportional increment once again identified iteratively to ensure that overall ownership rates are raised to meet projection year targets. The upshot of this approach is similar to that observed for the constant increments approach, except that the projected rapid increase in ownership rates amongst those households already identified as most likely to own a given appliance are exaggerated even further, due to the increment being proportional to the base year ownership probability. A fourth and final alternative considered was to increase the probability of owning a washing machine proportionally to the complement of this probability (one minus the probability of ownership), again scaled to raise overall ownership rates to projection year targets. This last approach leads to a relatively rapid increase in prevalence rates amongst initially low prevalence households, whilst prevalence amongst already high prevalence market sectors also increases, but at a much reduced rate, and without ever giving rise to complete market saturation.

In Yorkshire the market for dishwashers does not yet appear to be saturated, with only 12% of households owning a dishwasher in 1991. Further rapid growth in ownership amongst all household types is forecast, suggesting that a constant or proportional increment approach would be most appropriate for dishwashers. For reasons of theoretical credibility and computational efficiency, a constant increment approach is favoured. In particular, it was felt that the proportional increment approach exaggerates the undesirable effect of complete saturation at the top end of the market. In contrast, the market for washing machines in Yorkshire is already fairly saturated; in 1991 89 % of households owned one. This suggests that, for washing machines, an approach based upon the complement of the probability of ownership would be more appropriate. The results of applying these two models can be seen in figure 13.3.

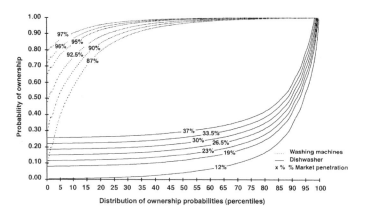

Figure 13.3. Market penetration models for washing machines and dishwashers

The forecast prevalence rates of washing machine ownership include households owning a washer-drier. In order to model the market penetration of washer-driers, it was simply assumed that the probability of a household owning a washing-machine actually owning instead a washer-drier was equivalent to the forecast proportion of households with a washer-drier divided by the proportion of households forecast to own a washing machine. As washing-machine owning households are, on average, those with a higher probability of owning a washing machine, this is equivalent to assuming that households with the greatest probability of owning a washing machine are also those most likely to own a washer-drier.

13.7.2 Changing Washing Machine and Dishwasher Efficiencies

Herrington (1996, Table A.1) suggests that in 1991 households owning a washing machine were likely to use their machine on average 4.5 times a week, at an average water consumption of 100 litres per cycle. By 2025 changes in appliance efficiency are forecast to lead to a 25% reduction in water use per wash cycle, resulting in an overall decrease of 4% in per capita water consumption. A similar reduction in overall per capita water consumption is predicted to arise as a result of improvements in dishwasher efficiency over the same period. In order to reflect these changing efficiencies over time, the imputed water consumption for households owning a washing machine and/or dishwasher were reduced accordingly.

Unfortunately, the Yorkshire Water Services DCM makes no distinction between households owning a washing machine and households owning a washer-drier. As the estimated level of washer-drier ownership at the time of the survey was relatively low (<10%), it was simply assumed that no household in the DCM owns a washer-drier. For all households imputed as owning a washer-drier in the base projection year (1991), initially estimated household water demand have been revised upward to take into account the additional demand associated with the drying cycle. This upward adjustment was calculated on a simple pro rata basis given the ratio between water use for the drying cycle and the washing cycle. For years other than the base year of the projection, the additional water use per wash due to the drying cycle was reduced proportionately to the forecast increase in efficiency of washing machine wash cycles.

Two alternative forecasts of future washer-drier efficiency are incorporated in the model. In both cases, the efficiency of the washing cycle of washer-driers is assumed to improve at the same rate as the washing cycle of conventional washing machines. The difference lies in the way in which the efficiency of the drying cycle is assumed to improve. In the first forecast the efficiency of the drying cycle is assumed to remain unchanged over time, following Herrington (1996, p147). However, this assumption is felt to be unrealistic, and in stark contrast to forecast improvements in washing machine and dishwasher efficiency over the same thirty-year period. In the second forecast, therefore, the efficiency of the drying cycle is assumed to improve at the same rate as the washing cycle. There is no evidence to support this alternative assumption, but the rate of improvement in efficiency assumed amounts to only a 25% reduction over 30 years, in comparison to a forecast 50% improvement in water efficiency by dishwashers over the same period.

13.7.3 Trend Increases in Other Micro-components

Herrington (1996) forecasts significant trend increases in three key non-appliance related micro-components of water demand. These are personal washing, garden watering and miscellaneous usage. In order to incorporate the forecast trend increases in these micro-components, imputed per capita water demand is revised upward in proportion to the forecast increase in each type of water usage. This assumes a similar proportional increase in water consumption across all household types, but no information is available to permit any alternative solution.

13.8 Illustrative Results for Selected Scenarios

Table 13.4 lists eleven alternative scenarios designed to explore the relative and cumulative impact of the various micro-component trends upon domestic water consumption. The forecast trends are all taken from Herrington (1996), with two exceptions. First, a set of higher forecast increases in appliance ownership are included, based upon extrapolations from a published time-series of appliance ownership rates (OPCS, 1993, Table 3.48). Second, as noted in section 13.7, the possibility that the efficiency of washer-driers might increase over the projection period is admitted. The scenarios range from Scenario 1, in which no micro-component of demand is assumed to change between 1991 and 2025, through to Scenario 11, in which every micro-component is assumed to change over time, in exactly the fashion forecast by Herrington. The remaining scenarios explore the relative importance of the various micro-components in determining overall demand.

Table 13.4. Alternative water demand scenarios

Micro-components	Scenarios										
	1	2	3	4	5	6	7	8	9	10	11
	Forecast increase in micro-component										
Appliance ownership											
Washing machine	−	+	++	−	+	++	−	−	−	+	+
Dishwasher	−	+	++	−	+	++	−	−	−	+	+
Washer-driers	−	+	+	+	+	+	−	−	−	+	+
Appliance efficiency											
Washing machines	−	−	−	+	+	+	−	−	−	+	+
Dishwashers	−	−	−	+	+	+	−	−	−	+	+
Washer-driers	−	−	−	+	+	+	−	−	−	+	−
Personal washing	−	−	−	−	−	−	+	−	−	+	+
Garden watering	−	−	−	−	−	−	−	+	−	+	+
Miscellaneous usage	−	−	−	−	−	−	−	−	+	+	+

− = *no increase from 1991*

+ = standard rate of increase, as forecast by Herrington (1996) *Climate change and the demand for water*, HMSO, London.

++ = higher rate of increase, extrapolated from time series published in the 1991 General Household Survey (OPCS, 1993, Table 3.48)

Clearly, future levels of domestic water consumption will be driven, not only by the trends outlined above but also, and perhaps more fundamentally, by changes in population size and composition. However, whilst the consideration of population change is important, and is the focus of current research, it remains instructive to consider what would be the impact of forecast changes in water consuming behaviour on current demand, should the population stay unchanged.

The forecasts of total domestic water consumption arising from the eleven scenarios are presented in table 13.5. To a great extent these findings mirror those reported by Herrington (1996) for the South East. On a positive note, forecast improvements in appliance efficiency are predicted to more than outweigh forecast increases in appliance ownership. Set against this, forecast increases in water consumption due to personal washing and garden watering are larger than those for changes in appliance ownership, even for the higher market penetration variants. The relative importance of the arbitrary increase in miscellaneous usage forecast by Herrington is also highlighted by the 5.7% increase in consumption forecast for scenario 9. Scenario 10 provides a 'best' forecast of domestic water demand by drawing together all of the separate micro-component trends. It differs from Scenario 11 only in the inclusion of a forecast improvement in washer-drier efficiency. In either case it can be seen that the forecast increase in domestic water consumption from 1991-2025 is approximately 24%. This increase in demand is attributable solely to trends in water consuming behaviour, induced by changes in wealth and lifestyle. Projected increases in population, including a significant rise in the total number of households, can only push this total higher.

Table 13.5. Variant water consumption projections

Scenario	Total domestic water consumption within region served by YWS Ltd.[1] (Scenario 1, 1991 = 100) Year						
	1991	1996	2001	2006	2011	2016	2025
1	100.0	100.0	100.0	100.0	100.0	100.0	100.0
2	100.7	102.0	102.4	103.2	104.0	104.7	105.8
3	100.7	102.0	103.0	103.9	104.8	105.7	107.0
4	100.7	99.5	98.2	97.1	96.0	96.2	96.2
5	100.7	100.2	98.8	97.8	96.8	97.1	97.4
6	100.7	100.2	99.1	98.2	97.1	97.5	98.0
7	100.0	101.6	103.2	105.1	106.9	108.6	112.0
8	100.0	101.0	101.9	103.1	104.2	105.4	107.4
9	100.0	100.6	101.1	102.1	103.1	104.1	105.7
10	100.7	103.4	105.0	108.3	111.2	115.6	123.3
11	100.7	103.5	105.2	108.6	111.7	116.2	124.0

[1] Results include households within the City of York supplied by the York Water Company.

In addition to the size of the forecast increase in demand, a key question for those planning and regulating water supplies, is the location of this forecast increase in demand. Given that number of residents is the main driver of domestic water consumption, it is perhaps no surprise to discover that, in the absence of population change, those wards with the greatest estimated water consumption in 1991 are also those forecast to experience the greatest absolute water consumption in 2025. However, the same is not true for the pattern of relative increase in water demand. This pattern, illustrated in figure 13.4, is identical whether mapped on a per capita or per household basis, as it has been necessary to assume that there are no economies of scale in water appliance usage; a counter-intuitive assumption that has been reluctantly adopted only in the absence of any concrete empirical evidence to contrary. Consequently, forecast variations in the relative increase in water consumption are attributable entirely to changes in water appliance ownership. The expectation is that, over time, ownership rates in poorer wards will catch-up slightly with those in more affluent wards, as the more affluent wards reach market saturation. Figure 13.4, therefore, is almost a map of relative affluence, with the poorest wards being those forecast to experience the greatest relative increase in total domestic water demand.

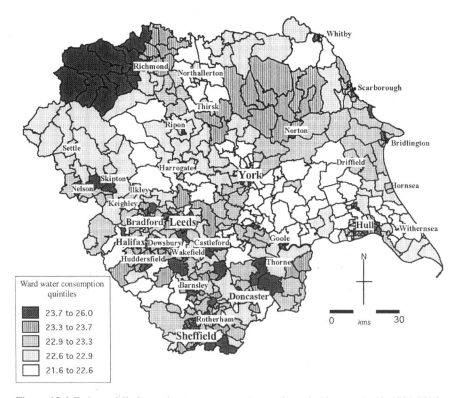

Figure 13.4. Estimated % change in water consumption per household: scenario 10, 1991-2025

13.9 Conclusions

Water industry providers and regulators increasingly require spatially detailed estimates of future water consumption, in order to best plan for future investment in water supply infrastructure. The current industry wide best practice involves the forecasting of regional domestic water consumption in the light of forecast trends in the various micro-components of water demand. This approach not only lacks spatial detail, but also presupposes detailed knowledge concerning household water consumption behaviour. Unfortunately, for reasons already noted, unbiased data on domestic water consumption are hard to come by. Static microsimulation offers a means of producing the required spatially detailed forecasts, for a variety of alternative demand scenarios, whilst incorporating forecast trends in the micro-components of household water demand and overcoming, at least in part, the problem of bias in water consumption data.

At the heart of the approach lie two key elements. First, a spatially detailed set of population microdata, of necessity synthetically estimated given the lack of publicly available alternatives: second, imputation. Imputation offers a way of overcoming the likely bias contained in Daily Consumption Monitor data as, even if the DCM data themselves are unrepresentative of households at large across the study area, the estimated water demand functions can still be used to estimate relatively unbiased demand functions for household 'types'. Importantly, the use of population microdata allows the incorporation of a wider variety of household level determinants in the imputation of water consumption than would be possible were any other approach adopted. The main criticism of such an imputation based approach is the implicit assumption that imputed household characteristics vary across the study region as a result of variations only in those household attributes captured in the imputation function. The validity of this assumption has already been considered. It is argued that, at the very least, the overall pattern of spatial demand is captured in the domestic consumption model. To the extent that the true spatial variation in imputed attributes is not captured, this will give rise to a homogenizing effect in estimated spatial patterns of demand. In any case regional averages will remain unaffected, offering alternative and, arguably, more robust estimates of regional domestic water consumption.

The domestic water consumption model presented in this chapter has allowed, for the first time, the estimation of the spatially detailed pattern of domestic water consumption in 1991 for unmetered households across the Yorkshire region. The model has also allowed the investigation of the likely impact of alternative forecast trends in the micro-components of water demand, both upon aggregate and spatially detailed patterns of demand. Three areas in particular are highlighted as requiring further research, in order to improve model forecasts. First, do larger households enjoy economies of scale when using washing machines and dishwashers? Second, are the forecast trends in personal washing, garden watering and miscellaneous usage really equally applicable to every household on a pro rata basis? This is perhaps a question in particular for the way in which trends in personal washing are forecast. Third, what will be the likely impact of future demographic changes? The trend towards decreasing household size, due to a combination of population ageing, increased household fission and changes in social behaviour means that the number of households is set to increase dramatically. As there are at least some limited (but currently unknown) economies of scale to be enjoyed by larger households, this trend can be expected to drive up even further future domestic water demand. Spatial variations in this additional demand will be driven not only by place-specific demographic trends but also by the differential impact of migration. Forecast changes in household numbers and population composition can be incorporated in static microsimulation models by reweighting the estimated base population microdata to agree with small area population projections, derived from models such as that outlined by Rees (1994).

Acknowledgements

All Census data and table layouts reported in this paper are crown copyright, as are all data from the General Household Survey. The Census Small Area Statistics are provided through the Census Dissemination Unit at the University of Manchester, with the support of ESRC/JISC/DENI 1991 Census of Population Programme. The Census Sample of Anonymised Records is provided through the Census Microdata Unit at the University of Manchester, also with the support of ESRC/JISC/DENI. Map boundary data were provided with the support of ESRC/JISC and uses boundary material that is Crown and Post Office copyright. General Household Survey data were made available through the Data Archive.

The research reported here is part of a wider research project into 'Model integration and development for sustainable water resource management and demand side control', funded jointly by the Engineering and Physical Research Council and Yorkshire Water Services Ltd. (award GR/K 93433), undertaken in collaboration with Dr G Mitchell, Professor A McDonald and Dr P Wattage from the University of Leeds. A debt of gratitude is also owed to Joanna Gomulka, of the London School of Economics, for her invaluable help in porting a sample reweighting algorithm into the water projection program suite. Any errors in thinking and calculation remain my own.

References

Beckman R.J., Baggerly K.A. and McKay M.D. (1996) Creating synthetic baseline populations, *Transportation Research A*, 30(6), 415-429.

Birkin M. and Clarke M. (1988) SYNTHESIS – a synthetic spatial information system for urban and regional analysis: methods and examples, *Environment and Planning A*, 20(12), 1645-1671.

Birkin M., Clarke G., Clarke M. and Wilson A. (1996) *Intelligent GIS: location decisions and strategic planning*, GeoInformation International, Cambridge.

Caldwell S.B. (1986) Broadening policy models: alternative strategies in G H Orcutt, J Merz and H Quinke [eds.] *Microanalytic models to support social and financial policy*, North-Holland, Amsterdam, 59-77.

Caldwell S.B. and Keister L.A. (1996) Wealth in America: family stock ownership and accumulation, 1960-1995, in Clarke G.P. [ed.] *Microsimulation for urban and regional policy analysis*, Pion, London, 88-116.

Caldwell S.B., Clarke G.P. and Kesiter L.A. (1998) Modelling regional changes in US household income and wealth: a research agenda, *Environment and Planning C*, 16, 707-722.

Chambers R.L. (1996) Robust case-weighting for multipurpose establishment surveys, *Journal of Official Statistics*, 12, 3-32.

Clarke G.P.[ed.] (1996) *Microsimulation for urban and regional policy analysis*, Pion, London.

Clarke G.P., Kashti A., McDonald A. and Williamson P. (1997) Estimating small area demand for water: a new methodology, *Journal of the Chartered Institution of Water and Environmental Management*, 11, 186-192.

Dale A., Middleton E. and Schofield T. (1995) New Earnings Survey variables added to the SARs, *SAR Newsletter 6*, Census Microdata Unit, University of Manchester.

Department of the Environment (1995) *Projections of households in England to 2016*, HMSO, London.

Duley C. (1989) A model for updating census-based household and population information for inter-censal years, unpublished Ph.D. thesis, School of Geography, University of Leeds

Duley C. and Rees P. (1991) Incorporating migration into simulation models, in Stillwell J. and Congdon P. [eds.] *Migration models: macro and micro approaches*, Belhaven Press, London, 228-261.

Eason R. (1996) Microsimulation of direct taxes and fiscal policy in the United Kingdom, Ch. 2 in A Harding [ed.] *Microsimulation and public policy*, North-Holland, London, 23-45.

Falkingham J. and Lessof C. (1992) Playing God: or LIFEMOD – the construction of a dynamic microsimulation model, Ch. 2 in in Hancock R. and Sutherland H. [eds.] *Microsimulation models for public policy analysis: new frontiers*, Suntory-Toyota International Centre for Economics and Related Discplines, London School of Economics and Political Science, London, 5-32.

Ghosh M. and Rao J.N.K. (1994) Small area estimation: an appraisal, *Statistical Science*, 9(1), 55-93

Gilbert N. and Doran J. [eds.] (1994) *Simulating societies*, UCL Press, London.

Gomulka J. (1992) Grossing-up revisited, in Hancock R. and Sutherland H. [eds.] *Microsimulation models for public policy analysis: new frontiers*, 121-131.

Hägerstrand T. (1952) *The propogation of innovation waves*, Lund Studies in Geography, B:4, Lund.

Hancock R., Mallender J. and Pudney S. (1992) Constructing a computer model for simulating the future distribution of Pensioners, incomes for Great Britain, Ch. 3 in Hancock R. and Sutherland H. [eds.] *Microsimulation models for public policy analysis: new frontiers*, Suntory-Toyota International Centre for Economics and Related Discplines, London School of Economics and Political Science, London, 33-66.

Harding A. (1993) *Lifetime income distribution and redistribution: applications of a microsimulation model*, North-Holland, London.

Herrington P. (1987) Water demand forecasting in OECD countries, OECD Environment Monograph no. 7, Paris, available only from OECD Environment Directorate, 2, Rue André-Pascal, 75775, Paris, Cedex 16 France.

Herrington P. (1996) *Climate change and the demand for water*, HMSO, London.

Merz J. (1994) Microdata adjustment by the minimum information loss principle, Discussion paper no. 10, Forschungsinstitut Freie Berufe, Universtät Lüneburg, Lüneburg, Germany.

Mills I. and Teague A. (1991) Editing and imputing data for the 1991 Census, *Population Trends*, 64, 30-37

Mitchell G. and Wattage P. (1998) Industrial and commercial water demand in the YWS Ltd. region: model development and forecasts to 2020 – Technical report, The Environment Centre, University of Leeds.

Mitchell R., Martin D. and Foody G.M. (1998) Unmixing aggregate data: estimating the social composition of enumeration districts, *Environment and Planning A*, 30(11), 1929-1941.

Nelissen J. (1998) Annual versus lifetime income redistribution by social security, *Journal of Public Economics*, 68(2), 223-249.

Office of Population Censuses and Surveys (1993) *General Household Survey 1991*, London, HMSO

Orcutt G.H. (1957) A new type of socio-economic system, *The review of economics and statistics*, 29, 116-123.

Pearson M., Rajah N. and Smith S. (1993) The distributional effects of different methods of charging households for water and sewerage services, Institute of Fiscal Studies, available only from the Office of Water Services, Centre City Tower, 7 Hill Street, Birmingham B5 4UA.

Rees P.H. (1994) Estimating and projecting the populations of urban communities, *Environment and Planning A*, 26, 1671-1697.

Simpson L. and Middleton L. (1998) Characteristics of census undercount: what do we already? Adjusting for census output undercount, One Number Census Research Workshop, Leeds, May 12-13.

Simpson S., Diamond I. and Tye R. (1994) What was the real population of local areas in mid-1991: population estimates by age and sex for wards, Estimating with Confidence Working Paper 10, Dept. of , Social Statistics, University of Southampton.

Szeto K.L. and Devlin N.J. (1996) The cost-effectiveness of mammography screening – evidence from a microsimulation model for New Zealand, *Health Policy*, 38(2), 101-115.

van Tongeren F.W. (1995) *Microsimulation modelling of the corporate firm*, Springer, Berlin.

Wanders A. (1998) Integrating aggregate and disaggregate migration data from the 1991 population census, paper presented at the Royal Geographical Society (with the Institute of British Geographers) annual conference, Guildford, January 1998.

Williamson P., Clarke G.P. and McDonald A.T. (1996) Estimating small-area demands for water with the use of microsimulation, in Clarke G.P. [ed.] (1996) *Microsimulation for urban and regional policy analysis*, Pion, London, 117-148.

Williamson P., Birkin M. and Rees P.H. (1998) The estimation of population microdata by using data from small area statistics and samples of anonymised records, *Environment and Planning A*, 30, 785-816.

Zhao Z. (1996) The demographic transition in Victorian England and changes in English kinship networks, *Continuity and change*, 11(2), 243-272.

14 Fuzzy Geodemographic Targeting

Linda See and Stan Openshaw

14.1 Introduction

Geodemographics is the analysis of spatially referenced demographic and lifestyle data, which has proved to be a major geographical marketing research tool of enormous commercial value. The rapid growth experienced by the geodemographics industry during the 1980s in both North America and Europe (Flowerdew and Goldstein, 1989) continues today at an exponential rate; projections of the market value of geodemographics for 1996 were £100 million, almost double that of 1995, with a far larger sum of money invested in software and decision support systems containing an embedded geodemographic component (Sleight, 1997).

As a marketing tool, geodemographic systems are used to target segments or subgroups of the population that display higher than average response rates when presented with the opportunity to purchase an existing or new type of product or service (Flowerdew and Goldstein, 1995; Watts, 1994). By linking a customer address list to a geodemographic classification, it is possible to build a geodemographic profile of the customer base. By matching the response rates of previous direct mailing campaigns or other promotional means with the area typologies provided by a geodemographic system, new products or services can be marketed by targeting prospective customers in high response areas, thereby improving the chances of a positive response over a purely random targeting method. Since a one percent response rate for a direct marketing campaign is considered adequate from a commercial perspective, any system that can add some intelligence to the targeting process, and improve upon what equates to a failure rate of 99%, will possess commercial value (Openshaw, 1989).

Although geodemographics has become an effective and significant marketing technology, the conventional approach has not evolved much since it was originally conceived. Openshaw (1989) notes that there are two areas of uncertainty or fuzziness associated with conventional geodemographics that still need to be addressed. First there is fuzziness present in the actual geodemographic classification. This uncertainty is also inherent in all clustering algorithms that impose a crisp boundary between classes. Conventional neighbourhood geodemographic classifications such as ACORN or MOSAIC are crisp because they ignore the fact that some of the small areas being classified may be fairly close in the multivariate taxonomic space of the classification to more than one residential neighbourhood type, but they require that each geographical zone belong to only one cluster. This is not a case of classification error but a reflection of the fuzzy nature of reality in which some areas may differ only by very small amounts in terms of their multivariate data profile yet can be assigned to different clusters in the classification. The second kind of fuzziness relates to the geographical aspects of neighbourhood. A neighbourhood effect is a geographical phenomenon that may well reflect sociological concepts of neighbourhood or locality, whereby for various reasons people who live near to each other tend to be similar and share some common behavioural characteristics despite other differences (Openshaw et al., 1995). The neighbourhood effect is very important in geodemographics but largely uncontrolled. It has two primary components: the accuracy of the allocation of addresses to the correct geodemographic cluster and the spatial extent over which it is assumed to apply. In conventional systems it is a reflection of the size or scale of the areas being classified and the number of clusters.

Marketers often have a reasonably good idea of the ideal profile of the segments of the society they think are the principal target groups for a specific product. However, whilst this may be expressed in a highly crisp form so that conventional geodemographics can be used, their knowledge is seldom that precise. Alternatively, marketers can perform some customer profiling exercise in an attempt to define the target groups but without necessarily knowing what precisely these target segments consist of or what the labels mean. The various sources and causes of uncertainty and imprecision in conventional geodemographics are not well understood and in practice are more or less completely ignored. Consider an example. Suppose residential area type 27 is selected for mailing. The questions you could answer via a fuzzy geodemographic system (as opposed to a conventional approach) would be: (i) which areas not in type 27 are very similar to those areas that are; (ii) which areas not in type 27 are very near on the map to those areas that are; (iii) which areas not in type 27 are both very similar and near to those areas that are in type 27. It was to handle these types of questions that Openshaw (1989) originally proposed the idea of a fuzzy geodemographic system.

The reasons for formulating a fuzzy approach to geodemographics are partly theoretical, i.e. what would fuzzy geodemographics entail, and partly practical, i.e. what benefits might it offer. It has taken almost a decade for such a system to be realised and tested. Feng and Flowerdew (1998a,b) re-stimulated interest in the topic and are to be congratulated on being the first to outline a practical methodology. The purpose of this chapter is to revisit Openshaw's (1989) original ideas behind fuzzy geodemographics, to review the version proposed by Feng and Flowerdew (1998a,b), and to provide a further real world test of a fuzzy geodemographic system.

14.2 Client Requirements

The client is a leading UK financial institute whose main interest is in the development of high performance systems for customer targetting of financial products, i.e. through mailshots and telephone sales, and for making informed business decisions about customer purchasing and behaviour. The client is faced with a difficult modelling problem that is characterised by: a lack of response data, since random targetting exercises are required to collect the data but can be very expensive, and the data collected from these exercises exhibit very low response rates (typically 0.1%); hence the databases for model calibration are information poor. The addition of fuzziness to the conventional geodemographic approach provides another avenue for investigation into the development of higher performance targetting systems.

14.3 A Conventional Geodemographic Approach

In conventional geodemographics, a national small area census or part census and part non-census database is created covering a wide range of demographic (age, lifestage), socioeconomic (occupation, social class, car ownership, employment status) and household (type and tenure, amenities) variables. Most variables are crude reflections of need (life stage, age) and disposable income. These 'small areas' have a linkage to lists of postal addresses; for example, in the UK the unit postcode provides the linking mechanism to postcodes, census enumeration districts (EDs), wards, and postcode sectors. Historically, the postcode-to-census linkage has been imprecise and error prone.

A multivariate classifier is then applied to group these data into a manageable number of clusters of different types of residential area. Note that these clusters would at one time have been called social areas or even neighbourhoods, being characterised by a common set of features. Also note that the relative distinctiveness of the clusters and the accuracy of the assignment of the data entities to clusters is highly variable. Some areas could conceivably be very close (in taxonomic space) to more than one cluster; others might be a long way away from all the clusters. Yet in both cases the all or nothing nature of the crisp assignment process forces complete membership in the "nearest" cluster regardless of either goodness of match or uniqueness.

The clusters are then assigned descriptive labels, which summarise their principal characteristics. The clusters are labelled with reference to their dominant or most striking features. These are used to create idealised profiles or pen portraits or thumb nail sketches of typical residents. However, these clusters are seldom (if ever) completely homogeneous but will contain a number of subdominant non conforming types of persons and household for which the average cluster label is quite misleading. This is an example of an ecological fallacy and its extent will vary from one cluster to the next.

Once the system is developed, it can used by an organisation for targeting via mail, telephone or other direct means, as outlined in figure 14.1. The first step involves the geodemographic profiling of the customer database in which a neighbourhood type or cluster from a geodemographic system is attached to each customer. It is then a simple calculation to determine the market penetration of a particular good or service in each neighbourhood type in relation to a base area such as a national or regional average. After completing the geodemographic profiling, the database can be used for mail and telephone targeting (figure 14.1 step 2), which involves selecting the most relevant set of clusters thought to have a higher potential response to a given product or service.

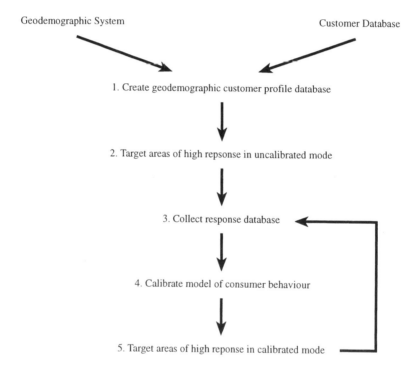

Figure 14.1. Use of a geodemographic system for targeting

These will correspond to the neighbourhood types that have a higher market penetration than the average. This task will also provide a better understanding of the client base; for example, financial institutes might want some general information about their gold credit card customers based on these area topologies.

The use of geodemographic profiling to target potential customers in this way is equivalent to employing an uncalibrated model that is based on the qualitative assumption that people living in an area characterised as type A will conform to the average for type A areas. Neighbourhood effects are involved here in that not only do "birds of a feather flock together" but they tend to behave in a similar way. However, this is a second type of ecological fallacy and the geographical distribution of the geodemographic types is also very poorly handled. But their geography is very interesting in that if there are 50,000 postcodes in geodemographic type A, then when mapped, these 50,000 areas may well appear as 2,000 separately identifiable regions; there will be parts of the map where several contiguous postcodes belong to the same type of area while in other parts of the map these types may be isolated. The problem is that neighbourhood effects are ill defined and implicit in the choice of data entities (i.e. unit postcodes or census Eds) and these are also affected by the number of residential area types that are defined. Despite these problems, if this type of approach is enough to yield a small increase in future response rates, it will prove to be more useful and therefore more commercially valuable than a random targeting of the population.

It is possible to stop at this stage and continue to use the geodemographic profiling as a way of identifying potential areas of high response in an uncalibrated manner, or the model can be calibrated (step 4) by collecting a response database based on previous targeting (step 3). Typically a previous mailing or telephone response database will be used to calculate response rates by geodemographic area type, and a gains chart, which plots the relative gains over a random targeting method, is then used to make future targeting selections in an iterative process of model updating and further targeting (step 5). However, the model calibration can be problematic due to the highly skewed nature of the response database; response rates are typically less than 1%. This may result in a response model that can only identify customers that will not respond and will ignore the very small percentage that might. Thus, data issues are clearly an important consideration in building the model.

14.4 Adding Fuzzy Logic to the Conventional Approach

The previous brief outline of conventional geodemographic systems highlights some principal areas of fuzziness and uncertainty, which include:

- fuzziness due to the classification method used
- fuzziness in the definition of geographical neighbourhood
- fuzziness in the selection of types of areas to target

The principal justification for developing a fuzzy geodemographic system is to try and tackle some of these areas of potentially significant uncertainty that might be addressed using fuzzy logic (Zadeh, 1965). Fuzzy logic is a generalisation of crisp Boolean logic, which no longer requires that a geographical zone belong exclusively to only one neighbourhood type. Instead it allows simultaneous membership in more than one cluster but to differing degrees; e.g., area A might belong to type A with a degree of membership of 0.75 and to type B with a degree of 0.25.

Adding fuzzy logic to the geodemographic process offers a number of potential benefits:

- it can deal with the vagueness resulting from conventional geodemographic classification methods;
- it provides an explicit mechanism for dealing with neighbourhood effects;
- it provides a modelling framework for building and calibrating a fuzzy targeting mechanism;
- it may improve the performance of geodemographics;
- it may breathe new life into old mailing lists; and
- fuzzy logic is becoming a very trendy technology so it can be used to market any resulting systems that make effective use of this technology.

The term fuzziness is preferred here rather than error because there is no way of measuring the errors, or if once measured, knowing how to remove them or reduce them. The presumption here is that a fuzzy approach would circumvent some of the problems and may provide the basis for a practical alternative technology to the conventional methods.

Figure 14.2 provides a diagrammatic representation of how fuzziness in both the classification and the geographical definition of neighbourhood can be represented. Conventional geodemographic systems populate the top left cell. The challenge for developers of fuzzy geodemographics is to develop targeting systems that can profitably and flexibly allow other cells in figure 14.2 to be targeted, to develop a means of selecting the most appropriate degrees of fuzziness, and then of persuading end-users that the ideas are not only interesting but have a practical value. The degree of difficulty in meeting these challenges should not be underestimated.

14.5 Developing a Fuzzy Geodemographic System

14.5.1 Identifying Fuzzy Neighbours of a Target Region

The easiest way of building a fuzzy geodemographic system is to add fuzzy logic to the uncalibrated model (figure 14.1 step 2) where one assumes that the purpose of the system is to find other areas that are similar or near to a set of targeted areas. This is useful because it provides a means of breathing new life into old mailing lists through the identification of potentially good prospects that have been "missed" by conventional geodemographics. This is the type of approach demonstrated by Feng and Flowerdew (1998a). They used a fuzzy k-means classifier of census ED data to obtain memberships of each ED belonging to each cluster in the classification.

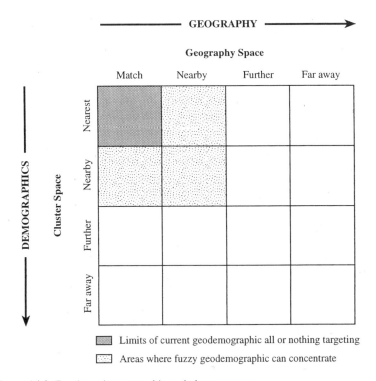

Figure 14.2. Fuzziness in geographic and cluster spaces

The fuzzy k-means algorithm (Dunn, 1974; Bezdek *et al.*, 1981) is a fuzzification of the crisp k-means classifier, which essentially replaces crisp boundaries with fuzzy ones and provides one method of coping with fuzziness in the demographic or cluster space. Iterative optimisation is used to minimise the fuzzy objective function for n observations and k clusters given by the following equation:

$$J(U,v) = \sum_{i=1}^{k} \sum_{j=1}^{n} (\mu_{ij})^f (d_{ij})^2 \tag{14.1}$$

which involves finding the best fuzzy partition matrix, U, that minimises the Euclidean distance, d_{ij}, between the observations and the cluster centers. μ_{ij} is the membership of an observation i in cluster j, and f is a fuzziness factor ranging between 1 and 2. A value of $f=1$ results in a crisp classification. As this value increases, the fuzziness of the classification increases until $\mu_{ij} = 1/k$ for all membership values. When this occurs, all or the majority of the observations will belong to each cluster equally, resulting in too much fuzziness. There is currently no established theory as to the optimal choice of this parameter (Bezdek *et al.*, 1984) since it is probably related in part to the actual data set. For the data set employed by Feng and Flowerdew (1998a), they found that a fuzzy classification exponent of 1.25 provided the fuzzy k-means algorithm with a sufficient means of identifying both a crisp classification (membership probabilities of above 0.75) and a 'near to' category (membership probabilities of greater than or equal to 0.25).

They then added what they term 'genuine neighbourhood effects'. They write "Neighbourhood effects reflect spatial interaction between contiguous areas" (p124) and they calculate a new membership function that adjusts the fuzzy classification probabilities. They state that: "The new membership value is taken as the average of the old membership value and the mean membership value of the neighbouring EDs weighted by distances between EDs as a decay measure" (p125). Hence:

$$\mu'_{ik} = \alpha \mu_{ik} + \beta \left[\sum_{k}^{m} \sum_{j}^{n} w_{ij} \mu_{jk} / \sum_{j}^{n} w_{ij} \right] \quad (14.2)$$

where:
μ_{ik} is the membership value provided by the fuzzy classifier
m is the number of clusters
n is the number of contiguous EDs to the ED of interest
w_{ij} is the distance weighting; note that if EDs i and j are non-contiguous then $w_{ij} = 0$ otherwise

$$w_{ij} = d_{ij}^{-a} \quad (14.3)$$

where d_{ij} is the geographical distance from i to j and a is an exponent set at -1. α and β are weights to represent the relative importance of demographic and geographical neighbourhood effects; note that $\alpha + \beta = 1.0$ and in their examples they use $\alpha=\beta=0.5$. Feng and Flowerdew (1998b) claim to test their method on real data but what they really do is demonstrate the existence of fuzziness. However, their approach does provide a way of potentially targeting a larger set of customers from an existing mailing list that might have been missed out via conventional means, and requires testing using real world data.

14.5.2 Building a Partly Fuzzy Geodemographic Response Model

It is important to remember that underlying geodemographics is the desire to identify the potential responsiveness of areas so that the most responsive ones can be targeted. First let us define the conventional response rate for any cluster j of a conventional classification as:

$$R_j = \sum_{i}^{n} C_{ij} / \sum_{i}^{n} P_{ij} \quad (14.4)$$

where:
- R_j is the response rate for the jth cluster,
- C_{ij} is the number of observed responders in a database or mailing in area i assigned to cluster j
- P_j is the population that could have responded in area i in cluster j, and
- n is the number of areas.

A fuzzy version is as follows:

$$R'_j = \sum_i^n \mu_{ij}(C_{ij} / P_{ij}) \tag{14.5}$$

which takes into account the fuzzy membership of the base areas with respect to the m clusters. Typically a user will sort the n areas by descending R_j values and plot the gains chart which shows for each j the cumulative P_j and C_j values. Some threshold will then be selected. The simplest fuzzy classification equivalent would be to use R'_j instead of R_j. However, this assumes that there are no neighbourhood effects. This formulation represents a fuzzy equivalent of a conventional geodemographic approach. The predicted fuzzy response rate for area i is

$$F_i = (\sum_j^m \mu_{ij} R'_j) P_i \tag{14.6}$$

As each area i can have a predicted response, the targeting mechanism would now sort the areas by F_i and then plot cumulative P_i against C_i.

A more general approach would be to include both fuzzy classification and geographical neighbourhood effects. The predicted fuzzy response for area i is now

$$Q_i = \alpha F_i + \beta \sum_k^n w_{ik} F_k / \sum_k^n w_{ik} \tag{14.7}$$

where w_{ik} is the weighting of the kth area with respect to its distance from i. The aim is to distinguish between those regions where there is a high response near to the area being considered and those where the predicted good response is an outlier and is not characteristic of the surrounding areas. To some large degree neighbourhood is a scale dependent phenomenon and different sizes of base unit may require very different spatial extents. So there is no good reason to assume that neighbourhood effects are purely a map contiguity phenomenon but there may well be either a relative or absolute distance cut-off. The w_{ik} values would be set to zero outside of this neighbourhood region and to some distance related weight within it. A relative distance cut-off would restrict the non-zero values to a kth nearest neighbour, whilst a distance threshold would emphasise physical distance. The former is more adaptive allowing larger distances in sparsely populated (or sparsely cliented areas), although distance decay may now be a matter of a different weight for each order of nearest neighbour.

14.5.3 Building a Fully Fuzzy Geodemographic Response Model

A fully fuzzy approach would seek to build a fuzzy response prediction model in which the inputs (fuzzy cluster membership and fuzzy distance) would be related to a fuzzy output variable (i.e. high, medium, low, poor responsiveness) via a set of fuzzy rules. The problem here is how to define these fuzzy rules, which would be of the following form:

IF zone A has a high degree of membership in cluster A
   THEN responsiveness of zone A will be high

IF zone A has a medium degree of membership in cluster A
   AND zone A is near to a zone with a high degree of membership in cluster A
      THEN responsiveness of zone A will be high

where cluster A is a potentially high responding cluster and both the membership in the cluster and the distance away from neighbouring zones are fuzzy quantities. However, one can see that the total number of possible rules governing the model is potentially quite large if there are rules for several clusters, and determining the fuzzy set definitions for distance is a non-trivial problem. With a previously collected response database it would be possible to find the rules and the membership function definitions of distance using an optimisation method such as a genetic algorithm, but this approach is left for further study as there appear to be far simpler alternatives worthy of investigation.

14.6 Telephone Targeting via a Fuzzy Geodemographic System

14.6.1 Creating the Fuzzy Classifications

The fuzzy k-means algorithm was applied to a set of UK wide postcode level factor scores for 8 independent variables from both the 1991 Census and other data sources. This data set was classified into 10, 25, 50 and 100 clusters, and three different values for the fuzziness factor were used: 1.10, 1.25 and 1.40. These different classifications will provide a range of results that will indicate the importance of the number of clusters used to develop the neighbourhood classification and how sensitive the fuzziness factor is to the data set. The computation time for clustering this data set – a matrix containing 1.6 million postcodes by 8 variables – increased exponentially as the number of clusters and the fuzziness factor were increased, ranging from 1 to 3 days on a fast UNIX workstation using optimised FORTRAN code.

14.6.2 Building a Partly Fuzzy Geodemographic Telephone Response Model

For developing the fuzzy geodemographic response model, a telephone response data set, for approximately 200,000 postcodes across the UK, was provided by the client. This data set was used to calculate the fuzzy response rates based on the memberships in each cluster as provided by the fuzzy classifications. A gains chart for each classification was then plotted, which indicates the advantage of using the system over a random targeting of the population. The gains charts for the 10, 25, 50 and 100 clusters with fuzziness factors of 1.10, 1.25 and 1.40 are provided in figures 14.3a to d. The straight line y=x that runs diagonally through each graph represents a random targeting of the population. The steeper the gains curve above this straight line, the greater the level of discrimination provided by the classification, and the greater the response above a random targeting of the population. Therefore, the aim of the exercise is to develop a geodemographic classification that produces the steepest possible gains curve, especially at the lower end of the graph, as this is the population that will be targeted.

All of the plots in figure 14.3 indicate that the fuzzy classifications provide some gains over a random targeting of the population. Increasing the number of clusters provides a slightly greater discrimination, i.e. a slightly steeper gains curve but not enough to aid in deciding the optimal number of clusters. Moreover, as the fuzziness factor increases, the steepness of the gains curve decreases, indicating that both 1.25 and 1.40 provide too much fuzziness. This is in contrast to an exponent of 1.25 used by Feng and Flowerdew (1998a,b) to classify a small subset of the UK, which implies that the much larger UK wide data set used in this example has more variation. If an even smaller fuzziness factor had been used, i.e., less than 1.10, this might have resulted in a steeper gains curve. Thus, some degree of experimentation with the fuzziness factor is required.

At this point the calibrated model would be used in a further targeting exercise to test how well the model would perform in future telephone targetings, where one would now expect response rates of greater than the usual 1%, based on how the fuzzy geodemographic models discriminated in the gains charts. This iterative cycle of model updating and continual targeting (figure 14.1 steps 3 to 5) is an exercise that needs to be carried out on a large number of different financial and other commercial targeting databases before one can really determine how well fuzzy geodemographic systems will perform in a commercial setting. This would involve finding private sector co-operation in providing both the data and the necessary commitment to participate in an iterative cycle of testing these types of models. This entire process would be further facilitated if developers of geodemographic systems began experimenting with the integration of fuzzy logic directly into their systems, and they should be encouraged to do so if the technology is to be considered seriously in the future.

14.7 Conclusions

This chapter argues that adding fuzzy logic to existing geodemographic systems represents one way of handling the uncertainty that is currently present in conventional systems, both with respect to classification errors and neighbourhood effects. It is also possible to take advantage of the full fuzzy logic modelling framework to develop rule-based fuzzy response models that can be used to predict the degree of expected response in a more general model of consumer behaviour. This type of approach offers the advantages of simplicity and the prospect of being able to combine a fuzzy approach with an optimisation method so that the fuzzy geodemographic systems can be calibrated or tuned for maximum performance. However, major problems of availability of practical commercial systems and user acceptability still need to be resolved. The term fuzzy immediately conjures up the spectre of extreme mathematical complexity; see for example, Zadeh (1965). However, as Openshaw and Openshaw (1997) explain, fuzzy logic is a wonderfully simple and easily applicable idea that can be expressed without any great methodological complexities. One solution might be to package the technology in an operational and commercially viable fuzzy geodemographic systems via a user-friendly mapping interface that removes some of the perceived complexity of this approach.

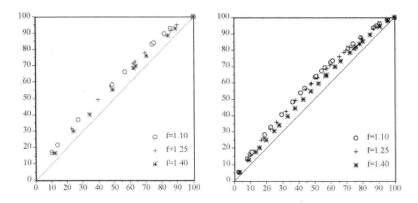

Figure 14.3. Gains charts for (a) 10 clusters and (b) 25 clusters

The task of estimating fuzzy sets, adjusting rules and changing weights can be left to invisible optimisation tools. The expert can look inside but the normal user need not bother. A good fuzzy targeting system should be understandable to anyone who was vaguely interested in how it worked and could be fine tuned to adapt its performance over time. It would also seek to recover its costs and therefore provide potentially substantial business benefits in the future.

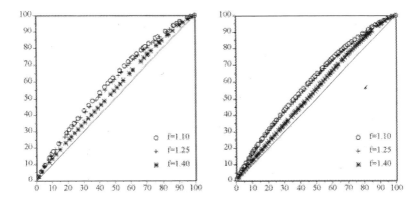

Figure 14.3. continued: Gains charts for (c) 50 clusters and (d) 100 clusters

References

Bezdek C.J. (1981) *Pattern recognition with fuzzy objective function algorithms*. Plenum Press, London

Bezdek C.J., Ehrlich R. and Full W. (1984) FCM: The Fuzzy c-means clustering algorithm, *Computer and Geosciences* 10, 191-203

Dunn J.C. (1974) A fuzzy relative of the ISODATA process and its use in detecting compact well separated clusters. *Journal of Cybernetics* 3, 32-57

Feng Z. and Flowerdew R. (1998a) Fuzzy geodemographics: A contribution from fuzzy clustering methods. In: Carver S. (Ed.): *Innovations in GIS 5*. Taylor and Francis, London, 119-127

Feng Z. and Flowerdew R. (1998b) The advantage of using fuzzy classification systems in geodemographic targeting. Paper presented at MRS Business Meets Academia Workshop, London

Flowerdew R. and Goldstein W. (1989) Geodemographics in practice: developments in North America, *Environment and Planning A 21*, 605-616

Openshaw S. (1989) Making geodemographics more sophisticated, *Journal of the Market Research Society 31*, 111-131

Openshaw S., Blake M. and Wymer, C. (1995) Using neurocomputing methods to classify britain's residential areas. In: Fisher P. (Ed.): *Innovations in GIS 2*. Taylor and Francis, London, 97-112.

Openshaw S. and Openshaw C. (1997) *Artificial intelligence in geography*, Wiley, Chichester.

Sleight P. (1997) *Targeting customers*. NTC Publications, Henley on Thames

Watts P. (1994) European geodemographics on the up, *GIS Europe 3*, 28-30

Zadeh L.A. (1965) Fuzzy sets, *Information and Control 8*, 338-353.

15 Analysing Access to Hospital Facilities with GIS

Martin Charlton, Stewart Fotheringham and Chris Brunsdon

15.1 Introduction

The location of a facility can be examined in the context of several classical frameworks in economic geography and operation research where the facility is located such that some objective function is minimised (ie. Love *et al.,* 1988; Ghosh and McLafferty, 1987; Wrigley, 1988; Fotheringham and O'Kelly, 1989). A frequently encountered objective, for example, is to find the optimal location of a new facility in terms of minimising the average time it takes individuals to travel to the nearest facility. That is, a new facility is added to an existing spatial distribution of facilities in order to achieve the maximum reduction in average travel times (the p-median problem). Other objective functions can of course be used to locate the new facility: it could for example, be located so that the maximum distance any individual has to travel is minimised (the minimax problem). Another slant on the problem is to simultaneously locate a set of facilities and to determine the allocation of demand to each of these facilities (the location-allocation problem). Still another is to model the choice of a facility by an individual as a probabilistic function of the attributes of each facility rather than as a deterministic one (a spatial interaction problem).

Although all these methods are commonly used to locate facilities having a profit motive, such as a retail outlet or a distribution centre or an industry being served with raw materials from a variety of sources, they can also be applied to the provision of health care services. One can view the location of health services in much the same way as that of a set of retail outlets in that patients need to be served by the facilities in the most efficient manner possible. This concept has led to the increasing use of standard retail location methods in the provision of health care facilities. Here, we examine the location of health care facilities in terms of the accessibility of patients to two hospitals in west central England, one in the market town of Shrewsbury, and one in the new town of Telford.

To date, the application of techniques for the location of facilities, health-related or otherwise, has typically depended on fairly crude measurements of distance or, in some more sophisticated studies, travel time. For a long period there has been a mismatch between the sophistication of the mathematical models used for facility location and the crudeness of the data, particularly that of distance, used to operationalise the models. In some cases, for example, straight-line distances were used between points to represent the spatial separation between demand and supply points. In other cases, rectangular or grid-based distances were used in an attempt to simulate spatial separation within urban areas. Here we describe how the effects of spatial separation, so vital to the successful implementation of facility location techniques, can be captured much more accurately through the use of Geographical Information Systems (GIS).

This study of health care provision thus provides some interesting facets of dealing with the generic issues of service provision and facility location from a GIS perspective. However, this is not the first time that the provision of health care facilities has received attention from geographers. For example, Mayhew and Leonardi (1982) examine considerations of equity and efficiency in health care systems, Hyman and Mayhew (1983) consider emergency services provision from a geometrical standpoint, Wilson *et al.,* (1990) examine resource allocation in health care provision, and Smallman-Raynor *et al.,* (1998) consider the optimal allocation of cancer treatment units.

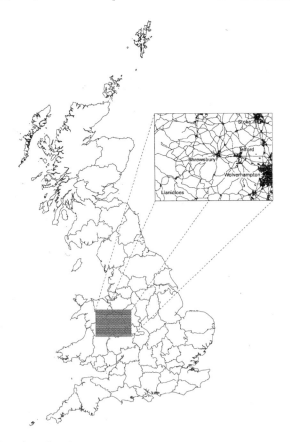

Figure 15.1. Location of study area

15.2 Client Needs

The general study area encompassing Shrewsbury and Telford is shown in figure 15.1 and in more detail in figure 15.2. The study area essentially includes the administrative county of Shropshire, with parts of the neighbouring counties of Cheshire, Staffordshire, West Midlands, Hereford and Worcestershire, Powys and Clwyd. Figure 15.2 shows the main roads and town in the study area to give some indication of scale. Shrewsbury is approximately in the middle of the study area, with Telford about half-way between Shrewsbury and the eastern border of the area. The hospitals of interest are also located in this plot. In Shrewsbury itself is the Royal Shrewsbury Hospital, whilst in Telford is the Princess Royal Hospital. There are other hospitals in Shropshire, for example in Ludlow and Whitchurch, but these are not included in the study as they do not have the patient facilities under scrutiny. Also of interest are hospitals in Stoke in Trent (North Staffordshire Hospital), Stafford (Staffordshire General Hospital), and Wolverhampton (New Cross Hospital).

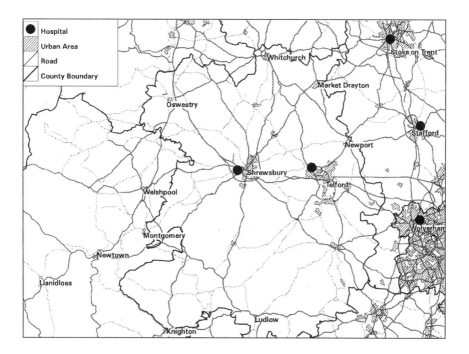

Figure 15.2. Urban areas and hospitals in the study area

The specific task set by the client was to determine the optimal allocation of health services between two 'competing' hospitals, one in Shrewsbury and one in Telford.

The services of interest deal with obstetrics, gynaecology, paediatrics, and neo-natal care. The question, however, is of general interest – the inefficient provision of medical facilities occurs in many parts of the county due to either shifts in the distribution of population or the building of new hospitals, or both. In some cases, inefficiencies exist when patients are poorly served by facilities and have to travel long distances for medical services. In other cases, inefficiencies exist because of the over-provision and duplication of services. In the specific situation examined here, the issue was to determine the optimal location of four different services that could each be offered at the two hospitals.

More specifically, the client requested the following tasks:

(i) a set of maps depicting the spatial distributions of fours sets of patient records and competing hospitals;
(ii) calculation of average travel times to the two hospital sites from four sets of patient records;
(iii) a set of maps depicting catchment areas for both hospitals based on shortest travel time from any point on the map;
(iv) a set of maps depicting the catchment area boundary between the hospitals at Shrewsbury and Telford and indifferent zones in which the travel time difference between the two hospitals was less than 5 minutes, less than 10 minutes, less than 15 minutes and less than 20 minutes;
(v) a set of maps depicting areas within certain travel times of each hospital. The times requested were 10, 20, 30 and 40 minutes;
(vi) an examination of the 'competition' faced from other nearby hospitals.

15.3. Data Issues

Patients using hospitals are routinely asked for identifying information including the address and postcode of their residence. The postcodes of patients using the facilities in Shrewsbury during one year provides a useful dataset with which to examine the spatial distribution of four groups of patients. The patients in each dataset were those who had been referred to a facility and who had undergone and completed a period of treatment in the hospital. In hospital parlance each patient is referred to as a 'Finished Consultant Episode' or FCE. The numbers of FCEs at each of four separate units within the hospital are shown in table 15.1.

Figure 15.3 shows the locations of the FCEs for the various specialities who visited the Royal Hospital Shrewsbury. The pattern for gynaecological and obstetric FCEs follows largely that of the distribution of the at-risk population. Notable is the distances some patients have to travel to reach the hospital. Newtown, in the southwest of the study area is some 50km from Shrewsbury, and Ludlow in the south of the study area is about 48km distant. Whitchurch, in the north of the study area, is 30km from

Shrewsbury. Shrewsbury and Telford are about 20km apart, joined by the A5/M54. The paediatric FCE distribution is somewhat different with an apparent lack of patients from Telford. This is because there is a paediatrics unit at the Princess Royal Hospital in Telford that caters for patients from the New Town. There are relatively few neo-natal cases (260). These are catered for at the Special Care Baby Unit at the Royal Shrewsbury Hospital that provides intensive care facilities for babies born prematurely or who require specialist attention over and above that normally provided in a maternity unit.

Table 15.1. FCEs by unit

Unit	Number	Coverage Name
Gynaecology	3095	patgyn
Obstetrics	3340	patobs
Paediatrics	1068	patpaed
Neo-Natal	260	patnn

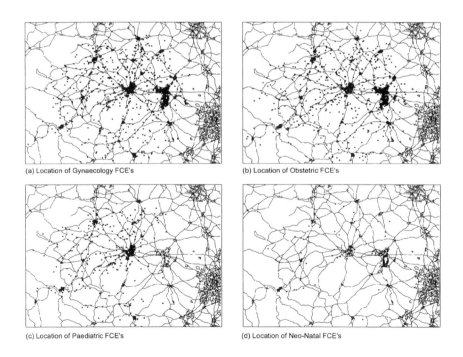

(a) Location of Gynaecology FCE's (b) Location of Obstetric FCE's
(c) Location of Paediatric FCE's (d) Location of Neo-Natal FCE's

Figure 15.3. Location of patients in each finished consultant episode (FCE) Type

15.4. Methodology

15.4.1 Data Preparation

From the spatial distributions of the four sets of FCEs we were able to examine the accessibility of patients to the two hospitals at Shrewsbury and Telford. There are many different measures of accessibility; here we decided to examine average travel times to each hospital for each of the patient types as this seemed most relevant to both hospitals and patients. We also considered the extent to which the potential catchment areas of each hospital might overlap. Typically, one might define isochrones around each hospital and count the number of patients in each of say 5-minute time bands. However, comparing the distributions is not easy, and a simple average was felt sufficient to give a helpful summary.

In order to compute mean travel times there are several necessary steps in the preparation and manipulation of the various pieces of source data. We decided to carry out the modelling using the GRID module in ARC/INFO. To compute the weighted mean distance on a lattice we require for each cell in the grid the number of FCEs originating in the cell and the minimum travel time distance from that cell to each of the two hospitals. The weighted mean is the sum of the product of the FCEs and the minimum distance in each cell, divided by the total number of FCEs. However, arriving at this point requires a series of data manipulations that are described below. Four ASCII files of the postcodes of the four types of FCEs were made available for analysis. Grid references were obtained by matching these against the Central Postcode Directory (Postzon, 1985). From the grid reference lists, four point coverages were created. Mapping these suggested that an appropriate extent for the study area would be a rectangular section of the West Midlands, 110km by 80km. The names of the coverages are shown in table 15.1. The FCEs are represented as point locations with no further attribute information. The grid references for both hospitals were obtained from 1:50000 maps, a 'representative' point being chosen roughly in the middle of the extent of each hospital's site. The shortest route algorithms in ARC/INFO will calculate a path between two nodes in a road network. One option would be to determine the nearest node to each FCE in the road network and then total the number of FCEs for each of the four patients types at each node. However, computing the shortest routes would entail several thousand invocations of the PATH command. This was therefore rejected as unwieldy. A useful function in the GRID module is *costdistance*, which will compute the least-cost route from a given set of source cells in the lattice to a set of destination cells. As input, it requires a grid containing the location of one or more source cells and a grid of cost-per-unit-distance in moving across any given cell. If we also have grids of the frequencies of FCEs, then the product of these two will eventually yield the weighted mean. To obtain a grid of the cost-per-unit-distance of moving through each cell we used the *linegrid* function, and some assumptions about average speeds on different types of roads, to convert from the vector representation of the network to a raster representation. A decision has to be made using this function about the raster size. Too small a raster size would result in lengthy computation times, and too coarse a raster would result in loss of accuracy in the calculations. After some experiment, it was decided that a square cell of 500m size would suffice.

15.4.2 Data Processing

The first step is to create one frequency grid for each FCE coverage. There are several routes to accomplishing this. One is to use *setwindow* to define the extent of the output grids, *setcell* to define the grid cell (raster) size, and *pointstats* with the *npoints* option to count the points in each grid cell. A slightly more cumbersome method is to define a vector grid using *generate*, then *intersect* each point coverage with the vector grid, count the number of occurences of each cell code with *frequency*, then *joinitem* to merge the frequency counts with the original vector grid attribute table, and finally *polygrid* to create a raster grid of frequencies. At the time the work was carried out, *pointstats* was not available, so the second method was chosen.

The initial step is the creation of the grid polygon coverage named **gridcov**. The lower left corner is located at 290000,270000 (metres east and north of the origin of the National Grid), with a cell size of 500 metres, 161 rows and 221 columns. In the ARC/INFO examples below the prompts and commentary from the software are printed in Courier type and the commands and their arguments are printed in **Courier Bold** type.

```
Arc: generate gridcov
Generate: grid
Fishnet Origin Coordinate (X,Y): 290000,270000
Y-Axis Coordinate (X,Y): 290000,280000
Cell Size (Width,Height): 500,500
Number of Rows, Columns: 161,221
Generate: quit
Arc: build gridcov poly
```

It is useful to draw the resulting grid, if only to check that the extent is that which is desired. There are several other coverages available including roads (road centrelines) and urban (urban area boundaries). The following code does this:

```
Arcplot: &station 9999
Arcplot: mape gridcov
Arcplot: arcs roads
Arcplot: polygonshades urban 3
Arcplot: pointmarkers patpaed 17
Arcplot: pointmarkers patgyn 18
Arcplot: pointmarkers patobs 19
Arcplot: pointmarkers patnn 20
Arcplot: quit
```

The *polygonshades* command here plots the urban area extents with a green fill, and the *pointmarkers* commands plot the four FCEs with differently coloured small squares.

To determine in which cell of our grid each FCE lies we intersect the four separate FCE coverages with the grid coverage. The intersection operation requires the determination of which of 35581 polygons each of the 15526 FCEs lies in. One method of speeding up the computations is to use the *index* command to create spatial indices for each coverage.

```
Arc: index gridcov poly
Arc: index patgyn point
Arc: index patobs point
Arc: index patpaed point
Arc: index patnn point
```

The four intersection operations can now take place. Any FCE locations which are outside the study area are removed from the output coverage. We also must inform ARC/INFO that the input coverage (and the output coverage) is a point coverage.

```
Arc: intersect patobs gridcov patobs_int point
Arc: intersect patgyn gridcov patgyn_int point
Arc: intersect patpaed gridcov patpaed_int point
Arc: intersect patnn gridcov patnn_int point
```

One result of the *intersect* command is to transfer the ID of the grid square in which a point lies to the attribute table for that point.

```
Arc: list patgyn_int.pat
  1
AREA            =   0.000
PERIMETER       =   0.000
PATGYN_INT#     =   1
PATGYN_INT-ID   =   *****
PATGYN#         =   705
PATGYN-ID       =   *****
GRIDCOV#        =   1229
GRIDCOV-ID      =   34378
```

In this case, the ID of the grid square in which the first gynaecological FCE lies is 34378. The next operation is to determine the frequency of FCEs in each grid square. This is done thus:

```
Arc: frequency patgyn_int.pat patgyn_int.frq
Enter Frequency item names:
================================================================
Enter the 1st item: gridcov-id
```

```
Enter the 2nd item: end
Enter Summary item names:
================================================================
Enter the 1st item: end
```

The input is the attribute table for the intersected coverage and the output is a table showing the frequency of occurrence in each grid square. A listing of the first few records of this table for the neo-natal FCEs is shown below. There was one FCE in grid square 1891, 3 in 1903 and so on. This information is then merged with the attribute table for the grid coverage.

```
Arc: list patnn_int.frq
Record         CASE#           FREQUENCY        GRIDCOV-ID
1              1               1                1891
2              2               3                1903
3              3               2                2112
4              4               1                2335
... ...... .. .. .. .. .. .. .. .. .. .. .. .. .. .. .. ... ...
```

The frequencies are stored in the attribute tables with an item name of *frequency*. As we have four tables to merge, we must change the name of this item in each table so that it is unique.

```
Arc: tables
Enter Command: select patnn_int.frq
465 Records Selected.
Enter Command: alter frequency
COLUMN ITEM NAME        WIDTH     OUTPUT     TYPE      N.DEC
  5 FREQUENCY           4         5          B         -
Item Name: countnn
Item Output Width: <return>
Item Type: <return>
Alternate Item Name: <return>
COLUMN ITEM NAME        WIDTH     OUTPUT     TYPE      N.DEC
  5 COUNTNN             4         5          B         -
Enter Command: q stop
```

In the example above we have altered the item name frequency to countnn. This is carried out for the other three frequency tables. We are then in a position to merge the frequency counts with the grid coverage attribute table.

```
Arc: joinitem gridcov.pat patnn_int.frq gridcov.pat ~
Arc: gridcov-id gridcov-id
```

This is carried out for the other three frequency tables as well. Inspection of the grid coverage attribute table shows that the merges have been successful and the counts are transferred to the grid coverage's attribute table. The added items are shown in italic in the listing below. Each grid square now has the frequencies of each FCE occurring within its boundary as attribute data.

```
Arc: items gridcov.pat
   COLUMN   ITEM NAME      WIDTH   OUTPUT   TYPE   N.DEC
   1        AREA           4       12       F      3
   5        PERIMETER      4       12       F      3
   9        GRIDCOV#       4       5        B      -
   13       GRIDCOV-ID     4       5        B      -
   17       COUNTNN        4       5        B      -
   21       COUNTOBS       4       5        B      -
   25       COUNTGYN       4       5        B      -
   29       CASE#          4       5        B      -
   33       COUNTPAED      4       5        B      -
```

The gynaecology FCE frequencies are stored in the item `countgyn`, those for obstetrics in `countobs`, those for paediatric in `countpaed`, and those for neo-natal in `countnn`. Now, the conversion to raster form can take place. Each raster takes the value that is contained within the count attribute for the polygon it represents. For the neo-natal data, the output grid will be named **gridnn**.

```
Arc: polygrid gridcov gridnn countnn
Converting polygons from gridcov to grid gridgyn
Cell Size (square cell): 500
Convert the Entire Coverage? (Y/N): y
Number of Rows = 161
Number of Columns = 221
```

The next task is to create the cost grid for *costdistance*. We start by making some assumptions about the speeds at which one might expect to travel along each different type of road. In the roads coverage, the ID of each arc is a code that indicates the nature of the road. There are essentially six types of road in the road coverage:

Motorway
Trunk Road
Dual Carriageway
A Road
B Road
Other Road

To reflect different driving conditions in urban and rural areas, we assigned different speeds to both forms of environment as well as to the different types of roads. The assumptions about road speeds are in the table below – the URBAN-SPEED item contains the urban speed assumptions, and the NORMAL-SPEED item contains those for rural areas.

```
Arc: list roads.lkp
```

Record	ROADS-ID	SYMBOL	URBAN-SPEED	DESCRIPTION	NORMAL-SPEED
1	130	1	15	Other Road	25
2	222	6	20	A Road	40
3	226	6	20	A Road	40
4	227	6	20	A Road	40
5	229	3	17	B Road	30
6	230	3	17	B Road	30
7	231	6	40	Dual Carriageway	60
8	232	3	40	Dual Carriageway	60
9	233	6	40	Dual Carriageway	60
10	235	8	50	Motorway	70
11	173291	6	30	Trunk Road	50
12	173292	6	30	Trunk Road	50
13	173295	6	30	Trunk Road	50

The speed assumptions are in miles per hour. In order to determine whether roads are urban or non-urban, we used the identity operation to add urban area feature codes from the urban area coverage to the arcs in the road coverage. The *intersect* operation is not appropriate since those arcs outside the urban areas would be removed from the output coverage (i.e. they do not fall in the geometric intersection of the two coverages).

```
Arc: identity roads urban roads_urb line
```

Once this is accomplished, we can add an item to the Arc Attribute Table, AAT, of the roads coverage to hold the passage speeds for each arc.

```
Arc: additem roads_urb.aat roads_urb.aat speed 4 5 b
Arc: additem roads_urb.aat roads_urb.aat time 4 12 f 3
```

The relevant road speeds are copied across using the INFO commands below.

```
Arc: info
ENTER USER NAME>ARC
ENTER COMMAND >SELECT ROADS_URB.AAT
ENTER COMMAND >RELATE ROADS.LKP BY ROADS-ID
ENTER COMMAND >RESELECT URBAN-ID = 0
ENTER COMMAND >CALCULATE SPEED = $1NORMAL-SPEED
ENTER COMMAND >NSELECT
ENTER COMMAND >CALCULATE SPEED = $1URBAN-SPEED
ENTER COMMAND >Q STOP
Arc:
```

The AAT is selected, and a temporary join is made between it and the road speed lookup table, ROADS.LKP, with the ROADS-ID as an item common in each table. We then select only non-urban arcs (their URBAN-ID is zero), and assign values from the NORMAL-SPEED item in the related table. The $1 indicates that this is the *first* related table. As INFO reads each record from the AAT, it checks the value of the ROADS-ID and finds the record with the same ROADS-ID in the lookup table. When a match has been found, the requisite speed is copied into the empty field for the record in the AAT. Then we swap the records currently selected with those currently not selected (using NSELECT) and repeat the operation for the urban arcs.

The final operation is to convert the arcs of the road coverage into a grid. However, we need one further set of assumptions. These deal with the situation where there are several road types in the same grid-square. As we are using speeds as the attribute for conversion, we must provide a table that shows how the speeds are to be resolved when several roads are competing for the same grid-square. In this case we decided to assign the fastest speed of road within a cell to that cell, the assumption being that, all else being equal, patients are more likely to travel by faster roads than slower ones!

```
Arc: list weight_table
   Record    CODE    WEIGHT
   1         15      1
   2         17      2
   3         20      3
   4         25      4
   5         30      5
   6         40      6
   7         50      7
   8         60      8
   9         70      9
```

The *linegrid* command is used to effect the conversion. We decided after some experiment that a grid size of 500 metres square was appropriate. It might be felt that this is somewhat crude but a trade-off was required between accuracy and speed of calculation. Grid cells for which there are no arcs passing through receive a special code, *nodata,* to indicate that they have had no value assigned.

```
Arc: linegrid roads_urb roadgrid speed # weight_table
Cell Size (square cell): 500
Convert the Entire Coverage(Y/N)?: y
Enter background value (NODATA | ZERO): nodata
Number of Rows = 162
Number of Columns = 221
```

The rest of the computations take place in the GRID module. The *costdistance* function requires that the grid cells contain the cost (time) of travelling one unit distance across the square. As our units are metres, these will be the time required to traverse 1 metre. We first convert the raster values that are currently in miles per hour to unit distance traverse times in minutes. Then, assuming a travel speed of 10 mph across all other grid cells, we convert the nodata cells to 0.00372 minutes per metre.

```
Grid: rdtmgrd = 1.00 / ( ( roadgrid * 1609 ) / 60.00 )
Grid: timegrid10 = con(isnull(rdtmgrd),0.00372,rdtmgrd)
```

The hospital locations required conversion to grid format and then these can be used as a source grid to create time grids for the two hospitals.

```
Grid: shrewgrid = pointgrid(shrewsbury,#,#,#,500)
Grid: telfgrid = pointgrid(telford,#,#,#,500)
Grid: shrewtime10 = costdistance(shrewgrid,timegrid10)
Grid: telftime10 = costdistance(telfgrid,timegrid10)
```

Table 15.2. Code for mean.aml

```
&args people dist
/* Initialise the counter variables
s = scalar(0)
sx = scalar(0)
sx2 = scalar(0)
/* Pass [1] : compute the weighted mean
docell
  s += %people%
  sx += %people% * %dist%
end
meanx = scalar(sx div s)
/* Pass [2] : compute sum of squared deviations from the mean
/* for calculation of variance, standard deviation
/* and standard error of the mean
docell
  sx2 += %people% * ( %dist% - meanx ) * ( %dist% - meanx )
end
varx = scalar(sx2 div s)
&type Mean value (minutes)
print meanx
&type Variance
print varx
&type Standard Deviation
print sqrt(varx)
stderr = scalar(sqrt(varx) div sqrt(s))
&type Standard Error of the Mean
print stderr
/* Calculate the upper and lower 95% confidence interval
upperx = scalar(meanx + ( 1.96 * stderr ))
lowerx = scalar(meanx - ( 1.96 * stderr ))
print upperx
print lowerx
&return
```

We experimented with assumptions of 15mph and 20mph as traverse speeds for the *nodata* cells to examine the sensitivity of the results to the different assumptions.

There is no single function in GRID to provide the weighted mean computation, so we created a macro, shown in table 15.2, which uses the Arc Macro Language (AML) to do the work. The formula for the weighted mean travel time to hospital j, is:

$$\bar{x}_j = \frac{\sum_{i=1}^{n} people_i traveltime_{ij}}{\sum_{i=1}^{n} people_i}$$

where *peoplei* represents the number of FCEs within grid cell i and *traveltimeij* represents the minimum travel time from grid cell i to hospital j.

The macro takes two arguments, the name of the grid containing the FCE frequencies, and the name of the accumulated cost grid. It is used thus:

```
Grid:   &run mean grdgyn shrewtime10
```

15.5. Results

The results for both hospitals for the four FCEs and 3 sets of 'non-road' speed assumptions are shown in table 15.3 below.

Table 15.3. Mean access times in minutes

Patient Type	Telford			Shrewsbury		
	10	15	20	10	15	20
Gynaecology	24.5	24.0	23.8	20.5	20.5	20.4
Obstetric	21.4	21.0	20.7	21.4	21.3	21.2
Paediatric	33.2	31.8	31.6	17.9	17.8	17.7
Neo-Natal	21.9	21.5	21.3	23.6	23.6	23.5

The labels **10 15 20** for each hospital refer to the assumed travel time in miles per hour of crossing grid cells containing no roads. We assumed a travel time of 10 mph for such cells but examined the sensitivity of the results to variations in this assumption by redoing the analysis using travel times of 15mph and 20 mph.

The hospital at Shrewsbury is clearly more accessible for gynaecology patients than is the hospital at Telford with average travel times being approximately four minutes less. There is little or no difference in travel times to the two hospitals for obstetric patients. There is clearly a huge difference in average travel times for paediatric patients with times to Shrewsbury being almost half those to Telford. However, this is because the FCE records are for patients using the hospital at Shrewsbury and there is a paediatrics unit already operating at Telford. Finally, average travel times for neo-natal patients are slightly higher to Shrewsbury than to Telford, although the differences are not great. These finding persist even with different definitions of travel times for cells having no digitised road system.

The travel time surfaces for both hospitals are shown in figure 15.4. Alternative assumptions concerning travel across the surface in areas where roads are not present in the roads coverage are also shown. In each case, the hospital is at the lowest point on the surface. Locations with a higher elevation require longer to reach the hospital. The influence of the different types of road can be clearly seen. A 'time valley' runs east-west across the surfaces – this corresponds to the location of the M54/A5/A458 running from north of Wolverhampton to Welshpool. Also clear on the Shrewsbury surface is the time valley due the A49 from Ludlow to Shrewsbury to Whitchurch (approximately south-north). The surfaces for the 10mph 'non-road' speed assumption are comparatively rougher than those for the 20mph assumption. Noticeable on all surfaces is a 'time escarpment' in the north-west corner of the study area – this also corresponds to high ground that lies on the Welsh/English border.

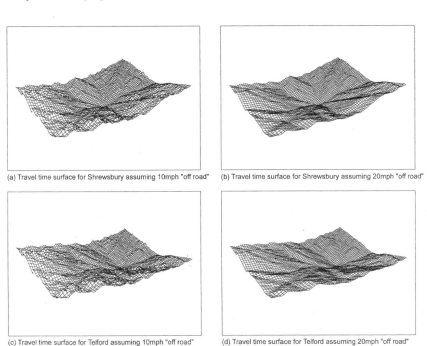

(a) Travel time surface for Shrewsbury assuming 10mph "off road" (b) Travel time surface for Shrewsbury assuming 20mph "off road"

(c) Travel time surface for Telford assuming 10mph "off road" (d) Travel time surface for Telford assuming 20mph "off road"

Figure 15.4. Travel time surfaces for Shrewsbury and Telford hospitals

Figure 15.5a shows the locations of the Royal Shrewsbury and Princess Royal Hospitals on the 10mph Shrewsbury travel time surface. The Royal Shrewsbury is located at the lowest point on the surface, with Telford lying in the M5 'time valley'. Figure 15.5b shows the same surface in a slightly different form. This time, a grayscale is draped across the surface, with lighter areas indicating locations for which the access time is longer that those for areas coloured dark. The influence of the different road types is relatively clear, with lower access times to roads with faster travel speeds. Figure 15.5c and 15.5d show the relationship between access speed and location of two of the FCE types (gynaecology and neo-natal), but draped over the Telford accessibility surface. The mean access time is then the sum of the individual access times divided by the number of patients in the FCE. As the data are aggregated to a grid, and the access times are represented as a grid with the same mesh locations, the actual calculation is of a weighted mean.

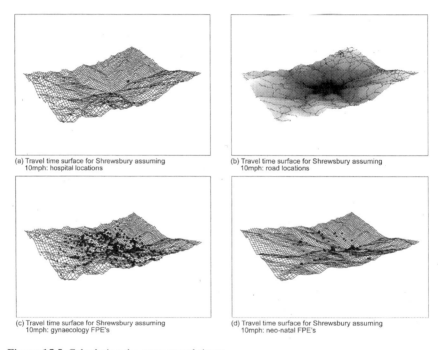

(a) Travel time surface for Shrewsbury assuming 10mph: hospital locations

(b) Travel time surface for Shrewsbury assuming 10mph: road locations

(c) Travel time surface for Shrewsbury assuming 10mph: gynaecology FPE's

(d) Travel time surface for Shrewsbury assuming 10mph: neo-natal FPE's

Figure 15.5. Calculating the mean travel times

There will be some locations for which travel times are similar to either hospital. If we subtract both travel time surfaces (ie shrewtime – telftime) there will be a group of cells with the value zero. These cells are equal in terms of travel time to either hospital; cells to the west of this line will be closer to Shrewsbury than Telford, and cells to the east of this line will be closer to Telford than Shrewsbury. Any patient on this line might be regarded as being *indifferent* in their choice of hospital. However, it might be that 5 minutes either way does not make much difference to an individual in choosing a hospital. That part of the surface which corresponds to 5 minutes difference or less we may term an indifference *zone*. We can examine various indifference zones, say under 5 minutes, 5 to 10 minutes, and so on, and count the number of patients within each zone. The extent of 5, 10, 15, and 20 minute indifference zones is shown on figure 15.6.

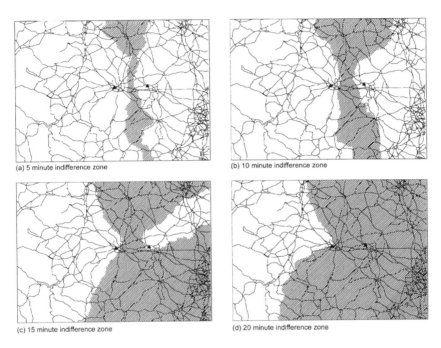

Figure 15.6. 5, 10, 15 and 20 minute indifference zones

Calculating the indifferences within the GIS is carried out as follows:

```
Grid: shtl10df = shrewtime10 - telftime10
Grid: shtl10dfpos = con(shtl10df < 0,- shtl10df,shtl10df)
Grid: shtldif5m = int(SHTL10DFPOS div 5 ) + 1
Grid: shtldifply = gridpoly(shtldif5m)

Arc: intersect patgyn shtldifply indif_gyn point
```

First we create the time difference surface shtl10df. Values that are positive are at locations where the time required to get to Shrewsbury is greater than that required to reach Telford. Negative values will occur at those locations where the opposite is the case. We use the *con* function to convert those locations with negative values to positive – using the *abs* function would have been somewhat more elegant. To create a grid of 5 minute indifference values we divide the 5 by 5, take the integer part, and add 1. A value of 1 in the resulting grid indicates that the location is in the 0-5 minute indifference band. To show the bands as hatched areas requires that we convert the surface to polygons, using *gridpoly*, then the maps can be shaded using *polygonshades* as appropriate. ARC/INFO includes an item name grid-code in the PAT of the coverage to contain the relevant surface value.

To calculate the proportions of an FCE type which lie in each zone, we need to intersect the point coverage for the FCE with the indifference zone coverage we have just created and use the *frequency* command to compute the frequencies in each grid-code zone. It indicates, for example that less than 7% of the FCEs lie within the 5 minute indifference zone but that 60% of the patients are located within the 15 minute indifference zone.

Table 15.4. Numbers of gynaecology patients in each indifference zone

FREQUENCY	INDIFFERENCE	PERCENT
196	0 - 5 minutes	6.33
452	0 - 10 minutes	14.60
1840	0 – 15 minutes	59.45
2501	0 – 20 minutes	80.81

It is often desirable to determine the extent of the area that can be reached within a certain time. The boundary of this area, a line that connects all locations which can be reached in the same time, is known as an *isochrone*. There are alternative approaches to the depiction of isochrones. An immediate reaction is to think that 'contours' could be drawn at say 10 minute intervals from zero. Then it is a simple matter to draw these 10 minute intervals on a map containing information on urban areas and the locations of patients. However, if it is required that an area corresponding to 0-10 minutes travel time, or 0-20 minutes travel time, is shaded, there is an easier route. First we create a grid in which the values are 1 for 0-10, 2 for 10-20 and so on. Then we polygonise this grid with *gridpoly* and select and shade as appropriate. The relevant commands are:

```
Grid: shriso10 = int(shrewtime10 div 10) + 1
...
Arc: gridpoly shriso10 shrisoply10
...
Arcplot: arcs roads
Arcplot: polygonshades urban 73
Arcplot: polygons urban
Arcplot: reselect shrisoply10 polys grid-code le 2
Arcplot: polygonshades shrisoply10 49
```

Figure 15.7 shows the extents of 10, 20, 30 and 40 minute drive areas for Shrewsbury. The same maps might also have shown the same drive times for Telford with a different hatch. The indifference zones would be then shaded with a cross hatch, although this would not allow us to compute the numbers of cases in the overlap area. Clearly we can also calculate the numbers of patients that lie in each distance band.

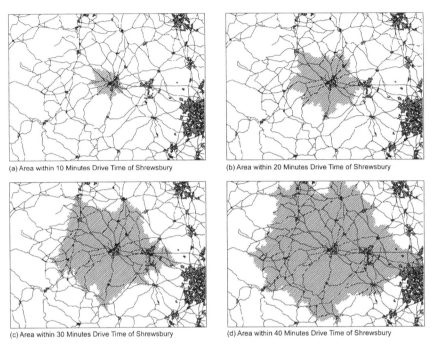

Figure 15.7. Extent of 10, 20, 30 and 40 minute drive time areas for Shrewsbury

There are three hospitals outside Shropshire that also have facilities similar to those at Shrewsbury. These are the hospitals in Stoke, Stafford, and Wolverhampton. It was useful to consider the overlap between the travel time areas for Shrewsbury and these other hospitals. First we have to create a travel time surface for the three non-Shropshire hospitals. Their locations are in a point coverage named nonshrophsp. A grid was created named nonshropgrd using po*intgrid*. This is then used as a source grid with *costdistance* and the resulting output grid nonshroptim contains shortest times to the nearest non-Shropshire hospital. The drive time areas are calculated using the trick of integer truncation used above. The resulting surface is then converted back to polygon form with gridpoly.

```
Grid: nonshropgrd = pointgrid(nonshrophsp,#,#,#,500)
Grid: nonshroptim = costdistance(nonshropgrd,timegrid20)
Grid: nonshropiso = int(nonshroptim / 10 ) + 1
Grid: nonshropply = gridpoly(nonshropiso)
```

Figure 15.8 shows the overlaps between the 10, 20, 30 and 40 minute drive time areas for the non-Shropshire hospitals and the equivalents for Shrewsbury. For the 20 minute plot, the relevant *arcplot* commands are:

```
Arcplot: arcs roadspolygonshades urban 61
Arcplot: polygons urban
Arcplot: reselect shrisoply10 polys grid-code le 2
Arcplot: polygonshades shrisoply10 49
Arcplot: reselect nonshropply polys grid-code le 2
Arcplot: polygonshades nonshropply 37
```

The results are rather interesting. There is no overlap between any of the 10 minute drive time areas. The 20 minute drive time area for Shrewsbury includes a small part of Telford, but the same drive time area for the non-Shropshire hospitals does not. The urban extent of Telford is within both the 30 minute drive time area for Shrewsbury and the non-Shropshire hospitals. In other words it is possible, if there were no facilities in Telford to reach either Shrewsbury or the nearest non-Shropshire hospital inside 30 minutes. The 40 minute drive time surface includes considerable overlap between the Shrewsbury hospital and those outside Shropshire.

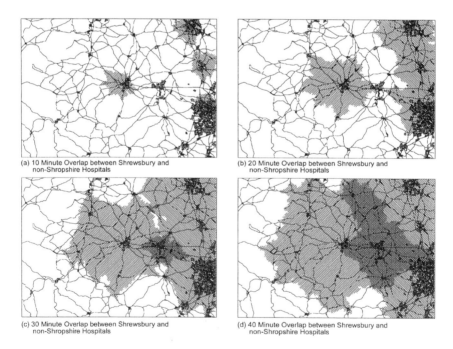

(a) 10 Minute Overlap between Shrewsbury and non-Shropshire Hospitals

(b) 20 Minute Overlap between Shrewsbury and non-Shropshire Hospitals

(c) 30 Minute Overlap between Shrewsbury and non-Shropshire Hospitals

(d) 40 Minute Overlap between Shrewsbury and non-Shropshire Hospitals

Figure 15.8. Extent of 10, 20, 30 and 40 overlap between Shrewsbury and Non-Shropshire hospitals

15.6. Conclusions

It could be argued that GIS is a relatively low grade technology for data manipulation. However, what this exercise has demonstrated is that a sequence of relatively simple *geometrical* operations on spatial data permits a greater understanding of the *geography* of a particular problem. The provision of health care facilities has implications for both the providers of such facilities, in that there are economic and organisational problems to be faced, and to the users of such facilities, who are naturally concerned about being able to get quickly to a hospital containing the appropriate unit. Given data on the locations of patients and potential sites, a series of spatial data manipulations can provide useful insights to assist in any decision process regarding the location or relocation of facilities and services. The problem is not just one that is germane to the provision of treatment and care facilities at hospitals. The rationalisation of the provision of general practitioner facilities is another area in which the suitability of alternative locations may be assessed via a series of simple GIS operations. The important point to note is that GIS does not replace a decision, it merely helps in providing information that enables that decision to take place.

References

Fotheringham A.S. and O'Kelly M.E. (1989) *Spatial Interaction Models: Formulations and Applications* Kluwer: Dordrecht.

Ghosh A. and McLafferty S.L. (1987) *Location Strategies for Retail and Service Firms* Lexington: Lexington, MA.

Hyman G.M. and Mayhew L.D. (1983) On the geometry of emergency service medical provision in cities, *Environment and Planning A*, 15(12), 1669-1690

Love R.F., Morris J.G. and Wesolowsky G.O. (1988) *Facilities Location: Models and Methods* North-Holland: New York.

Mayhew L.D. and Leonardi G. (1982) Equity, efficiency and accessibility in urban and regional health care systems, *Environment and Planning A*, 14(11), 1479-1507

Postzon (1985) *The Postcode Address File Digest*, The Post Office, London

Smallman-Raynor M.R., Muir K.R., and Smith S.J. (1998) The geographical assignment of cancer units: patient accessibility as an optimal allocation problem, *Public Health*, 112(6), 379-383

Wilson R.M., Gibberd R.W. and Mayhew L.D. (1990) Resource allocation and planning in health care systems, *Environment and Planning A*, 22(12), 1657-1665

Wrigley N. (ed.) (1988) *Store Choice, Store Location and Market Analysis* Routledge: London.

16 GIS and Large-scale Linear Programming: Evolution of a Spatial Decision Support System for Land Use Management

Richard L. Church

16.1 Introduction

Forest management and planning is a major task faced by private timber companies and public agencies. As an example, the U.S. Forest Service is responsible for managing over 190 million acres of forests and grasslands in the U.S., an area that is almost twice the size of Germany. Thus, it is not surprising that the U.S. Forest Service uses many types of analysis tools including the use of input-output models to assess the impact of forest operations on local and regional economies. The most common approach to plan activities for a forested region over decades involves decomposing the decision making process into three hierarchically-defined components: strategic, tactical, and operational (Weintraub and Cholaky, 1991; Nelson, Broadie, and Sessions, 1991). Each level of management involves the development of decision support models, either optimization or simulation. These models aid in exploring alternatives and generating tradeoffs among objectives. Strategic models help identify overall targets for harvesting and conservation activities over a long planning horizon and over large tracts of land. Tactical models help specify where the activities should be accomplished in order to meet strategic targets. Finally, operational models help translate tactical level decisions into detailed operations plans. This type of hierarchical decomposition has been used in forest planning by governmental agencies and by private industry (Church, Murray and Barber, 1994).

At each level of forest planning and management, GIS is commonly used. GIS has been used by many public agencies and private companies in forest operations as an aid in mapping activities and resources, tracking inventory by spatial units, specifying operations by spatial units, merging information from satellite imagery, and storing and supplying data for spatial operations models (see for example Carroll *et al.*, 1995). Special versions of general-purpose GIS systems have been developed and marketed for forest management applications.

16.2 Client's Needs

This chapter describes the evolution of an approach designed to assist in tactical level planning for the U.S. Forest Service, based upon the integration of GIS methods, optimization models, and software tools. The development began with the realization that a major spatial analysis problem existed in tactical level planning. Like many planning problems, the problem definition evolved over a period of years from a simple, straightforward statement, to a complex and fuzzy definition requiring significant amounts of spatial data. The same type of evolution occurred in the needs for sophisticated mapping and data manipulation. Software requirements evolved too from simple mapping functions and model integration to more complex models and mapping issues. The development path started in 1992 and continues as this chapter is written. As is true in many GIS applications, it is impossible to foresee all of the problems that will arise. The main objective of this chapter is to present this evolution, along with the imposed constraints, providing a good example in modelling and GIS in a large governmental agency involving land management and planning. A GIS is defined as a software system capable of storing and retrieving geographical data for the purposes of mapping, analyzing and modelling spatial data. Early GIS systems were based upon the use of raster or cell based data storage units. Such systems were easy to develop and provide a basis from which to easily integrate spatial models, like watershed runoff models, viewshed models, and cellular automata process models. Vector-based GIS can store data associated with points, lines, and irregular shaped areas or polygons. The U.S. Forest Service has used geographical information systems utilizing both types of data structures, vector and raster. Spatial data concerning stands, watersheds, sensitive habitats, etc. is most often entered and stored as vector data. For management purposes, stand data (i.e. small polygonal areas) are aggregated in two different ways: 1) into larger contiguous spatial units called analysis areas, and 2) into sets of stands, each containing the same type of vegetation and age-class. For the first type of aggregation, each planning unit represents a number of spatially connected stands of different strata and age classes, whereas in the second type of aggregation, the set of stands represent a spatially discontinuous group of areas across a large zone (or perhaps an entire ranger district of a National Forest).

16.3 Model Design

Although a number of strategic models have been developed and used by the U.S. Forest Service, the most widely applied is the set of models called FORPLAN (Johnson and Stuart, 1987; Bare and Field, 1987; Barber, 1986; Kent et al., 1991) and the successor model set called Spectrum. These models are based upon the second type of aggregation, namely stands of the same strata and age-class across a wide area or zone of the forest. To optimize forest activities, primarily harvesting, the coordinated allocation of choices (CAC) model is typically used. The basic form of the CAC model involves two decisions: 1) the choice of an operating regime for each zone, and 2) the choice of activities for each strata-age class within the zone, consistent with the operating regime of the zone. Zone decisions are represented by the Y decision variables and strata activities are represented by the X decision variables. The linear programming formulation of the CAC model as specified in Church, Murray, and Barber (1999) is given below:

$$\text{Maximize } Z = \sum_z \sum_m \sum_n c_{zmn} Y_{zmn} + \sum_s \sum_t \sum_{i \in U_p} \sum_{k \in W_{p_i t}} \hat{c}_{sp_i k} \hat{X}_{sp_i k} \quad (16.1)$$

Subject to:

$$\sum_m \sum_n Y_{zmn} = 1 \quad \forall z \quad (16.2)$$

$$-\sum_z \sum_m \sum_n a_{zmnspt} Y_{zmn} + \sum_{i \in U_p} \sum_{k \in W_{p_i t}} \hat{X}_{sp_i k} - T_{sp(t-1)t} + T_{spt(t+1)} = 0 \quad \forall s, p, t \quad (16.3)$$

$$0 \leq Y_{zmn} \leq 1 \quad \forall z, m, n \quad (16.4)$$

$$0 \leq \hat{X}_{sp_i k} \quad \forall s, k, p_i \quad (16.5)$$

$$0 \leq T_{sp(t-1)t} \quad \forall s, p, 1 < t < T \quad (16.6)$$

where:
- s = index of analysis areas (or strata);
- z = index of management zones in the forest;
- m = index of coordinated allocation of choices;
- n = index of timing choices for zones;
- p = index of prescription sets;
- t = index of time periods;
- k = index of timing choice associated with a prescription on an area;
- p_i = prescription i of the prescription set indexed as p;
- $\hat{X}_{sp_i k}$ = the acres allocated to timing choice k of prescription i in the set p defined for analysis area s;
- Y_{zmn} = the proportion of zone z allocated to timing choice n of the CAC m;
- c_{zmn} = the contribution to the objective function of timing choice n of the CAC m for zone z;
- $\hat{c}_{sp_i k}$ = the per acre contribution to the objective function of timing choice k of prescriptions i in set p defined for analysis area s;
- U_p = the set of prescriptions in set p;
- a_{zmnspt} = acres made available in time period t to prescription set p for analysis areas if timing choice n of CAC m is chosen in zone z;

$W_{p,t}$ = the set of timing choices k of the prescription i in set p that have their first management action in period t;

$T_{sp(t-1)t}$ = the number of acres made available for but not allocated to analysis area s under prescription set p in period $t-1$.

The objective of the CAC model is to maximize the value or return associated with both individual activities and zoning decisions. Zoning provides complimentary functions in adjacent areas. For example, if a zone is designated wilderness, then harvest activities may be prevented entirely. Values may be measured in terms of the value of protecting valuable habitat in a zone or in terms of the present value of harvests. For an industrial application, the objective function may represent the present value of net revenues, and zoning decisions may represent staging production areas for road building, etc. (Church, *et al.*, 1999). The CAC model is based on two structural constraints. Constraint (16.2) allocates each zone to a management strategy. Constraint (16.3) translates zone management and timing decisions to analysis areas. Acreage in an analysis area cannot be assigned to a prescription and timing choice more than once in equation (16.2). Solutions to this model specify levels of activities for each decade on each strata for up to fifteen decades. Activities or prescriptions represent treatment strategies like clear cutting, thinning in a specific decade followed by clear cutting 40 years later, etc. Each prescription is an operation and timing choice, which provides desired outputs. Since acreage can be assigned to only one prescription and timing choice, it is easy to track that acreage in terms of its state in any decade (e.g., acreage of some strata/age-class that is clear cut in the first decade is only ten years old in the second decade). The CAC model can also contain restrictions on the total amount of acres subject to harvesting or thinning activities in a given zone per decade (not given in above formulation). Such constraints are used to limit sediment loads onto nearby streams, prevent visual impacts (e.g., keep the activities in a given viewshed to be less than ten percent of the viewshed area in each decade), prevent too much of oldgrowth from being harvested, etc. Also, constraints may be included which provide for at least a minimum amount of harvesting activities in a given zone (e.g., provide appropriate amounts of openings for forage areas for elk), or provide timber for economic development and stabilization.

Additional constraints written at the zonal level are used to represent US Forest Service standards and guidelines for activities. Such standards help to prevent damage to streams due to sediment, provide for scenic vistas without an overwhelming visual blight caused by such activities, promote suitable habitat for desired species, and reduce risk of fire. Strategic models like the CAC model contains thousands of variables and constraints, even when the number of zones is small. The reason for this is that the number of potential prescription and timing choices for a given strata can be quite large. If the number of zones needs to be kept relatively small, then the size of such zones can be quite large. In fact, the size of such zones is significantly larger than the size of the planning units which are used in managing the standards and guidelines when activities are planned on the ground. Herein lies the rub, is it possible that a strategic solution violates one or more guidelines when tested at the level of spatial detail used in applying the standards and guidelines? The problem is that the strategic model specifies activities on strata which might be unevenly scattered across a large zone. Further, just which stands of a given strata should be treated in a given decade, when say only 25% of the area of that strata is to be treated in that zone? That is, no specific spatial arrangement is specified in the output of the strategic model.

To test standards and guidelines accurately at the planning unit level, a ground-based plan is necessary. This is the objective of the tactical model, namely to translate a strategic plan into activity levels of planning units called analysis areas (approximately 2000-6000 acres). Such analysis areas represent an aggregation of stands as described above, but are much smaller in area than a zone (20,000-60,000 acres). Constraints which represent standards and guidelines are invoked either on individual analysis areas or a set of neighbouring analysis areas, essentially at the same level as applied at the operational level. Personnel in the Forest Service dubbed this the spatial disaggregation process (SDP). Astute personnel suspected that strategic solutions may not always yield a viable solution when invoking standards and guideline constraints at the analysis area level as compared to the zonal level as in the CAC model. But, they had no easy way to check their suspicions. This was the desired goal of SDP: can we identify whether a strategic solution is feasible at the tactical level? Further, can we visualize the solution in terms of whether any flexibility is left in the standards and guideline constraints? As an attempt to address this problem our project was initially called Visualization of the Implementation Process (VIP).

16.4. Visualization of the Implementation Process (VIP)

The VIP research program began as an attempt to provide a GIS based tool which can provide:

- A tactical level model which attempts to translate a strategic model output into a ground-based plan for the next decade, subject to standards and guideline constraints.
- A user interface that could be used to explore flexibility in a given solution and explore alternatives
- A map display depicting the activity levels and available slack in the constraints for each analysis area
- A solution routine that is fast, allowing the analyst to test a number of different scenarios, each one representing changes in constraint levels, harvest targets, etc.

The major concern was for the current decade. Forest analysts were concerned whether forest plans could be implemented without difficulty starting now, let alone worry about the feasibility of activities starting decades from now. Consequently, the initial VIP design focused on a one- time period analysis.

The major question to be resolved is whether a solution to FORPLAN or Spectrum meets the standards and guideline constraints at the analysis area level as compared to the zonal level. More specifically, can the activity levels prescribed by the strategic models be mapped to specific areas (i.e. analysis areas) for operational planning and meet all of the constraints that will be required at that level? The simplest and most straight-forward approach to this problem is to prorate all of the activities evenly across the landscape according to the amounts of different strata and the levels of assigned activities. To demonstrate how this done, we need to introduce some notation (see Church and Barber, 1992 and Church et al., 1999):

θ_{zsp_ik} = the number of acres of strata s in zone z that are to be treated using prescription i in set p with timing choice k;

$\theta^u_{zsp_ik}$ = the number of acres of strata s in sub-unit u of zone z that are to be treated using prescription i in set p with timing choice k;

a_{zs} = the number of acres of strata s in zone z;

a^u_{zs} = the number of acres of strata s in zone z that are in sub-unit u.

Assume that each zone is divided into a set of sub-units, called analysis areas or planning units. We want to estimate the extent to which activities are to be assigned to each analysis area. Within the zone at the strategic level, activities are determined for each type of strata (and associated age-class). If 50% of that specific strata falls within a given analysis area, then 50% of the activity associated with that strata should be assigned to that zone. This can be represented mathematically as:

$$\theta^u_{zsp_ik} = \theta_{zsp_ik} \frac{a^u_{zs}}{a_{zs}} \qquad (16.7)$$

As an example, if 20% of strata s falls within sub-unit u, then 20% of each treatment activity for that type of strata is assigned to sub-unit u. From this assignment of prescriptive activities, we can then compute just which period the activity begins as:

$$H^u_{zsp_it} = \alpha^t_{zsp_ik} \theta^u_{zsp_ik} \qquad (16.8)$$

where:

$H^u_{zsp_it}$ = the acres treated in strata s in sub-unit u of zone z with prescription i in set p in period t;

$\alpha^t_{zsp_ik} = \begin{cases} 1, & \text{if prescription } i \text{ in set } p \text{ and timing choice } k \text{ begins in period } t \\ 0, & \text{otherwise.} \end{cases}$

Suppose that treatment activity represents thinning, then $H^u_{zsp,t}$ represents the number of acres of strata s of sub-unit u in zone z that are thinned in period t. With this proportioning of activities to sub-units, we can now translate all activities for strata to the sub-units level. If each one of these activities produces an impact value, it is then possible to sum across the activities in the unit multiplied by their impact levels and generate an estimate of impact at the sub-unit level. For example, let us suppose that the impact was sediment generation. We can then estimate the level of sediment generation in a given sub-unit as:

$$\sum_s \sum_p \sum_{i \in U_p} \beta_{sp} H^u_{zsp_i 1} \tag{16.9}$$

where:
β_{sp} = the amount of sediment produced by the prescriptive action p on an acre of strata s.

If a standard and guideline condition for sediment generation is established as an upper limit for sub-unit u, then we can test the feasibility of the proportioned solution by testing whether the following inequality holds:

$$\sum_s \sum_p \sum_{i \in U_p} \beta_{sp} H^u_{zsp_i 1} \leq \Gamma_u \tag{16.10}$$

where:
Γ_u = the upper limit on sediment produced within sub-unit u;

In general, we can establish this type of condition for any standard and guideline constraint in the following manner:

$$\sum_s \sum_p \sum_{i \in U_p} \beta_{spv} H^u_{zsp_i 1} \leq \Gamma_{uv} \tag{16.11}$$

where
Γ_{uv} = the upper limit for the impact on sub-unit u for standard or guideline v;
β_{spv} = the amount of impact type v produced by the prescriptive action p on an acre of area s.

For each guideline or standard, we can structure the above type of constraint and test the feasibility of simply prorating all activities to the sub-unit level. If all such conditions are met at the planning unit (sub-unit) level, then the strategic model solution is easily translated into individual tracks of land for operations planning. If, however, some or all of these conditions are not met then it is not clear whether any ground based solution can meet the strategic model-based targets and activities. The degree to which any such condition is violated can be easily calculated and expressed in terms of a percentage. Thus, if a guideline limited a level of impact to 100 and the calculated level of impact was 110, then such a condition was violated by 10%. Just as we can calculate levels of violation in terms of percentages, we can also calculate the level of slack, whenever the constraint is met. Thus, if the activity produced an impact level of 90 and the upper threshold was 100, then we compute a 10% slack. Slack is considered a good thing in forest management, as it depicts flexibility by not having pushed activities to an upper limit. Forest service personnel wanted to see a map of analysis areas, coloured by the degree of slack or violation. This was depicted by colouring an area as pale yellow, when the constraint was just met; green when slack was present; red when a violation occurred. Light green was used to represent small percentages of positive slack. Dark green was used to represent large percentage levels of slack. Shades of red were used to represent percentage ranges of constraint violation. The greatest levels of violation were represented as dark red. A quick glance at the map was all that was needed to identify plan flexibility, problem areas, or major issues of infeasibility. Personnel termed a plan 'dead', if they saw 'red.' A VIP presentation of a 'proportioned' FORPLAN solution is given in figure 16.1 for the Ochoco National Forest.

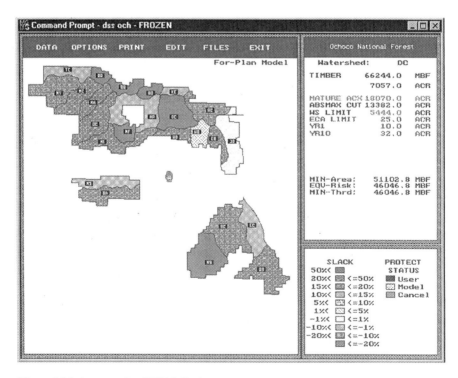

Figure 16.1. An example of VIP 1 display

In virtually all applications, U.S. Forest Service personnel found that a simple prorated set of activities to sub-units, from the general forest-wide strategic solution, involved considerable threshold violation. Identifying the existence of a problem (in translating strategic activities to operational units) was a major start, but what was equally necessary was an approach to determine just what was feasible. To address this issue, we can use the above mathematical expressions in a more flexible context. First, note that:

$$H^u_{zsp_it} = \alpha^t_{zsp_ik}\theta^u_{zsp_ik} = \alpha^t_{zsp_ik}\theta_{zsp_ik}\frac{a^u_{zs}}{a_{zs}} \quad (16.12)$$

The general standard and guideline constraint given above can then be rewritten by making a simple substitution as:

$$\sum_s\sum_p\sum_{i\in U_p}\sum_k \beta_{spv}\alpha^1_{zsp_ik}\theta_{zsp_ik}\frac{a^u_{zs}}{a_{zs}} \leq \Gamma_{uv} \quad (16.13)$$

If we consider that activities must differ from a simple proportioning scheme based on the ratio of areas, a^u_{zs} and a_{zs}, we can then define this fraction as a variable rather than a fixed quantity:

$X^u_{zsp_ik}$ = the fraction of analysis area s in sub-basin u of zone z that is assigned prescription i in set p with timing choice k.

Converting the fixed proportioning process to a flexible proportioning process yields a new version of the general standard and guideline constraint:

$$\sum_s\sum_p\sum_{i\in U_p}\sum_k \beta_{spv}\alpha^1_{zsp_ik}\theta_{zsp_ik}X^u_{zsp_ik} \leq \Gamma_{uv} \quad (16.14)$$

We can now use this construct to define a general proportioning model (Church, *et al.*, 1999):

Bridging Analysis Model A (BAM-A)

$$\text{Minimize } Z = \sum_z\sum_s\sum_i\sum_k\sum_u \left(S^u_{zsp_ik} + R^u_{zsp_ik}\right) \quad (16.15)$$

Subject to:

$$\sum_s\sum_p\sum_{i\in U_p}\sum_k \beta_{spv}\alpha^1_{zsp_ik}\theta_{zsp_ik}X^u_{zsp_ik} \leq \Gamma_{uv} \quad \forall u,v \quad (16.16)$$

$$\sum \theta^u_{zsp_ik} X^u_{zsp_ik} \leq \theta_{zsp_ik} \quad \forall z, s, p_i, k \tag{16.17}$$

$$\sum_p \sum_{i \in U_p} \sum_u \sum_k X^u_{zsp_ik} \leq 1 \quad \forall z, s \tag{16.18}$$

$$X^u_{zsp_ik} - R^u_{zsp_ik} + S^u_{zsp_ik} = \frac{a^u_{zs}}{a_{zs}} \quad \forall z, u, s, k, p_i \tag{16.19}$$

$$0 \leq X^u_{zsp_ik} \leq 1 \quad \forall z, u, s, k, p_i \tag{16.20}$$

$$0 \leq R^u_{zsp_ik} \tag{16.21}$$

$$0 \leq S^u_{zsp_ik} \tag{16.22}$$

where:

$R^u_{zsp_ik}$ = amount that assigned activity exceeds area proportion;

$S^u_{zsp_ik}$ = amount that assigned activity is under area proportion.

The objective of BAM-A is to allocate treatment/prescriptions in such a manner that all conditions are met and that the allocation is as close as possible to the simple proportioning scheme. BAM-A attempts to fit as close as possible the simple proportioning scheme to the landscape, but makes changes in order to maintain implementation constraints. Constraints (16.16) ensure that all translated activities meet the standards and guideline conditions structured at the sub-unit level. Constraints (16.17) maintain that the acreage of a given analysis area in a zone that is to be treated under prescription p and timing choice k is assigned to acreage among the sub-units of that zone within the strata. This means that BAM-A will not substitute acreage use on one strata with another type of strata, nor will it substitute treatment of acreage of the same strata over different zones (Church, et al., 1999). Constraint (16.18) maintains that acreage within a given type in a sub-unit is not over assigned. Constraint (16.19) defines the proportional characteristics of an area for treatment translation. Constraint (16.20) imposes non-negativity conditions and bounds on decision variables.

BAM-A links or bridges a strategic solution to smaller management areas and attempts to maintain the integrity of the strategic solution while disaggregating it to individual analysis areas that maintain environmental constraints. BAM-A is, however, a rather restricted model that allows little flexibility in substituting activities in meeting strategic outputs or goals. It is designed so that all activities on a given strata remain on that strata, rather than substituting activities on various strata in order to efficiently meet the strategic model output targets. We can structure a similar model to BAM-A which allows great flexibility in determining the type and location of the activities as long as the sum of the outputs reaches the level of the outputs given in the strategic model. We can call this model BAM-B. Forest Analysts chose BAM-B for development.

The one time period BAM-B model was defined in such a manner that it would minimize the amount of land treated in order to meet the strategic targets or goals. This model was nicknamed the Min_Area model. It was recognized early on that a greedy style algorithm could be used to find an optimal solution to the one time period Min_Area model. A second complementary model called Eqv_Risk model was also defined. This model attempted to spread all activities as evenly as possible in order to maintain an equivalent amount of slack in the most constraining threshold condition in each planning unit. Min_Area would concentrate activities as much as possible, whereas Eqv_Risk would spread activities out as much as possible. The Eqv_Risk model could be solved by a simple search procedure which identified the highest slack percentage, such that all constraints and activities were feasible. These two models were also included in the first version of VIP software.

16.5 Software Development

The original prototype for VIP was written in LISP and integrated in Windows 3.0. Since Windows 3.0 was new and the LISP product contained many bugs associated with MS Windows, it was nearly impossible to handle heap memory. Usually, the prototype system, after operating for ten minutes, would require rebooting. Even though the prototype did provide all of the needed functionality, the system could not be considered a production tool. This lead to a complete restructuring of the code into a Turbo Pascal, DOS application. This system provided essentially the same functionality, but relied on a proven, bug-free compiler. Forest Service personnel then used this system to test forest plans.

An example map display of the VIP system is given in figure 16.1. The interface was simple but allowed for a number of actions. First, if one clicked on a unit with the mouse, you could see a detail window of data associated with the unit (upper right portion of the screen). Second, a model could be selected and executed by selecting an appropriate 'Options' menu item. Third, the display was a colour coded map of the analysis areas where the colour represented the remaining slack associated with the tightest constraints in a unit. The colour key is given in the bottom right portion of the display. The figure gives a display of the first decade results of cumulative activities associated with a FORPLAN solution on the Ochoco National Forest (USA). The prorated FORPLAN solution is the default baseline solution for VIP. Most of the time such a solution violates many of the Standards and Guidelines at the analysis area level. This can be seen for each analysis area that is coloured in red or a shade of red. Strategic solutions were often found to violate a number of constraints written at the analysis level or set level. This means that such solutions are not immediately implementable, and require re-balancing using the Min_Area or Eqv_Risk models. These results were invaluable to the analysts, because it demonstrated a major problem for the short run.

As the VIP system was used new elements were requested. Of these new functions, there were three basic categories: 1) model enhancements, 2) interface and visualization enhancements, and 3) algorithm enhancements. In fact, as use of the models expanded a steady stream of interface enhancements was suggested. First, users needed a system to digitize analysis areas, as Forest Service GIS staff were in short supply and could not support this need in a timely manner. This need was met by developing a special purpose digitizer program that could be used by virtually all staff using a small, inexpensive digitizer and a PC. Labels and label location on the map could also be edited within VIP by use of the mouse. It was also possible to edit model targets and set specific levels of activity in a given planning unit and optimize the remaining activities.

The models were originally implemented to handle one type of treatment, which was defined as 'equivalent clear-cut acres.' Although analysts debated about the value of converting each type of activity into an equivalent amount of clear-cutting activity, most believed that this was sufficiently accurate enough to estimate the impact of various operations and test standard and guideline conditions at the analysis area/ unit level. The first version of VIP also utilized what was termed composite strata, which represented the average of all strata in a planning unit.

After analyzing the first decade in a number of forest plans for a number of National Forests in the western United States, questions were raised as to what would happen in subsequent decades. This meant that the models needed to be enhanced by adding multiple time periods. To do this required abandoning the special purpose algorithms, as they could not be easily expanded to solve multiple time period Min_Area and Eqv_Risk problems optimally. We acquired a Turbo Pascal routine which could solve modest sized linear programming problems and embedded the routine in VIP to solve the multiple time period problems. We also added the capability to easily visualize results on the map by quickly moving back and forth in time. This major change led to the release of VIP version 2.0. Since there were no royalties associated with the LP solver, we could distribute the system without cost to any forest service office.

The development and evolution of the VIP system is depicted in the upper portion of figure 16.2. There are three columns, one for system name, one for basic features, and one for software environment. The VIP prototype was developed in LISP. Unfortunately, it was not stable enough to use as an applications program. The first release version was developed in Turbo Pascal. All working versions of VIP remained as DOS applications using Turbo Pascal. The multi-time period version was released as VIP version 2.0. This system is still fully operational as a DOS application.

Use of the system began to tax the problem size dimensions supported in VIP 2.0. There was also a memory barrier that was reached within the DOS application, that prevented further expansion of the system functionality. Essentially many tricks like the use of overlays had been used to the maximum extent, and now it was necessary to consider making a major change in development platforms. To meet the growing needs for both functionality and problem size, we redeveloped the system in C++ as a MS Windows application. At this time it was decided to change the name of the

GIS and Large-scale Linear Programming 317

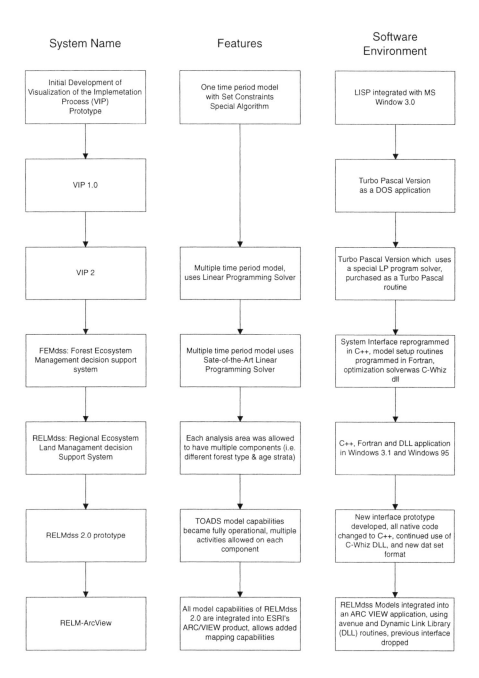

Figure 16.2. Evolution of VIP to RELM- Arc/View

decision support system to FEMdss, Forest Ecosystem Management-decision support system. This system utilized C++ programs for the user interface, FORTRAN dynamic linked library routines for model setup, and a state-of-the-art LP solver called C-Whiz by Ketron Corporation. Since the C-Whiz solver was being used as the FORPLAN solver, many analysts had C-Whiz and could then use FEMdss. FEMdss was a short-lived system as it underwent substantial extensions shortly after it was finished. The name was changed to RELMdss, Regional Ecosystem and Land Management decision support system.

Just as analysts wanted to expand their analysis from one time period to multiple time periods, many requested the option of using multiple strata types in a given planning area. Such elements of a planning unit were called components. RELMdss had the added capability of handling both components and time. A hierarchical capability was also added to RELMdss so that it would be possible to solve lower levels with ever increasing data and spatial resolution. The system allowed links to be established so that the modelling lat onle level could be assigned seamlessly to the next lower level (Church et al., 1994).

Forest analysts began to adopt an approach which targeted desired outcomes over some planning horizon. These targets were called desired future conditions. For example, it would be nice to increase the amount of oldgrowth by 10% or the number of acres devoted to 'park-like open stands' in a National Forest. To accomplish this type of analysis in RELMdss, additional model forms were developed to optimize desired future conditions in conjunction with previously developed model elements.

Many modelling and analysis systems are developed incrementally over time, with an evolution of functions, techniques, visualization, etc. This is due to several general factors. First, software and hardware capabilities have changed significantly over the past two decades. Twenty years ago, virtually all operations and planning models were executed on mainframes or minicomputers. As workstations and PCs evolved in their capabilities, much of the decision support system technology evolved on such hardware platforms as well. The vision of what was needed and what could be accomplished in land management modelling lelvolved within the forest service personnel ranks as well. Some of the changes that occurred in the VIP-RELM systems were a direct response to changes in the planning process that have taken place in the U.S. Forest Service. The point to be made, though, is that such systems do evolve in an incremental fashion and there are often few opportunities to wipe the slate clean and start from scratch given the history of information and understanding. This opportunity did occur when it was requested that the system handle alternate forms of treatment activity.

The models that were already solved within the VIP-RELMdss systems contain thousands of variables and thousands of constraints, without introducing any additional modelling complexity like alternate treatments. Since alternate treatment/prescriptions on separate components within each analysis area increased the level of complexity in terms of modelling and vlilsualization by almost an order of magnitude, we took the opportunity to design from scratch a new data base structure, model formulations, and interface. The new models have the ability to place constraints at each level and time period involving: global forest, analysis area, components within an analysis level, treatment alternatives, component-treatment alternatives at each level, and all of these conditions on sets of analysis areas. Thus, it is possible to solve a BAM-B style model where clear cutting of oldgrowth across a forest is limited to a set number of acres: or, the amount of a given component subject to some type of activity is limited. Such conditions can now be included in addition to the threshold constraints which represent standard and guideline constraints. The first version of this new expanded model set and visualization interface was called the RELMdss 2.0 prototype.

Originally, the VIP and RELMdss interfaces were developed outside of GIS. Data from GIS was packaged in a form that could be used by VIP/RELM, FORPLAN, etc. As the interface contained many mapping functions which duplicate those found in GIS systems, it was only natural to consider integrating the models into an existing GIS. With the introduction of ARC/View 3.0, this possibility became a serious consideration. Since ARC/View is now widely available among forest service offices, it is now possible to distribute an ARC/View application to forest service offices without needing to purchase additional software. Ironically, the cost of integrating RELM into an ARC/View application was borne by the U.S. Bureau of Land Management.

RELMdss now exists as an Arc View application. It is a fully integrated system in that it effectively uses the capability of the GIS software, the optimization routines, and expanded BAM-B models. Further development is underway to provide alternate solution methodologies for the current model set.

16.6. Conclusions

One of the largest natural landscapes on earth is forest lands. Many of these have been managed as a natural resource by industry and governments. In the United States alone, the U.S. Forest Service manages more than 190 million acres of forest and grassland. To manage such large tracts of land, large scale optimization has been all but required. Strategic level decisions involve devising forest plans and goals for a long planning horizon (e.g., 150 years) and at a high level of data aggregation. Unfortunately, strategic plans may not be feasible when attempting to identify just where such activity should be allocated. Tactical level modelling helps to bridge solutions reached at a strategic level with the details and scale of operational planning. This chapter presents the development of a large-scale modelling system for forest management which attempts to bridge strategic goals and operational requirements. We present several 'tactical-level' planning models that have been developed as a part of a research effort supported by the U.S. Forest Service. These models have been utilized in land management and planning by the Forest Service and have been implemented into a spatial decision support system. This system has evolved over time, starting with a system called VIP and evolving to an ArcView GIS application called RELM-ArcView. The evolution of this decision support system follows closely with the evolution of software tools and PC hardware. The design of this spatial decision support system utilizes a model set that has evolved as analysts responded to changes in both the organization and its perception of ecosystem management requirements.

Acknowledgements

I would like to acknowledge the assistance of US Forest Service personnel from regions 5 and 6, especially Klaus Barber, James Merzinich, Richard Dyrland, and Malcom Kirby as well as Duane Dippon of the Bureau of Land Management. I would also like to thank Phil Aune from the Pacific Southwest Research Station for support in this cooperative research project. I want to acknowledge Dr. David Lanter for the initial design of the interface in LISP. My gratitude goes to Scott Loban who was instrumental in developing a working version in Turbo Pascal. My sincere appreciation for many hard hours of work to Alan Murray who developed the Fortran drivers and interface with C-Whiz, Michael Figueroa who developed the C++ version of the interface and the experimental prototype, B.J. Okin who developed the C++ optimization drivers, and Rusty Dotson who integrated the models into ArcView GIS. All were instrumental in making this an interesting and rewarding research project.

References

Barber K. (1986) Large FORPLAN models: an exercise in folly, In *Lessons from using FORPLAN*, USDA Forest Service, Land Management Planning, Washington D.C.

Bare B.B. and Field R. (1987) An evaluation of FORPLAN from an operations research perspective, USDA Forest Service General Technical Report RM-140, Fort Collins, CO.

Carroll B., Landrum V. and Pious L. (1995) Timber harvest scheduling with adjacency constraints: using ARC/INFO to make Forplan realistic in *1995 ESRI User Conference Proceedings*, Redlands, California, USA (http://www.esri.com/base/common/userconf/proc95/prochome.html).

Church R. and Barber K. (1992). Dissaggregating forest management plans to treatment areas, *Stand Inventory Technologies 92*, 287-295, Portland, Oregon.

Church R., Lanter D. and Loban S. (1992) VIP: a spatial decision support system for the US Forest Service, Working paper, Department of Geography, University of California, Santa Barbara, CA.

Church R., Murray A. and Barber K. (1994) Designing a hierarchical planning model for USDA Forest Service planning, in *Proceedings of the 1994 Symposium on Systems Analysis and Forest Resources*, 401-409.

Church R., Murray A., and Barber K. (1999) Forest Planning at the Tactical Level Working paper (available from the first author).

Johnson K.N. and Stuart T. (1987) *FORPLAN Version 2: Mathematical Programmer's Guide*, Land Management Planning Systems Section, US Forest Service, Washington, D.C.

Kent B., Bare B.B., Field R. and Bradley G. (1991) Natural resource land management planning using large-scale linear programs: the USDA Forest Service experience with FORPLAN. Operations Research 39, 13-27.

Nelson J., Brodie J.D. and Sessions J. (1991) Integrating short-term, area-based logging plans with long-term harvest schedules, Forest Science 37, 101-122.

Weintraub A. and Cholaky A. (1991) A hierarchical approach to forest planning, *Forest Science 37*, 439-460.

17 Crime Pattern Analysis, Spatial Targeting and GIS: The Development of New Approaches for use in Evaluating Community Safety Initiatives

Alex Hirschfield, David Yarwood and Kate Bowers

17.1 Introduction: GIS in Regional Science

A Geographical Information System is a system of hardware, software and procedures designed to support the capture, management, manipulation, analysis, modelling and display of spatially referenced information. Such systems enable links to be established and relationships to be analysed between different phenomena by cross referencing data sets drawn from a range of sources (e.g., data on demography, infrastructure, land use, environmental quality, health, crime and disorder, social cohesion, etc).

In theory, GIS would be an ideal tool for visualising and exploring many of the research questions of interest to regional scientists. Indeed, Griffith has observed how 'regional scientists, like geographers, harbour a spatial perspective and show that space matters' (Griffith, 1999, p.25). In practice, however, the number of GIS applications in mainstream regional science is limited despite the fact that there are a large number of potential applications. For example, the combination of mathematical programming models and GIS can provide an aid to decision makers especially when different scenarios are tested and the spatial implications identified (see chapter 16 for example).

Hanink and Cromley (1998) used GIS to derive land use maps produced using a scoring system which identified the suitability of parcels of land for different functions. The added value of the GIS was in its ability to visualise the output from an optimal land assignment model. GIS has also featured in other research concerned with the development of multi-criteria evaluation procedures in land use modelling (Carver 1991, Pereira and Duckstein 1993). More recently, the use of GIS to facilitate natural resource evaluation using the technique of travel cost analysis is a further example of a relevant application (Brainard, Lovett and Bateman, 1999).

Research in the use of fractals for modelling urban growth has also drawn upon the functionality of GIS to test theoretical models of changing urban morphologies (Batty and Longley, 1994). More recent research in this area has involved the integration of satellite imagery data and socio-economic information within a GIS to create appropriate data sets for the more rigorous testing and evaluation of such models (Longley and Mesev, 1997).

Attempts have been made to integrate capabilities for spatial statistical analysis into GIS. Perhaps the best example is SPACESTAT, a software package which can be used to test for spatial autocorrelation and to undertake spatial multivariate analysis (including spatial regression) and other procedures on geo-referenced data sets (Anselin, 1992). Work in this area has been taken forward by others with interests in both methods of spatial analysis and GIS (Fotheringham and Rogerson, 1993; Bailey and Gatrell, 1995).

The use of GIS in business and service planning has been well established. Longley and Clarke (1995) recognised the importance of planning relating to the management of budgets in industry and the allocation of resources in the service sector. They explain how in recent years businesses and service organisations have been looking to GIS to assist in the planning process. Case studies are given regarding the type and quality of information that is required for high accuracy GIS analysis (e.g., Martin and Longley, 1995), the way in which GIS is used commercially to identify target customers (e.g., Batey and Brown, 1995) and the evolution of specific software solutions to address geographical problems identified by the user (e.g., Clarke and Clarke, 1995).

Focusing on the use of GIS for problem solving in the service sector, a number of GIS applications have been developed in the health field. Examples include applications concerned with:

- the re-configuration of health services;
- accessibility to health services;
- service delivery analysis and planning;
- health needs assessment;
- epidemiological studies;
- resource allocation and the definition of priority areas.

Gatrell *et al.,* (1991) used GIS to explore the extent to which spatial clustering could be observed in the incidence of Motor Neurone disease in north west England. Research on the Wirral in Merseyside examined the spatial distribution of community based health services by the types of service provided and the location and social characteristics of the patients who used them. The study identified catchment areas for different services and highlighted spatial mismatches between existing health care provision and patterns of need (Hirschfield, Brown and Bundred, 1995). Other studies have used GIS to examine spatial variations in the provision of community pharmacies and their proximity to each other and to primary health care services to inform locational decision making (Hirschfield, Wolfson and Swetman, 1994).

Another field in which GIS has proved useful is in the analysis of crime data. Within this field there is substantial variety in the types of task to which GIS technology is applied. For instance, it has been used to assist in the identification of the residential location of serial offenders (e.g., Canter and Larkin, 1993), to assess the extent to which people have a heightened fear of crime in particular areas (Brunsdon, 1997) and to identify areas of high crime for use in the allocation of resources to combat crime in a local authority and police service context (e.g., Chainey, 1997).

This chapter will focus in more detail on the application of GIS in the field of crime. It will begin by outlining some of the methods used in the identification of areas of high crime and go on to describe an application developed by the authors to streamline the process of 'hotspot' identification.

17.2 Client Needs

Many researchers have found clear evidence of patterns in crime data (Sherman *et al.*, 1989, Cohen and Felson 1979, Wilson 1987, Mayhew, *et al.*, 1993). These include tendencies for incidents to occur in the same areas (spatial clustering), or at certain times of the day, or to feature specific modus operandi, affect particular types of property, or to be perpetrated against victims with similar demographic and social characteristics. Discernible patterns in the distribution of crimes emerge precisely because most of them are not unique random events but rather share a number of common characteristics or features in terms of when they occur, where they occur and who are the victims. Various theories have been advanced to explain such patterns. They include the routine activities theory which explains crime (mainly that against the person) in terms of the convergence in space and time of motivated offenders, suitable targets and the absence of 'capable guardians' against crime (Cohen and Felson 1979) and social disorganisation theory which attributes crime in selected neighbourhoods to problems of family disruption and a breakdown of social cohesion (Shaw and McKay 1969, Sampson and Groves, 1989, Bursik 1988). Other theories explain crime patterns in terms of the intersection of criminal opportunities with offenders' motivation, mobility and perceptions of target areas (Brantingham and Brantingham 1991), segregation effects (Braithwaite 1979), ghettoisation (Wilson 1987) and the impact of 'bordering effects' particularly the proximity of poor areas to affluent neighbourhoods (Hirschfield, Brown and Bowers, 1995, Bowers and Hirschfield, 1999).

Spatial patterns of crime should not be viewed in isolation to the functions of different areas. Wikstrom (1991) stresses how particular areas may be devoted to different types of land use (residential development, retailing, industry, leisure, open space) and how the activities and population profile of an area may vary according to the day or time of day (e.g., city centres on weekday mornings and Saturday nights).

The targeting of crime prevention strategies on the basis of detailed information about the local circumstances and context of crime can be facilitated greatly by information systems capable of handling spatially-referenced crime and incident data and cross-referencing them with contextual information on land use, infrastructure and demographic and social conditions (Ekblom 1988). This is where applications involving the use GIS have the most to offer. Good intelligence in this area is also essential for evaluating the impact of crime prevention programmes as well as being an effective aid in the planning of police operations against crime.

There are a number of precedents in the development of GIS applications for crime pattern analysis. These have ranged from relatively simple PC-based systems for simple forms of spatial search, map query and pin-mapping visualisation of crimes (Hirschfield, Brown and Todd, 1995) through to more sophisticated, workstation mounted, GIS products which were able to support a wider range of functions and analytical capability (Cross and Openshaw, 1991). However, in the past, even the more elaborate forms of workstation mounted GIS were incapable of supporting the range of spatial analysis and pattern identification functions needed by crime pattern analysts due to a lack of appropriate tools. All too often the user was required to do much of the really hard work in trying to distill patterns, in particular, deciding the locations on which to focus attention, selecting crime categories, time periods and other characteristics of crimes. Some progress towards automating these procedures was made in the development, by the Centre for Research on Crime, Policing and the Community at the University of Newcastle, of prototype applications such as ARC/Crime (Cross and Openshaw, 1991) and the Crime Pattern Analysis System or CPAS adopted by Northumbria Police (Openshaw *et al.*, 1991).

The arrival of more powerful PCs alongside enhanced GIS products in recent years now provides an opportunity to develop relatively sophisticated strategic force-wide GIS crime pattern analysis applications which are affordable, portable and deliverable.

A GIS-based crime pattern analysis system was developed by the authors as part of an ESRC-funded research project which examined relationships between the spatial concentration of disadvantage and levels of crime on Merseyside in north west England (Hirschfield, Bowers and Brown 1995). A diverse range of data sets were brought together during the course of the project. This included geographically-referenced information from the Merseyside Police command and control system recording every telephone call made to the police over a three year period (2 million records), recorded crime incidents comprising details of the location of the crime and that of the victim and information on offenders including their residential location. These data were cross-referenced with information on demography, social conditions, residential area type (geodemographics) and land use to explore how spatial patterns of crime allegations and actual crimes related to the underlying socio-economic environment. Particular attention was placed on the extent to which different types of crime tend to cluster in disadvantaged areas and whether crime risks in terms of being a victim of crime or an offender, increase where affluent areas border directly or are proximate to disadvantaged areas.

During the course of the ESRC research project, the research team were commissioned by Merseyside Police Authority to evaluate the Safer Merseyside Partnership (SMP) initiative. This is a nine-year programme funded by the Single Regeneration Budget which aims to reduce levels of crime through a combination of physical crime prevention measures or 'target hardening' (e.g., window locks, alarms, closed-circuit television) and social action (e.g., youth diversion schemes). An intensive programme of research monitoring and evaluation is now underway using the techniques developed during the research project.

The combined research and evaluation agendas have placed heavy demands on the available technology. Both activities have relied upon GIS-based applications to deliver the following:

- the identification of areas with a high spatial concentration of actual crimes and crime allegations usually referred to as 'hot spots';
- the identification of cases of repeat victimisation (e.g., addresses which have been burgled more than once within a year) and their spatial distribution;
- the identification of socio-economic conditions within areas of high crime, their immediate hinterlands and other user-defined areas.

In order to meet these requirements, the research team utilised a major GIS package in conjunction with an existing crime pattern detection program developed by criminologists in the United States and two software applications developed in-house by the research team; one for producing demographic and land use profiles for crime 'hot spots' and the other for identifying repeat burglaries (Hirschfield and Bowers 1997a, Johnson, Bowers and Hirschfield, 1997).

The definition of 'hot spots' was carried out using the Spatial and Temporal Analysis of Crime Software (STAC) which was developed in Chicago (Illinois Criminal Justice Information Authority, 1994). STAC produces hot spots using the following procedure. Firstly, it reads in computer files which contain grid references of crimes falling within a rectangular region of interest; the co-ordinates of the region itself and parameters which define the spacing and radius of circles in which the number of crimes are calculated. With these inputs STAC uses automatic search procedures to identify discernible spatial clusters of criminal activity or 'hot spots.' STAC summarises the boundaries of the defined clusters as standard deviational ellipses (Ebdon, 1988) which it exports as a file of grid co-ordinates.

A practical use of this technique has been in the identification of 'Youths Causing Annoyance' (YCA) hot spots. Eleven SMP-funded youth diversion initiatives are currently operational, each with their own well-defined target area. The GIS system has been used to produce hot spot maps that can be used as a guide to detached youth workers in defining where alleged problems are located within their areas. It is hoped that in each case a re-run of the analysis after the initiative will show a decrease in the levels of YCA calls and a change in the location of hot spots.

Since one of the priorities of the SMP was to tackle problems of repeat victimisation, information on the number and location of domestic properties that had been burgled more than once within a year was essential. Although seemingly a straightforward task, identifying the number and location of repeats was one of the most difficult to undertake. Yet this information is crucial, not only for identifying how to respond to burglary and where to target limited resources, but also, for monitoring the impact of measures (e.g., target hardening) in reducing crime and possible side effects (e.g., displacement).

Software was therefore developed which used address checking procedures to identify addresses at which more than one incident of residential burglary had occurred in a period of one year. Unfortunately, data relating to residential burglary from police records contains many inputting errors and inconsistencies. Therefore the software included 'fuzzy matching' facilities and manual intervention procedures to ensure that as many repeat victimisations as possible were identified (Johnson et al., 1997).

This information has been of great importance to the SMP. The residential properties that are given a target hardening grant (which can be money to provide window locks or a burglar alarm) are selected using demographic information relating to the age and employment status of the occupiers, information relating to whether or not the property is within the schemes' area of operation (European Union Objective One priority areas) and the level of victimisation suffered by the occupants. This information had been collected manually and with the danger of a certain level of bias, since much of the information relating to levels of victimisation was collected from the prospective grant recipients. The repeats programme has enabled the SMP to identify vulnerable properties independently of the occupants, since it is widely acknowledged that vulnerability increases as the number of victimisations suffered increases. GIS methods have also enabled the identification of repeat victimisations that have occurred in Objective One areas. This method also has the added advantage that computerised records can be kept on the properties that received grant aid, which facilitates monitoring and evaluation. In 1997, this work was extended to enable repeat crimes affecting small businesses and vulnerable public buildings (e.g., schools and hospitals) to be identified (Bowers, Hirschfield and Johnson, 1998).

It is useful to have the ability to identify areas of high crime especially for operational policing. However, often further insights into conditions found within crime 'hot spots' are needed which may provide clues which can lead us to an explanation of why such an area exists where it exists. For this reason a social, demographic and land-use profiling tool, referred to as the 'The Profiler' was developed by the research team for producing demographic, land use and crime risk profiles for areas of high crime (Hirschfield and Bowers 1997a). The Profiler produces contextual information such as the level of disadvantage in an area, demographics of the area's residential population, information relating to the level of unemployment in the area, the number of clubs, pubs or schools in the area and the level of many different types of crime in the area.

The programs and procedures described in this section are stand alone applications which produce results that can be visualised through a GIS and in some cases subjected to further analyses using GIS functionalities. The procedures have been very fruitful both from the point of view of research and in terms of their more practical applications. However, in order to streamline the monitoring and evaluation process and to cut down on the processing time of many of the requests for information made by Partnership members, the authors have created a fully integrated system, which brings together the elements of the analyses described above into a single seamless procedure which will do much to facilitate spatial query and crime pattern analysis in general.

17.3 Enhancing Crime Pattern Analysis Techniques

The remainder of this chapter discusses how the hot spot delineation and area profiling tools described above were integrated into a single GIS application. In the process of developing the latter, further consideration was given to the effectiveness of the methods adopted by STAC to delineate areas of high crime. As a result, a series of alternative approaches were developed in order to create an improved set of GIS tools for crime pattern analysis. These include the development of methods for depicting variations in crime rates across areas using small regular polygons (crime surfaces) and the creation of iterative searching techniques for defining crime clusters.

The enhancement of the existing stand-alone programs evolved largely through a process of trail and error. Just as the first program was generated to remedy some of the limitations of STAC and 'The Profiler' subsequent programs were written to refine some aspects of these programs or to explore the potential for new areas of research which become possible with an integrated GIS system.

A number of weaknesses were identified in STAC. Some of these stemmed from the fact that the software was run as a separate stand alone program without a mapping front end. This caused delays in the speed with which hot spots could be visualised since outputs from STAC had to be processed and imported into a GIS for display and analysis. This imposed constraints on the number of analyses which could be undertaken by the investigator and limited the scope for interacting with the underlying data base to search for evidence of crime displacement and other processes.

There were also drawbacks in running the Profiler separately from both STAC and the GIS. This was not only time-consuming but also limited the ability of the analyst to compare easily and speedily the demographic, social and land use characteristics of hot spots derived on the basis of alternative criteria (e.g., for different types of crime, by time of day, according to the age, gender and other characteristics of the victims of crime). The first wave of enhancements sought to remedy these shortcomings by bringing together STAC and the Profiler into a single GIS application.

17.3.1 An Integrated Hot Spot and Profiling System

The arrangement which has been described of using different systems to perform tasks and transferring files between them (e.g., the interactions between GIS, STAC and the Profiler) is defined as the loose-coupling approach. Input data are generated in GIS, ellipses are generated externally and are eventually imported back into GIS for display and analysis. The programs achieve their purpose but only after a significant amount of technical effort, which for many, particularly policy makers and senior police officers, would prove far too cumbersome and impracticable. What decision makers require are simple-to-use software tools which generate a range of easily digestible outputs – maps, graphs, statistics – which can be used to inform decision making, particularly in relation to the targeting of resources.

To streamline the process it was decided, from the outset, to build a customised application around a commercial GIS system. A program using existing software routines was likely to run faster, be prototyped more speedily, be more flexible and easier to modify with changing requirements and be readily available to a wide set of users.

Mapinfo's programming language, MapBasic, gives the developer access to a range of in-built spatial analysis tools which forego the need to sort and process grid references in arrays vis-à-vis the original STAC program. Conceptually this made the task of programming STAC's algorithms much easier. The program literally overlays a grid of points and the GIS calculates which of these fall within a defined 'search area' using its spatial 'Square Lattice' (SQL) functionality. A buffer function is applied to each grid point to generate a circle, and spatial SQL is used to determine the circles with the highest numbers of crimes. i.e. hot areas. Overlapping hot areas are merged to form hot clusters by buffering them with a distance of zero metres. This spatial analysis routine replaces STAC's time consuming algorithm which looks at each crime in turn to determine if it falls within more than one hot cluster.

Trigonometric functions are then applied to the crimes within each cluster to generate a standard deviational ellipse (Ebdon, 1988). The ellipse is centred on the mean centre, and is oriented in the direction of maximum dispersion. The boundary is two standard deviations from the centre of the ellipse, and so if crimes are normally distributed, about 95% of observations should fall therein. The ellipse is drawn automatically on the screen, making the manual import of grid co-ordinates from STAC redundant.

Steps were taken to integrate the Profiler, which had been a free-standing FORTRAN Program, into Mapinfo to provide the investigator with an option to generate, automatically, a socio-demographic profile for each of the ellipses. The GIS intersects the ellipses with each layer in a MapInfo workspace and exports crime and socio-demographic data which has been apportioned to the area of the ellipse directly into a commercial spreadsheet. It is fairly straightforward to produce customised demographic, social and land use profiles for each crime hot spot using the spreadsheet's formulae, formatting and graph functions. The complete profile can be saved as a template for every subsequent analysis. The workspace and template are prompted for at the beginning of the MapInfo Program, meaning that once the program has been launched the analyst can expect maps, graphs and statistics to be generated automatically in a matter of seconds.

These developments are a considerable improvement on the loose coupling approach because they allow the user to delineate crime hot spots and to produce profiles of them interactively. This gives the user greater speed and flexibility in the definition of hot spots providing scope for exploring the spatial manifestation of incidents according to crime type, time of occurrence, modus operandi, social characteristics of victims, social characteristics and location of known offenders and other variables. They also facilitate the search for changes in crime patterns in response to situational crime prevention initiatives such as the programme of security improvements to domestic dwellings (the 'target hardening' initiative) being implemented by the SMP,

Crime Pattern Analysis, Spatial Targeting and GIS   331

social diversion schemes (e.g., detached youth work targeted on known juvenile disturbance hot spots) and other schemes impacting upon the geography of crime and disorder.

One of the perceived benefits of hot spots is their ability to identify spatial patterns from the mass of detail which appears on a map when crime incidents are mapped as a series of points. Figure 17.1 shows a point map of all the incidents of burglary that occurred in a one year period in the City of Liverpool. Because of the very large number of crimes depicted (>8,400) it is very difficult to disentangle this map visually in order to identify areas within the district that suffer from higher than average levels of residential burglary. This is due to the fact that the map leaves the interpreter swamped with information and unable to distill significant patterns. An interesting observation that can be made from the point map is that it depicts areas within Liverpool that are residential in their nature. It can, therefore, be seen as a de facto population density and land-use classification map. Most residential burglaries will only occur in residential areas, so it is important that any crime pattern analysis tool that is used on the information highlights purely residential areas as regions of high crime or hot spots.

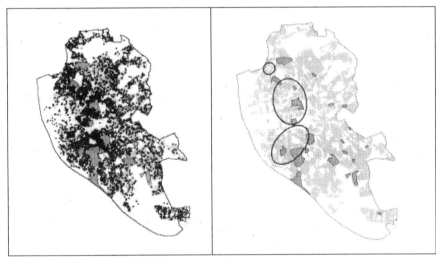

Figure 17.1. Burglaries against dwellings in Liverpool

Figure 17.2. 'STAC' burglary hotspots in Liverpool

The points in figure 17.1 were submitted to the hotspotting procedure and the results can be seen in figure 17.2. It is apparent from this map that the program has produced 3 ellipses which symbolise residential burglary hot spots in Liverpool. The most southerly hot spot covers the districts of Smithdown and Kensington, the smaller hot spot in the North covers the area of Walton and the hot spot in the middle covers the districts of Breckfield and Anfield. Figure 17.2 shows that generally, these hot spots do cover residential areas (the grey shading signifies such areas) but also encircle some areas of non-residential land use and park land.

Although the hot spotting procedure is far more efficient when fully integrated into a GIS, a number of problems remain in using standard deviational ellipses derived solely from the distribution of individual crime incidents to define areas of high crime. The question that must be asked is whether a high density of point locations is an indicator of an area with a high level of crime.

The aim in STAC, as with other traditional point pattern analysis, is to identify areas where levels of crime are significantly higher than elsewhere. There is also an implicit assumption that by using these techniques to identify such areas one has captured the underlying socio-economic/environmental influences that may be accounting for the pattern. The paradox is that by simply looking at the geometry of points one ignores these underlying characteristics. One could only claim that hot spots based on counts also depict higher than average crime rates where the population of an area was distributed evenly throughout space. However, such an even distribution of population is unlikely to be the case. It is more likely that the hot spots are delineating areas which have a large number of residents, or a high population density as well as areas with a high number of crimes occurring. This could mean that the crime rate within Liverpool's burglary hot spots is no greater than the average for the City of Liverpool. In short, the absence of a denominator in the derivation of hot spots (e.g., total population, number of properties) results in crime clusters being identified without any reference to the population at risk.

Other drawbacks with the approach in general and with STAC in particular include:

- the inability to establish the statistical significance of the ellipses by comparing an observed number of incidents with an expected frequency;
- the inability to identify from the ellipse any variations in levels of crime within the hot spot;
- the wide variations in the number and size of hot spots from one analysis to another;
- the lack of correspondence between the shape of the hot spots and underlying patterns of land use.

The inherent weaknesses in the hot spot approach reinforce the need to examine alternative methods for delineating areas of high crime which take into account the characteristics of the underlying population. The second refinement which was pursued achieved this by isolating spatial units with higher than expected rates of crime.

17.3.2 Derivation of Crime Surfaces

An alternative means of identifying areas with high levels of burglary is by producing burglary rates for particular geographical areas and producing a thematic shaded map which symbolises the rates. The areas that suffer from a high rate of burglary could be labeled as 'hot areas'. One of the problems with mapping crime rates is identifying which boundaries should be used to produce the rates. Census Enumeration Districts (EDs) are often chosen as the unit of mapping for rates, since it is easy to obtain population denominators for such areas. Although widely adopted in practice, it is unlikely that using EDs will be the most revealing way of displaying crime rate information, since EDs were not created specifically with crime pattern analysis in mind. Perhaps a more relevant geography for the City of Liverpool would be that of police beats. Beats with high burglary rates are shown in figure 17.3. The figure shows that the beats with high crime rates form three distinct areas. This is interesting since the results could have shown a far more fragmented picture. It is apparent that the high crime beats tend to cluster together. It is also apparent that the high crime beats are varied and irregular in size and shape. It is interesting to look at the way in which the high crime beats relate to the hot spots from figure 17.2. The immediate impression that is gained is that the hot spots fail to pick up some of the areas in which the burglary rate is very high. In particular, there are areas of high crime just south west of the Smithdown hot spot and north west of the Walton hot spot.

Inevitably, the spatial pattern of high crime will change depending upon the units that are selected for analysis. The beats are used for operational policing, but there is a question as to whether they are the most appropriate spatial units for identifying areas of high crime. Whilst it may be possible to use crime pattern analysis to redefine them, the way in which this should be done is by no means clear.

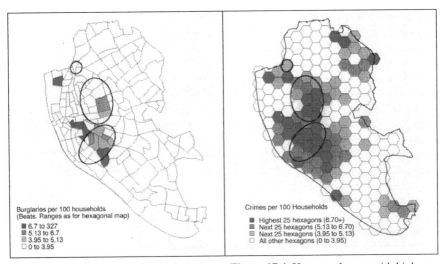

Figure 17.3. Police beats with high burglary rates in Liverpool

Figure 17.4. Hexagonal areas with high burglary rates in Liverpool

If figures 17.3 and 17.2 are compared, it becomes apparent that several of the beats that show high levels of crime are areas of park land. This highlights another possible limitation of the use of beats. The high crime areas may just be identifying areas where there are very few residential buildings that happen to have been burgled more often than average. This information may be useful to the academic, who is striving to identify why certain areas suffer high crime rates, but such information might not be of much practical use for operational policing, since it does not indicate the areas in which the greatest concentration of burglaries is occurring.

A more flexible approach to producing a high crime picture on an areal-level basis is to eliminate the problem of units being constructed for different purposes and being irregular in size and shape by producing arbitrarily defined, but regular, units of analysis.

The second software enhancement attempts to address some of the shortcomings in using fixed administrative boundaries by generating a tessellated surface of hexagons covering the entire area under study. Both individual level crime data (which constitute the numerators in crime rates) and underlying population and land use data (the denominators) are then generated for each hexagon using a combination of data aggregation and data apportionment techniques.

The starting point is a triangular grid of points at 50m spacing. By coding the grid with x and y values any size of hexagons in steps of 50m can be specified by the user. This saves appreciable disk space. The hexagons are produced by buffering these points. Normally a GIS would buffer a point with fifty or more segments to give the illusion of a circle. By specifying just six segments, hexagons are created effortlessly. The number of crimes falling inside each hexagon is determined using GIS's point-in-polygon routines. An estimate of the number of households within each hexagon is determined by using the GIS's overlay functionality to apportion data held against digitised EDs. Only the residential parts of each ED have been digitised to improve the accuracy of this apportionment. The numbers of crimes per 100 households is calculated by the GIS and displayed as a thematic map which resembles a patchwork quilt. The default bands are based upon the standard deviation of the crime rates such that hexagons with more than two standard deviations above the mean form the highest crime category. The user can interactively change the bands to identify, for example, areas surpassing a pre-defined crime rate.

A rerun of the high crime rate analysis using a lattice of adjacent hexagons is shown in figure 17.4. A fairly different picture emerges to that in figure 17.3. Areas within the boundaries of the two larger hot spots that had fairly high levels of burglary in figure 17.3 are classed as areas with even higher levels of burglary in figure 17.4. In fact, the hexagon-based analysis seems to coincide with the results of the hot spot analysis to a greater degree than the beat-level analysis. This shows that a great deal of thought should be put into the area of analysis in all crime pattern analysis exercises. Interestingly, the hexagon-based analysis also shows a certain level of clustering of areas of high crime. In fact, it indicates that the two larger hot spots that have been produced might be delineating some regions of a single, large 'hot area'. It would be possible using a 'buffer' command in MapInfo to integrate this irregular cluster of highest burglary rate hexagons into a single hot area, which could then be investigated in its own right. Lastly, it is apparent that the hexagon-based crime rate analysis produces more high burglary areas than the equivalent beat-based analysis. This may be one indication that the hexagonal areas bring out more information than the police beats.

Apart from presenting a more detailed and, it could be argued, easier to understand picture of crime across a study area, the hexagonal pattern lends itself to more sophisticated analysis. One of the limitations of STAC is that a crime cluster is summarised by a single geographical object, a standard deviational ellipse, while the real picture suggests variation across the surface. Rather than produce a socio-demographic and land use profile of the area which falls within an ellipse uniformly there would seem to be some merit in focusing attention on those areas where crime rates are highest.

The hexagonal areas have the advantage that they are not restricted to ED-based geography, although, EDs are the smallest units for which Census data are routinely available. The MapInfo system includes an apportionment programme which enables Census information to be estimated for the hexagonal areas. Using this apportionment technique, population and land use profiles can be produced for each hexagonal area. This information can be used to create a database which can be used to explore relationships between levels of crime and socio-economic conditions through ecological statistical analyses based on this new geography.

Table 17.1 shows the results of such an analysis. Burglary rates for each of the hexagonal areas in Liverpool have been correlated with four socio-economic factors apportioned to the ED from Census information. These four factors are:

- the percentage of households without a car;
- the percentage of the population who are recent migrants;
- the percentage of households living in non self-contained accommodation;
- the percentage of persons aged 16-24 that are unemployed.

Table 17.1. Relationships between levels of crime and socio-economic conditions through ecological statistical analyses based on hexagons.

	% Households without a Car	% Population Recent Migrants	% Households Living in Non Self-contained Accommodation	% Youths 16-24 Unemployed
Burglary per 100 households	**0.1274** *0.0106*	**0.3240** *0.000*	**0.3509** *0.000*	**0.0632** *0.424*

Notes; Figures in bold are the Pearson's Correlation Coefficients. Figures in italics are the two-tailed significance values of the Coefficients.

It can be seen from table 17.1 that recent migrants and households in non self contained accommodation are significantly and positively correlated with the burglary rate. In contrast, youth unemployment and households without a car are not significantly correlated with the burglary rate. This implies that a lack of suitable guardianship and the presence of easy targets due to a transient population which is not very socially cohesive and open access housing have greater implications on the level of burglary in an area than the level of disadvantage and unemployment.

17.3.3 A Prototype Criminal Geographic Analysis Machine

The methods described above, particularly the integrated hot spot and profiling system, give the user more control in exploring crime patterns and receiving valuable information to help explain observed phenomena. However, it is also helpful to the investigator, particularly to practitioners with limited time, if some of the effort in detecting crime patterns can be reduced through automation. There is the further problem with the deductive approach, which places the investigator in the 'driving seat' as to whether or not the crime patterns which emerge are real or simply depicting selection bias. Our research has shown that the spacing of the grid and size of circles of hexagons can generate completely different descriptions of the same crime data; the Modifiable Areal Unit Problem (MAUP). The characteristics of this problem have been discussed at length elsewhere (Openshaw 1984).

To investigate these concerns an attempt has been made at implementing a Geographical Analysis Machine (GAM) for crime analysis. In the late 1980s, Openshaw introduced the technique to investigate whether there were clusters of leukaemia cases around power stations (Openshaw and Craft 1991). In simple terms, for circles of a given radius, their GAM ranked the observed number of leukaemia incidents among a large number of randomly generated values. If the observed value was greater than say, 99, random values, the circle was classed as a significant crime hotspot. To address the MAUP the analysis was repeated for many different sized circles. It was argued that real clusters would be manifest at all spatial scales which would be obvious when all the significant circles were superimposed on the same map.

The research reported in this chapter is based on a variant of the original GAM system. The GAM starts by creating a grid of overlapping circles with a minimum specified radius and separation. The actual number of crimes within each circle is found using a point-in-polygon procedure. The number of households in each circle is then found using an ED apportionment method. Rates can then be produced for each circle. The GAM then increases the radius of the overlapping circles by a stepped amount and repeats the process. Once a maximum radius size has been processed the GAM increases the separation of the circles and repeats.

Using this GAM procedure results in the derivation of crime rates for circles of many different radii and levels of separation. Since rates are being used, all of these circles can be ranked according to their crime levels. To visualise the circles with exceptionally high crime rates, those within the top 5% and those within the top 1% of all circles were identified for mapping.

Figure 17.5 shows the results of the GAM analysis on the domestic dwelling incidents. The map shows circles with 5 different sized radii generated on grids ranging from 120m to 1000m at 20m intervals. It is apparent from figure 17.5 that there are certain areas in which high crime circles of different radii overlap. It is these locations that are stable hot spots since they remain irrespective of the areal unit being used.

The differences between figure 17.5a and figure 17.5b demonstrate that the criteria used for selecting high crime areas can make quite a difference to the resulting picture. There is obviously a region in west Liverpool that has a particular problem with domestic burglary since many contiguous and overlapping circles remain when the circles within the top 1% for burglary rates are plotted.

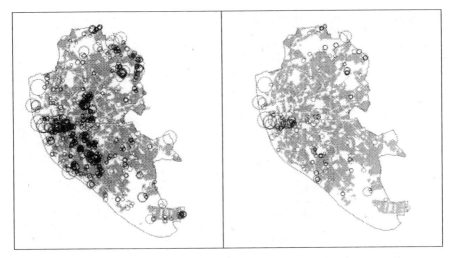

Figure 17.5a. High burglary areas from GAM analysis (top 5%)

Figure 17.5b. High burglary areas from GAM analysis (top 1%)

The results from the GAM can be compared with both the STAC and hexagonal area-based analyses. The GAM circles, especially those in figure 17.5a, coincide, to some extent, with the STAC hot spot shown in figure 17.2. However, in the case of the Breckfield/Anfield hot spot, there is a large area to the west which has no significance according to the GAM analysis. The STAC approach appears to be too rigid in its definition of hot spots as ellipses. Furthermore, there is a large area in the west of Liverpool, which has high burglary rates according to the GAM but is not identified by the STAC analysis.

The results produced by the GAM appear to accord more closely with those which emerge from the hexagonal area-based analysis since the areas of greatest density and significance in figure 17.5 are very similar in shape to the 'hottest' areas identified by the crime surface analysis shown in figure 17.4. The GAM and the hexagonal area-based analysis also both highlight the area of high crime in the north east of Liverpool; an area that has not been picked up by the STAC analysis.

This preliminary visual inspection helps to confirm that the pattern of clustering indicated by the hexagons (figure 17.4) may, after all, represent 'real' clusters (i.e. clusters that are stable on different spatial scales). This finding is reassuring for future research as the hexagon map was generated in about 30 seconds compared with 3 hours for the GAM analysis.

More in-depth GAM analyses can also pick up on statistically significant areas of high crime; such a rigorous method of identifying problems would give the academic confidence in pursuing possible causes or explanations for such 'hot' areas. This is currently being explored by the authors. GAM is also far less subjective than many of the other techniques of crime pattern analysis and it is a possible way forward for helping to define the level of resolution at which further empirical analyses should take place.

17.4 Conclusions and the Need for Further Research

This chapter has taken the same set of burglary dwelling data, applied three different methods of crime pattern analysis and generated three different outputs. Other researchers may have generated yet more results by applying their own analytical techniques. It is virtually impossible to judge which is the correct answer – the one which best describes the occurrence of crime clusters. It could be argued that the best judges of the techniques and outputs are the intended users of the products. The police officer may well consider the ellipses to be the superior method because of its emphasis on the volume of crime which is his/her concern. STAC, for example, is used by a variety of law enforcement agencies in the United States. A policy maker, faced with harsh decisions about the allocation of scarce resources may welcome the patchwork quilt display which shows how one part of an area compares with another. The academic, who is most concerned about data quality, statistical significance and methodology might favour the GAM approach.

As researchers into crime analysis we have a collective responsibility to improve the overall reliability and accuracy of methods for non-specialist users. Neither the police nor policy makers in local government are likely to have the time, knowledge or inclination to fine tune the methods, especially if the results appeal to them. Yet we have seen the dangers of choosing some arbitrary spatial scale for mapping.

The methods described, though currently standalone, should complement each other in the longer term. Attention will need to be turned to comprehending the outputs from the GAM in order to determine at what spatial scales particular types of crimes operate. These could be used to select automatically the appropriate grid separations and radii for the ellipse and hexagonal analyses, removing any human intervention. The GAM analysis, while helping to corroborate the description of crime clusters identified by other means, still does not quite prescribe what should be the definitive radius for conducting cluster analysis on burglaries. Research needs to establish at what point the GAM can indicate that 'enough is enough' and return an answer. Openshaw *et al.*, (1991) noted that the GAM was primarily a visualisation aid. Their GAM was groundbreaking in its attempt to remove subjectivity from the methodology; now there is a need for techniques to add objectivity to the interpretation of the results.

The way in which GAM has been used in the above analysis can be seen as relatively simplistic, in that, areas of high crime are defined, not in relation to well-established risk factors, but purely in terms of their crime rates. However, much is known about underlying factors which account for high levels of crime. If this information was used, it would significantly change the nature of the maps by depicting areas where levels of crime exceed those predicted on the basis of known risk factors. These are areas with inexplicable high crime given the known risk factors, within which potential causal processes need to be examined more closely. In such areas problems might be exacerbated by special circumstances or unknown correlates of crime.

The enhancements which have been carried out will assist the evaluation team by speeding up and improving the delineation of high crime areas to which scarce resources should be targeted and the derivation of full social, demographic and land use profiles describing such areas.

There are also likely to be benefits to Merseyside Police who are currently implementing a strategy to devolve crime pattern analysis capabilities to Local Research and Intelligence Officers in individual police stations. The strategy is being supported by the SMP's purchase of 23 copies of Mapinfo which are being supplied with customised interfaces for querying and mapping crime records. A series of discussions has already been held to explore how the research/evaluation team might assist the police with these developments by sharing with them their recent developments in crime pattern analysis.

Further ways of improving and refining the approaches discussed in this chapter will continue to be investigated by the authors. Particular attention will be focused on how systems might be further developed to predict the location of crime clusters given the necessary and sufficient conditions for their formation. This work will build upon pioneering studies using neural networks (Olligschlaeger 1996) and other approaches.

Acknowledgements

This analysis utilised the Spatial and Temporal Analysis of Crime Software developed by the Illinois Criminal Justice Information Authority. Any opinions, findings and conclusions of recommendations expressed in this publication are those of the authors and do not necessarily reflect the views of the Illinois Criminal Justice Information Authority. The authors also wish to thank Merseyside Police for supplying data sets used in the research.

References

Anselin L. (1992) *SpaceStat: a program for the analysis of spatial data.* National Center for Geographic Information and Analysis, University of California, Santa Barbara.

Bailey T. and Gatrell A. (1995) *Interactive spatial data analysis.* Longman. London.

Batey P. and Brown P. (1995) From human ecology to customer targeting: the evolution of geodemographics in Longley P. and Clarke G.(eds) (1995*) GIS for Business and Service Planning*, GeoInformation International, Cambridge.

Batty M. and Longley P. (1994) *Fractal cities: a geometry of form and function,* Academic Press, London and San Diego.

Bowers K. and Hirschfield A. (1999) Exploring links between crime and disadvantage in north west England: an analysis using geographical information systems, *International Journal of Geographical Information Science,13(2)*,159-184.

Bowers K. Hirschfield A. and Johnson S. (1998) Victimisation Revisited: A case study on non-residential repeat burglary on Merseyside, *British Journal of Criminology, 38, (3)*, 429-452.

Braithwaite J. (1979) *Inequality, crime and public policy*, Routledge and Kegan Paul, London.

Brainard J., Lovett A., and Bateman I.(1999) Integrating geographical information systems into travel cost analysis and benefit transfer, *International Journal of Geographical Information Science, 13(3)*, 227-246.

Brantingham P.J. and Brantingham P.L. (1991) *Environmental Criminology*, Waveland, Prospect Heights.

Bursik R.J. (1986) Ecological stability and the dynamics of delinquency in Reiss Jr A.J. and Tomry M. (eds) *Communities and crime*, University of Chicago Press, Chicago.

Brunsdon C. (1997) The geography of the fear of crime – Some Empirical Findings, Proceedings of the Crime and Health Data Analysis Using GIS Workshop, Sheffield, UK.

Canter D. and Larkin P. (1993) The environmental range of serial rapists, *Journal of Environmental Psychology*, 13, 63-69

Carver S. (1991) Integrating multi-criteria evaluation with GIS, *International Journal of Geographical Information Systems, 5*, 321-339.

Chainey S. (1997) Crime mapping in the London Borough of Brent, Proceedings of the Crime and Health Data Analysis Using GIS Workshop, Sheffield, UK.

Clarke G.P. and Clarke M. (1995) The development and benefits of customised spatial decision support systems in Longley P. and Clarke G.P. (eds) (1995*) GIS for Business and Service Planning,* GeoInformation International, Cambridge.

Cohen L.E. and Felson M. (1979) Social change and crime rate trends: A routine activity approach, *American Sociological Review , 44*, 588-608.

Cross A. and Openshaw S. (1991): Crime pattern analysis: The development of ARC/CRIME', paper presented to the 3rd national conference and exhibition of the Association of Geographic Information (AGI), Birmingham, October 1990.

Ebdon P. (1988): *Statistics in Geography*, Blackwell, Oxford.

Ekblom P. (1988): Getting the best out of crime analysis, Home Office Crime Prevention Unit Paper 10, London: Home Office.

Fotheringham A.S. and Rogerson P. (1993)(eds.) *Spatial analysis and GIS.* Taylor and Francis, London.

Gatrell A., Mitchell J., Gibson H. and Diggle P. (1991) Tests for spatial clustering in epidemiology: with special reference to Motor Neurone disease, in Clifford Rose F. (ed), *New Evidence in MND/ALS Research*, Smith-Gordon, London.

Griffith D. (1999) Statistical and mathematical sources of regional science theory: Map pattern analysis as an example, *Papers in Regional Science, 78*, 21-45.

Hanink D. and Cromley R. (1998) Land use allocation in the absence of complete market values, *Journal of Regional Science, 38, 3*, 465-480.

Hirschfield A. and Bowers K.J. (1997a) The development of a social, demographic and land use Profiler for areas of high crime, *British Journal of Criminology, 37(1)*,103-120.

Hirschfield A. and Bowers K.J. (1997b) The effect of social cohesion on levels of recorded crime in disadvantaged areas, *Urban Studies, 34(8),* 1275-1295

Hirschfield A. Brown P.J.B. and Bowers K.J. (1995) Exploring relations between crime and disadvantage on Merseyside, *European Journal on Criminal Policy and Research, 3(3),* 93-112.

Hirschfield A., Brown P.J.B and Bundred P. (1995) The spatial analysis of community health services on Wirral using geographic information systems, *Journal of the Operational Research Society, 46,* 147-159

Hirschfield A., Brown P.J.B. and Todd P. (1995) GIS and the analysis of spatially-referenced crime data: experiences on Merseyside, *International Journal of Geographical Information Systems, 9(2),*191-210.

Hirschfield A., Wolfson D. and Swetman S. (1994) The location of community pharmacies: A rational approach using Geographic Information Systems, *International Journal of Pharmacy Practice, 3, (1),* 42-52.

Illinois Criminal Justice Information Authority (1994): What is STAC?, STAC News, 2(1), Winter 1994, p12

Johnson S.D., Bowers K. J. and Hirschfield A. (1997) New insights into the spatial and temporal distribution of repeat victimisation, *British Journal of Criminology, 37(2),*224-241.

Longley P. and Clarke G.P. (eds) (1995*) GIS for Business and Service Planning*, GeoInformation International, Cambridge.

Longley P. and Mesev V. (1997) Beyond analogue models: space filling and density measurement of an urban settlement, *Papers in Regional Science, 76, 4,* 409-4 27.

Martin D. and Longley P. (1995) Data sources and their geographical integration in Longley P. and Clarke G.P. (eds) (1995) *GIS for Business and Service Planning*, GeoInformation International, Cambridge.

Mayhew P. Maung N.A. and Mirlees-Black C. (1993) *The 1992 British Crime Survey*, Home Office Research Study 132, London: HMSO.

Olligschlaeger A.M. (1996) Artificial neural networks and crime mapping. Paper presented to the annual conference of the American Society of Criminology Chicago, USA, November 1996.

Openshaw S. (1984) The Modifiable Areal Unit Problem, *Concepts and Techniques in Modern Geography (Catmog) 38.* Geo-Abstracts, Norwich.

Openshaw S. and Craft A. (1991) Using Geographical Analysis Machines to search for evidence of clusters and clustering in childhood leukaemia and non-Hodgkin lymphomas in Britain, in Draper G. (ed.) *The geographical epidemiology of childhood leukaemia and non-Hodgkin lymphomas in Great Britain.* 1966-83 Studies in Medical and Population, 53, OPCS, HMSO, London, 109-26.

Openshaw S., Waugh D., Cross A., Brunsdon C., and Lillie J. (1991) A CrimePattern Analysis System for sub-divisional use, Police Requirement Support Unit, HomeOffice, 4-8.

Pereira J. and Duckstein L. (1993) A multi-criteria decision making approach to GIS-based land suitability evaluation, *International Journal of Geographical Information Systems, 7,* 407-424.

Sampson R.J. and Groves W.B. (1989) Community structure and crime: Testing Social Disorganisation Theory, *American Journal of Sociology, 94(4),* 774-802.

Shaw C. and McKay H. (1969) *Juvenile Delinquency and Urban Areas Chicago*, University of Chicago Press, Chicago.

Sherman L.W., Gartin P.R., and Buerger M.E. (1989) Hot spots of predatory crime: routine activities and the criminology of place, *Criminology*, *27*, 27-5 5.

Wikstrom P-O. H. (1991) *Urban crime, criminals and victims: The Swedish experience in an Anglo-American comparative perspective,* Springer-Verlag, New York.

Wilson W.J. (1987) *The Truly Disadvantaged,* University of Chicago Press, Chicago.

Subject Index

A

Absolute error 250-252
Accessibility 147, 165, 170, 172, 174–176, 190, 196, 198, 199, 202, 203, 283, 288, 298, 324
ACORN 270
Agglomeration 44, 102, 113, 139, 146
Agriculture 44, 52, 60, 73, 96
Alignment 143
Arc Macro Language (AML) 296
ARC/INFO 217, 218, 288–290, 300
ASDA 2, 140
ATMs 143, 239
Attractiveness 61, 138, 144, 146, 148, 150, 232

B

Banking 22–25, 52, 60, 96, 228
Bi-proportional matrix adjustment 47
Blue book of national accounts 47
Bordering effects 325
Brand preference 144–146
BRE 116, 117, 126
British Aerospace 2, 6, 38
British Airports Authority (BAA) 40
Buffering 183, 184, 330, 334
Business statistics office 47

C

CACI 142
Calibration 4, 137, 147, 148, 155, 162, 176, 189, 192, 271, 273
 see also Goodness-of-fit

Car ownership 166, 169, 174, 271
Categorical data 97
CATS household travel survey 19
CEEC-4 62, 68, 72–75
Central statistical office 47
Centre for land use and built form studies 161
CFA 47
CGE 6, 7, 61
 see also Multiregional computable general equilibrium model
Chicago 2, 6, 11–14, 16, 17–20, 26, 27, 30, 32, 34, 35, 327
Choice models 7, 97, 98, 100–104, 113, 161
 See also Discrete choice modelling
Cobb-Douglas 161, 171
Cohort-survival model 116
Combinatorial optimisation 249
Commodity flow accounts 47
Community charge 54, 55
Competing destination model 139
Confidentiality 7, 2, 211, 249, 250
Constant increment approach 260
Construction 21, 22, 29, 48, 50–52, 57, 58, 60, 75, 80, 89, 90, 96, 168, 181, 187, 204, 205, 216, 223, 259
Consultancy 2–4, 39, 115, 134, 159, 174, 212, 215, 221
Consumption profile 45
Consumption theory 98
CORSIM 244

CRESR 116
Crime 7, 323–339
Cross-hauling 27

D

Data reliability 256
DELTA 7, 59, 160–167, 169, 170, 171, 173, 174, 176, 177, 186, 187
Direct marketing 269
Directories 141
Discrete choice modelling 97, 98, 103, 104, 113, 160, 161
Distance deterrence 147, 148, 175
DoE 116, 118, 119, 120, 125, 126, 253
DSC 7, 164, 186
Dynamic models 91, 118, 121, 139, 148, 163, 165, 189, 193-195, 206

E

Earnings 47, 54
Eastern Europe 7, 61, 70, 71–74, 76, 232
Economics 1, 2, 163, 211, 239
EIRASS 105
Elasticity 68, 69, 76, 143, 144, 169, 197
Emigration 124, 128
Energy and water 44, 60
Entropy 137, 138
Entropy modelling/maximisation 160, 161
Environmental impact assessment 37
Environmental resources management for the regions 164
ESSO 2, 140
European internal market 63
European Union 63, 117, 328

F

Facility location 283, 284
Family expenditure survey 46, 47, 84, 143
Fertility 121, 124, 128, 132, 238
FHSAs 126
Financial services 139, 141, 142, 228, 239, 240
Fiscal impacts 38, 52, 55, 57
Fischer-Johansson model 61, 62

Floorspace 106, 108, 138, 139, 148, 168, 169, 170, 171, 172, 181, 184, 185
Forecasting 7, 14, 37, 82, 243–246, 265
Forest planning 305
Forestry and fishing 44, 60
FORPLAN 306, 310, 312, 315, 318, 319
Fractals 323
Fuzziness 269–279
Fuzzy logic 273, 274, 279, 280
Fuzzy matching 327

G

GAM 336–339
Gatwick 44
GDP 63, 64, 70, 82–84, 92–94
Genetic algorithms 139, 234
  see also Neural nets
Geodemographic profiling 247, 248, 272, 273
Geodemographics 7, 115, 248, 269–271, 274, 276, 326
Geographical information system (GIS) 7, 137, 155, 211, 212, 215, 217, 220, 221, 283, 284, 299, 303, 305, 306, 316, 319, 320, 323, 324-330, 332, 334, 339
Geography 1, 2, 6, 7, 20, 98, 115, 116, 126, 132, 134, 139, 223 160, 163, 233, 238, 273, 283, 303, 331, 333, 335
Gini index 74, 75
GMAP 7, 116, 121, 139, 140, 146, 223, 225, 226, 231, 239, 240
Goodness-of-fit 109, 250
  see also Calibration
Gravity trade model 64, 71
Greater Manchester Strategy Planning Model (GMSPM) 164

H

Headship rates 120, 123, 126, 129–131
Health services 22–25, 283, 285, 324
Heathrow Airport 6, 39, 40, 44, 46–49, 50–52, 55–57
Heathrow employment survey 48, 50
Heckscher-Ohlin model 74

Hot spots 327–329, 330–334, 337
Household consumption 16, 41, 44, 47, 55, 258
Household projection 116–123, 126, 128–131, 134, 252
Housing 32, 45, 46, 54, 81, 82, 88, 162,164, 170–173, 175, 182–186, 193, 195, 196, 206, 336

I

Iceberg model 67
Immigration 124, 128
In-commuting 45, 46
In-migration 91
Income multiplier 14-16, 30, 31
Independence from irrelevant alternatives property 101
Indicators 5, 82, 83, 97, 98, 104, 105, 128, 196, 202
  see also Performance indicators
Inductive modelling 139
Industrial gross output 42, 46
Information system 106, 121–126, 134, 137, 325
Inland Revenue 244
Input-output models 6, 7, 11, 12, 14–17, 19, 37, 41, 43, 44, 47, 48, 50, 52, 57, 79, 80, 83, 84, 86–88, 90, 161, 175, 305
Intelligence 140, 269, 325, 339
International passenger survey 128
International Study Group on Land-Use/Transport Interaction (ISGULTI) 161
International trade flows 68
Internet 11, 228
Interregional commodity flows 67
Investment model 175
IRPUD 162, 163
Isochrones 288, 300
ITS 164, 186

J

Japan 15
Journey-to-work 12, 14, 16, 26, 57, 225, 216

K

K-means classification 274–276, 278

L

Labour markets 81, 166, 169, 173
Labour requirements matrix 52
Land-use modelling 159, 160, 176
Land-use/transport models 160
Leontief model 14, 15, 22, 37, 41, 50, 52, 79, 80
Lifestyle 142, 263, 269
LILT 162
Liverpool 6, 38, 39, 43, 44, 49, 50, 54–57, 161, 185, 330–333, 335, 337
Local labour market 38, 40, 56
Location quotient 17, 47
London Chamber of Commerce and Industry (LCCI) 39
Long-term unemployed 44–46, 54, 58
Lowry model 160

M

Manchester 2, 7, 38, 164, 166, 169, 177, 178, 183, 184, 186, 187, 227, 266
MapInfo 330, 334, 335, 339
Marital status 117, 120, 121, 123, 126, 129, 130, 131, 243, 250, 251
Market penetration 150, 259, 260, 263, 272
Market share 61, 97, 98, 101, 102, 104, 105, 109, 112, 115, 140, 148, 238
Martin Centre 161, 162, 163, 167, 175
Maximum likelihood procedure 108
MEPLAN 161, 162, 171, 174, 189
Merseyside 2, 7, 38, 39, 43, 44–47, 54–58, 164, 184, 227, 324, 326, 339
Merseyside Police 2, 7, 326, 339
Metrolink 177, 179
Micro-economic theory 160
Microdata 244, 245, 246, 248, 249, 250, 251, 252, 253, 255, 256, 265, 266
Microsimulation model 7, 243, 244, 245, 246, 247, 248, 249, 255, 259, 265, 266
Migration 4, 39, 84, 91, 92, 116, 117, 120–126, 128, 129, 130–132, 134,

166, 169, 170, 175, 184, 185, 189, 190, 195, 196, 243, 244, 266
Minimax problem 283
Missing data 148, 256
Miyazawa framework 11, 12, 34
Modifiable Areal Unit Problem (MAUP) 336
Mortality 121, 124, 128, 132
MOSAIC 270
Multinomial logit model 100, 102-105, 107, 109, 110, 112, 113, 161, 190, 192, 196, 203
Multinomial probit model 100
Multiregional computable general equilibrium model 62, 64, 65, 74, 76, 79, 80
MVA 7, 117, 164, 174, 186

N

National insurance contributions 54
National table 44, 47
Neighbourhood effects 273, 274, 276, 277, 280
Nested logit model 102, 103, 105, 107, 109, 110, 112
Neural nets 139
  see also Genetic algorithms
New earnings survey 47
NHS 117, 122, 123, 126
NHSCR 122, 123, 126
NISRA 116
NOMIS 47
Northumbria Police 326
NUTS 62, 73

O

OPCS 116, 117, 119, 120, 126, 253, 262
Optimisation 196, 211, 217, 224, 226, 228, 231, 232, 234, 238-240, 249, 250, 275, 278, 280

P

P-median problem 283
Parameters 46, 55-57, 63, 68, 69, 90, 99, 100, 103, 104, 108, 110, 125, 138, 147, 148, 169, 176, 177, 192–199, 201, 232, 276, 327
Parking policies 105, 106, 110, 112, 113
Pen portraits 271
Performance indicators 5
  see also Indicators
Planning 4, 7, 38, 40, 50, 82, 84, 97, 104, 105, 109, 115, 117, 119, 120, 121, 132, 134, 140, 160, 163, 164, 167, 168, 174, 181–183, 186, 197–199, 203, 204, 221, 223–225, 239, 244, 245, 258, 264, 305, 306, 308, 309–311, 315, 316, 318, 319, 324, 325
Planning policy guidance 120
Population estimates 117, 120, 125, 127
Population geography 115, 132, 134
Population projection 115–123, 125, 128–130, 131, 134, 246, 266

R

Random utility theory 98
RAS technique 47
Re-branding 148, 150
Redistricting procedures 215
Regional economic model 166
Regional economics applications laboratory 14
Regional planning guidance 119
Regional science 1, 2, 4–6, 61, 76, 98, 138, 155, 160, 211, 223, 228, 240, 243–245, 323
Regional Trade Coefficients 17
Rent adjustment 170
Rents 139, 162, 168, 170, 171, 183, 184, 185, 186
ReRO 115, 118
Residence-work trips 33
Response rates 269, 271, 273, 279
Retailing 7, 12, 22–25, 83, 96, 106, 112, 139–142, 144, 145, 150, 164, 181, 223, 228, 240, 244, 283
Revealed choice models 103
Reweighting 244, 245, 246, 249, 253, 266
Ripple effects 20, 26

Subject Index   347

S

Safer Merseyside Partnership (SMP) 326
Sainsbury 2, 140, 144
Sample of anonymised records 174, 250, 266
Sensitivity analysis 55, 56, 76
Sequential estimation 104
Severn Trent Water 248
Sheffield Hallam University 116
Shopping behaviour 7, 98, 103, 105, 106, 109, 113
Shopping trips 31, 34, 108, 113, 197, 231
Simulated annealing 214, 215, 219, 221
Single Regeneration Budget 326
Social accounting system 6, 11, 12
Social disorganisation theory 325
Social security payments 52, 54, 55
Socio-economic Impact Assessment 37, 39, 40, 57
SPACESTAT 324
Spatial demographic analysis 115
Spatial interaction models 4, 6, 7, 16, 61, 62, 76, 97, 98, 113, 137, 138, 140, 141, 147, 155, 223, 228, 276, 283
Standard industrial classification 44, 96
Stated choice modelling 103
Stone-Geary 161
Store loyalty cards 141
Suburbanisation 121
Survival ratios 124
Susquehanna river basin 37
Switching points 214, 215
Synthetic population 245, 248, 249, 253, 255, 256, 259

T

Thorn EMI 140
Town and Country Planning Association 119
Toyota 2, 82, 140
Trade flows 20–22, 27, 61–64, 68, 70–72, 83, 86
Trade permits 73
Trans-Pennine model 166

Transport modelling 161, 163, 176, 186, 189
TRANUS 161, 162, 171, 174, 189
Trip matrix 173, 192
TRRL 161, 162

U

Under-enumeration 245, 252, 253
Unemployed workers 42, 45, 52, 54
Unemployment benefits 46, 47
University of Dortmund 162
University of Leeds 5, 7, 139, 140, 223, 266
University of Liverpool 39, 161
University of Newcastle 326
Urban models 7, 160
User interface 309, 318

W

W.H.Smith 140
Wage rate 47, 49, 74, 81, 84, 88, 90, 91
Water consumption/demand 243-250, 252–259, 261–266
Welsh Office 116

Y

YHRPC 119
Yorkshire Water 2, 7, 245, 249, 254–257, 261, 266

Figures

2.1	Four-fold division of Chicago	13
4.1	Impact of the catching up scenario on regional welfare in Austria	75
5.1	Regional consumption (money flows)	87
7.1	Boundaries of the local authorities of Yorkshire and Humberside, 1993	120
7.2	Net migration balances, Yorkshire and Humberside 1981-1992	122
7.3	The structure of the West Yorkshire model	127
7.4	Projection of marital status factors for males 30-34, females 75-79 in Bradford	130
7.5	Projected changes in population of Bradford wards under differant scenarios 1991-2001	133
8.1	Car data in Madrid	141
8.2	Adjustments to the car data	141
8.3	Demand estimations in NW England	142
8.4	The alignment process	143
8.5	Observed demand at ATM branches in West Yorkshire	144
8.6	Brand preference for groceries in Yorkshire and London	145
8.7	Scatter plot between observed and predicted centre revenues for the Yorkshire centres containing WH Smith Group stores	145
8.8	Scatter plot between observed and predicted centre revenues (using new attractiveness term), for the Yorkshire centres containing WH Smith Group stores	146
8.9	Outlet attractiveness for the petrol market	146
8.10	Set-up a business scenario	149
8.11	Open a new outlet	151
8.12	Close an existing outlet	151
8.13	Change the attributes of an outlet	152
8.14	Simulate the impact of rebranding	152
8.15	An example business scenario	153
8.16	Impact evaluation ... store sales	153
8.17	Impact evaluation ... brand totals	154
8.18	Impact evaluation ... the detail	154
9.1	Operation of the model over time	165

9.2	Component sub-models of GMSPM	166
9.3	Component sub-models: Trans-Pennine application	167
9.4	Development sub-model: total floorspace proposed	169
9.5	Development sub-model: allocation of development to zones	169
9.6	Location sub-model: households' utility of consumption	171
9.7	Location sub-model: households' change in utility of location	172
9.8	Location sub-model: households' location choice	172
9.9	Employment status sub-model: demand for workers by socio-economic group	173
9.10	GMSPM: other centres	179
9.11	GMSPM: transport results (i)	180
9.12	GMSPM: transport results (ii)	180
9.13	TPCM: areas of restricted development	181
9.14	TPCM: percentage change in housing supply	182
9.15	TPCM: percentage change in population	183
9.16	TPCM: changes in employment	184
9.17	TPCM: population of Sheffield inner city	185
9.18	TPCM: housing rents in Sheffield inner city	186
10.1	Modal split function	196
10.2	Procedure of the STASA model in general	199
10.3	Flow chart of the transport model	200
10.4	Flow chart of the population model	200
10.5	Resistance function individual transport	202
10.6	Resistance function public transport	202
12.1	Ideal solution to territory allocation problem	230
12.2	Description of the genetic dealer location algorithm	233-234
12.3a	Ford network – existing dealers	236
12.3b	Ford network – optimal network	236
12.4a	Nissan network – existing dealers	237
12.4b	Nissan network – optimal network	237
13.1	Estimated ward water consumption: scenario 1, 1991	257
13.2	Estimated water consumption per household: scenario 1, 1991	258
13.3	Market penetration models for washing machines and dishwashers	260
13.4	Estimated % change in water consumption per household	264
14.1	Use of a geodemographic system for targeting	272
14.2	Fuzziness in geographic and cluster spaces	275
14.3	Gain charts for a) 10 clusters b) 25 clusters c) 50 clusters d) 100 clusters	280-281
15.1	Location of study area	284
15.2	Urban areas and hospitals in the study area	285
15.3	Location of patients in each finished consultant episode type	287
15.4	Travel time surfaces for Shrewsbury and Telford hospitals	297
15.5	Calculating the mean travel times	298
15.6	5,10,15 and 20 minute indifference zones	299

15.7	Extent of 10,20,30 and 40 minute drive time areas for Shrewsbury	301
15.8	Extent of 10,20,30 and 40 overlap between Shrewsbury and Non-Shropshire hospitals	302
16.1	An example of VIP 1 display	312
16.2	Evolution of VIP to RELM Arc/View	317
17.1	Burglaries against dwellings in Liverpool	331
17.2	'STAC' burglary hot spots in Liverpool	331
17.3	Police beats with high burglary rates in Liverpool	333
17.4	Hexagonal areas with high burglary rates in Liverpool	333
17.5a	High burglary areas from GAM analysis (top 5%)	337
17.5b	High burglary areas from GAM analsyis (top 1%)	337

Tables

1.1	Major sponsors of applied regional science described in this book	3
2.1	Employment by region based on place of work, 1995	13
2.2	Comparison of interrelational income multipliers between 1960 and 1985	16
2.3	Aggregate trade flows between the regions	21
2.4	Origins of in-flows as percentage of total	21
2.5	Destination of outflows as a percentage of total	21
2.6	Impact multipliers for aggregate trade flows	22
2.7	Construction impacts by region	22
2.8	Employment impacts associated with an injection of $2 million	23
2.9	Net impacts by sector and region	25
2.10	Journey to work flows	26
2.11	Destination flows from each region as a percentage of the total outflow	26
2.12	Origin flows from each region as a percentage of the total inflow	27
2.13	Trade balance in work trips	27
2.14	Gross income flows between the regions	28
2.15	Destination of income originating in work performed in each region	28
2.16	Origin of income brought into each region	28
2.17	Construction income impacts	29
2.18	Durable manufacturing income impacts	29
2.19	Trade income impacts	30
2.20	Income flows on household basis	30
2.21	Interregional income multipliers for the four regions	31
2.22	Projects under construction for the Pullman district within region 2	32
2.23	Impact analysis of the proposed projects: total output	33
3.1	Summary of employment impacts of Heathrow Airport	51
3.2	Employment impacts of Heathrow in 2016 with terminal 5	53
3.3	Fiscal impacts of Liverpool Airport in Greater Merseyside resulting from airport operations in 2005	55
3.4	Outcome of sensitivity testing: heathrow study consumer expenditure	56
3.5	Outcome of sensitivity testing: Heathrow study local recruitment	56
4.1	The gravity trade model of bilateral trade flows	64

4.2	Estimates of the gravity trade model for six product classes in 1995	71
4.3	Austria's long-run trade potential with the CEEG-4 for six product categories	72
4.4	Impact of the catching up scenario on Austria at the national level	74
4.5	Impact of the long-run trade scenario on sectoral production in Austria	75
5.1	High sterling scenario – differences in GDP from base	93
6.1	Effects of the parking policy scenarios	111
7.1	Population projections for 2001, local authority areas of Yorkshire and Humberside	129
7.2	Household projections for 2006, local authority areas of Yorkshire and Humberside	131
8.1	Observed versus predicted sales volumes at selected petrol stations in Sicily	149
9.1	Sub-models in DELTA applications	168
9.2	GMSPM: population by area in 2011	178
9.3	GMSPM: employment by area in 2011	178
10.1	Statistical tests for the estimation of the STASA transport model	201
10.2	Traffic effects in the study area	205
12.1	Comparison statistics for European dealer networks	225
12.2	CMA partition of the UK	226-227
12.3	Oncology fieldforce territory summary statistics	231
12.4	Comparison of outcomes under IRP and MBF	233
12.5	Comparison of GA, IRP and brute force methods	234
12.6	Full network IRP run	235
12.7	Rationalised network IRP run	235
13.1	The micro-components of domestic water consumption	247
13.2	Small area census tabulations used as constraints in microdata estimation	250
13.3	Evaluation of estimated population microdata for an 'average fitting' enumeration district	251-252
13.4	Alternative water demand scenarios	262
13.5	Variant water consumption projections	263
15.1	FCEs by unit	287
15.2	Code for mean.aml	295
15.3	Mean access time in minutes	296
15.4	Number of gynaecology patients in each indifference zone	300
17.1	Relationships between levels of crime and socio-economic conditions through ecological statistical analyses based on hexagons	335

Author Index

Anselin L. 324
Aoyama Y. 189
Arentze T. 105
Armington P. 67
Azis I. 6
Baggerly K. 244
Bailey T. 324
Bailly A. 1,2,5
Balassa B. 63
Barber K. 305, 306, 307, 308, 310, 313, 314, 318
Barker T. 7,79
Barnes T. 1
Bare B. 306
Bart P. 231
Bateman I. 323
Batey P. 6, 41, 42, 48, 56, 324
Batty M. 160, 189, 323
Bauwens B. 63
Beckman R. 244
Ben-Akiva T. 97,98,103
Benoit D. 137, 147, 155
Berechman J. 189
Bertuglia C. 5
Bezdek C. 275, 276
Birkin M. 4, 7, 140, 141, 231, 232, 244, 249, 250, 252
Blair P. 47
Blake M. 270
Bly P. 161,189
Boden P. 117,126
Bon R. 18
Bonner E. 37, 41, 43

Borgers A. 102
Bowdry M. 214
Bowers K. 7, 137, 325, 326, 327, 328
Boyce D. 5,14,189
Boyle P. 116
Bradley G. 306
Brainard J. 323
Braithwaite J. 325
Brantingham P.J. 325
Brantingham P.L. 325
Breheny M. 1, 119
Brocker J. 61, 67, 73
Brodie J. 305
Brown D. 52
Brown P. 324, 325, 326
Brundred P. 324
Brunsden C. 7, 137, 324, 326, 338
Buerger M. 325
Burniaux J. 76
Bursik R. 325
Butler D. 215
Caldwell S. 244
Canter D. 324
Capron D. 117
Cardwell W. 139
Carroll B. 305
Carver S. 323
Castro L. 128
Chadwick A. 37,50
Chadwick G. 160
Chainey S. 324
Chambers R. 253
Champion T. 116

Chapman J. 37, 41, 43
Chase R. 37
Charlton M. 7, 137
Cholaky A. 305
Church R. 7, 305, 307, 308, 310, 313, 314, 318
Clarke G. 4, 5, 7, 97, 137, 139, 140, 144, 146, 147, 148, 155, 223, 232, 244, 245, 249, 252, 324
Clarke M. 4, 5, 7, 97, 139, 140, 141, 144, 148, 223, 231, 232, 244, 252, 324
Coccari R. 52
Coffey W. 1, 2, 5
Cohen L. 325
Cole S. 3, 4
Collins S. 70
Conway R. 80
Corner I. 117
Craft A. 336
Cromley R. 323
Cross A. 326, 338
Crouchley R. 139
Culf R. 7
Dale A. 244
de la Barra T. 167
Devlin N. 243
Dewhurst J. 41
Diamond I. 127, 252
Diggle P. 324
Dinwiddy C. 65
Diplock G. 139
Dixon P. 80
Doran J. 244
Drennan M. 6
Duckstein L. 323
Duley C. 243, 252
Dunn J. 275
Eason R. 244
Ebdon S. 327, 330
Ehrlich R. 275, 276
Ekblom P. 325
Eyre H. 146
Falkingham J. 243
Felson M. 325

Feng Z. 270, 274, 276, 279
Field R. 306
Fik T. 139
Fingleton B. 7
Fischer M. 6, 61, 62, 70, 138, 139
Flowerdew R. 269, 270, 274, 276, 279
Foody G. 244
Fotheringham A. 7, 116, 137, 139, 283, 324
Full W. 275, 276
Gallagher C. 52
Gartin P. 325
Gatrell A. 324
George F. 231, 232
Getis A. 138, 139
Ghosh A. 231, 283
Gibberd R. 284
Gibson H. 324
Gilbert N. 244
Glasson J. 37, 50
Gobel G. 203
Goldstein W. 253
Goldstone H. 37
Golledge R. 100
Gomulka J. 253
Gordon I. 175
Gore A. 120, 123, 126, 129, 131
Griffith D. 323
Groves W. 325
Gruetzmann K. 7
Guldmann J. 139
Guo J. 11-12, 20
Haag G. 7, 194
Hagerstrand T. 243
Hall P. 119
Hamilton C. 62, 64, 70
Hamilton H. 37
Hancock R. 243
Hanink D. 323
Harche F. 231
Harding A. 243
Harrigan F. 80
Harris B. 139, 189, 223, 231
Henderson D. 47

Hensher D. 103
Herrington P. 246, 247, 253, 259, 261-263
Hewings G. 6, 11-12, 14, 15, 17, 20, 43, 80
Hirschfield A. 7, 137, 324, 325, 326, 327, 328
Hirte G. 68
Hlavackova-Schindler K. 139
Homenidou K. 7
Hunt J. 161,171
Hutchinson B. 189
Hyman G. 284
Isard W. 6, 79
Israilevich P. 11-12, 14, 17, 20
Isserman A. 1
Jager-Roschko O. 61
Jensen R. 3, 80
Johansson B. 61, 62, 70
Johnston K. 306
Johnston R. 1
Johnston S. 327, 328
Joun R. 80
Kain J. 189
Kantorvich Y. 139
Kashti A. 249, 252
Keister L. 244
Kent B. 306
King A. 63
Knudsen D. 139
Langley R. 139
Lanter D. 305
Larkin P. 324
Lee D. 4,160
Leistritz F. 37
Leonardi G. 284
Leontief W. 14, 15, 37, 41, 79, 80
Lerman S. 97, 98, 103
Lessof C. 243
Lewney R. 7
Lillie J. 326, 338
Longley P. 323, 324
Louviere J. 103
Love R. 283
Lovett A. 323
Lowry I. 160
Luce R. 98, 99
Lunn D. 127
McDonald A. 249, 252
McFadden D. 104
McGregor P. 80
McKay H. 325
McKay M. 244
McLafferty S. 283
McLean I. 215
McLoughlin J. 160
Mackett R. 162
Macmillan B. 7, 212, 215, 218, 222
Madden M. 6, 41, 42, 43, 56, 161
Mahidhara R. 14, 17
Mallender J. 243
Marsh I. 63
Martin D. 244, 324
Martin J. 76
Maung N. 325
Mayhew L. 284
Mayhew P. 325
Merz J. 253
Mesev V. 323
Middleton E. 244
Middleton L. 127, 244
Miernyk W. 37, 41, 43, 52
Miller R. 6, 47
Milliman J. 37
Mills I. 254
Mirlees-Black C. 325
Mitchell G. 245
Mitchell J. 324
Mitchell R. 244
Miyazawa K. 11, 12, 14, 15, 16, 19, 34
Molho I. 175
Morris J. 283
Morris P. 37
Moses L. 80
Muir K. 284
Mulligan G. 139
Murdock S. 37

Murray A. 305, 307, 308, 310, 313, 314, 318
Nelissen J. 243
Nelson J. 305
Nijkamp P. 4, 67, 80, 189
O'Kelly M. 139, 283
Okuyama Y. 6, 15
Olligschlaeger A. 338
Oosterhaven J. 41, 43
Openshaw C. 280
Openshaw S. 7, 139, 141, 214, 269, 270, 280, 326, 336, 338
Oppewal H. 7, 105
Orcutt G. 243, 244
Ottensmann J. 144
Pacione M. 2, 144
Paulley N. 161, 189
Pearson M. 248
Pereira J. 323
Peterson W. 79
Phibbs P. 41, 43
Phiri P. 139
Pierce T. 211, 215, 218
Pious L. 305
Plane D. 1, 2
Polenske K. 17, 18, 80
Pooler J. 139, 144
Postzon A. 288
Pudney S. 243
Pugh A. 37
Rajah N. 248
Rammer C. 62, 70
Rao L. 214
Radcliffe N. 231
Rees J. 2
Rees P. 7, 116, 117, 121, 123, 125, 126, 127, 146, 243, 244, 249, 250, 266
Reismann M. 139
Rietveld P. 67, 80
Rijk F. 139
Roberts E. 37
Rodrik D. 70
Rodwin L. 1
Rogers A. 128

Rogerson P. 324
Rosenau P. 2
Round J. 47, 80
Roy J. 139
Saltzman S. 6
Sampson R. 325
Samuelson P. 67
Sayer R. 1, 4, 160
Schindler G. 11-12, 14, 17, 20
Schneider M. 6, 73
Schofield G. 56
Schofield T. 244
Schumann J. 67
See L. 7, 141
Scn A. 139
Sessions J. 305
Shellhammer K. 37, 41, 43, 52
Sheppard E. 1
Sherman L. 325
Simmonds D. 7, 159, 161, 171
Simpson S. 127, 244, 252
Shaw C. 325
Shoven J. 65, 68
Sleight P. 269
Small K. 189
Smallman-Raynor M. 284
Smith M. 231
Smith S. 248
Smith S.J. 284
Smith T. 139
Snickers F. 67, 80
Sonis M. 6, 11-12, 14, 15, 17, 20
Speiss K. 116
Spencer A. 144
Stillwell J. 7, 116, 117, 120, 123, 125, 126, 129, 131
Stone R. 79
Strout A. 79
Stuart T. 306
Swetman S. 324
Szeto K. 243
Taylor P. 147
Teague A. 254
Teal F. 65

Teitz M. 231
Therival R. 37, 50
Thomson J. 41
Thorbecke E. 6
Timmermans H. 7, 100, 102, 103, 105
Todd P. 326
Tokarick S. 63
Tongeren F. 243
Toyomane N. 18
Train K. 104
Tye R. 252
Tyne W. 104
Van de Gaag N. 117
Van der Mensbrugghe D. 76
Van der Waerden P. 102, 105
Van Geenhuizen M. 4
Van Imoff E. 116, 117
Van Wissen L. 116, 117
Varian H. 65, 66
Vickerman R. 1
Vorst A. 139
Wanders A. 244
Wattage P. 245
Watts P. 269
Waugh D. 326, 338
Webster F. 161, 189
Weeks M. 41, 48
Weidlich W. 194
Weintraub A. 305
Wegener M. 162, 189
Wesolowsky G. 283
West G. 3
Westin L. 70
Whalley J. 65, 68
Wiegard W. 68
Wikstrom P. 325
Williamson P. 7, 244, 249, 250, 252
Wilson A. 4, 5, 125, 137, 139, 140, 141, 160, 189, 223, 231, 232, 244
Wilson R. 284
Wilson T. 117
Wilson W. 325
Wineman W. 52
Winters L. 62, 64, 70
Wolfson D. 324

Wrigley N. 97-98, 283
Wymer C. 270
Yarwood D. 7, 137
Yeung B. 68
Zadah L. 273, 280
Zellner A. 37
Zhao Z. 243

Contributors

Terry Barker	Cambridge Econometrics, Covent Garden, Cambridge CB1 2HS, UK
Peter Batey	Department of Civic Design, University of Liverpool Liverpool L69 3BX, UK
Mark Birkin	GMAP Ltd, 15 Blenheim Terrace, Leeds LS2 9HN, UK
Kate Bowers	Department of Civic Design, University of Liverpool Liverpool L69 3BX, UK
Chris Brunsdon	Department of Geography, University of Newcastle Newcastle NE1 7RU, UK
Martin Charlton	Department of Geography, University of Newcastle Newcastle NE1 7RU, UK
Richard Church	Department of Geography, University of California Santa Barbara, California 93106, USA
Graham Clarke	School of Geography, University of Leeds Leeds LS2 9JT, UK
Martin Clarke	GMAP Ltd, 15 Blenheim Terrace, Leeds LS2 9HN, UK
Richard Culf	GMAP Ltd, 15 Blenheim Terrace, Leeds LS2 9HN, UK
Bernie Fingleton	Department of Land Economy, University of Cambridge, Cambridge CB3 9EP, UK
Manfred Fischer	Department of Economic and Social Geography Vienna University of Economics and Business Administration, A-1090, Vienna, Austria
A. Stewart Fotheringham	Department of Geography, University of Newcastle Newcastle NE1 7RU, UK
Kathrin Gruetzmann	Steinbeis Transfer Centre Applied Systems Analysis (STASA), Rotwiesenstr. 22D-70599 Stuttgart, Germany

362 Contributors

Gunter Haag	Steinbeis Transfer Centre Applied Systems Analysis (STASA), Rotwiesenstr. 22D-70599 Stuttgart, Germany
Geoff Hewings	Regional Economics Applications Laboratory, University of Illinois, 607 St. Mathews, Urbana, Illinois, 61801, USA
Alex Hirschfield	Department of Civic Design, University of Liverpool Liverpool L69 3BX, UK
K.Homenidou	Cambridge Econometrics, Covent Garden, Cambridge CB1 2HS, UK
Richard Lewney	Cambridge Econometrics, Covent Garden, Cambridge CB1 2HS, UK
Bill Macmillan	School of Geography & Hertford College, University of Oxford, Oxford OX1 3BW, UK
Moss Madden (deceased)	Department of Civic Design, University of Liverpool Liverpool L69 3BX, UK
Stan Openshaw	School of Geography, University of Leeds Leeds LS2 9JT, UK
Harmen Oppewal	School of Management Studies, University of Surrey Guildford, GU2 5XH, UK
Yasuhide Okuyama	Regional Economics Applications Laboratory University of Illinois, 607 St. Mathews, Urbana Illinois, 61801, USA
Phil Rees	School of Geography, University of Leeds, Leeds LS2 9JT, UK
Martin Schneider	Department of Economic and Social Geography Vienna University of Economics and Business Administration, A-1090, Vienna, Austria
Linda See	School of Geography, University of Leeds, Leeds LS2 9JT, UK
David Simmonds	David Simmonds Consultancy, 10 Jesus Lane Cambridge CB5 8BA, UK
Michael Sonis	Regional Economics Applications Laboratory University of Illinois, 607 St. Mathews, Urbana Illinois, 61801, USA and Bar Ilan University, Israel.
John Stillwell	School of Geography, University of Leeds Leeds LS2 9JT, UK
Harry Timmermans	Department of Architecture and Urban Planning Eindhoven University of Technology, 5600 MB Eindhoven, Netherlands

Paul Williamson   Department of Geography, University of Liverpool
                       Liverpool L69 3BX, UK

David Yarwood   Department of Civic Design, University of Liverpool
                       Liverpool L69 3BX, UK

Druck: Strauss Offsetdruck, Mörlenbach
Verarbeitung: Schäffer, Grünstadt

New from Springer

J. Bröcker, H. Herrmann (Eds.)

Spatial Change and Interregional Flows in the Integrating Europe

Essays in Honour of Karin Peschel

Within the broad frame of regional research in an international perspective, the contributions of this volume present new theoretical, methodological and empirical results as well as political strategies for the following topics:
- economic integration in the Baltic rim,
- innovation and regional growth,
- economic integration, trade and migration,
- transport infrastructure and the regions.

2001. XII, 267 pp. 39 figs., 26 tabs. (Contributions to Economics) Softcover * **DM 90**; £ 53; FF 339; Lit 99.400; öS 657; sFr 82 ISBN 3-7908-1344-3

L. Hedegaard, B. Lindström (Eds.)

The NEBI Yearbook 2000

North European and Baltic Sea Integration

The NEBI Yearbook 2000 aims to provide a balanced picture of both the integrationist opportunities and disintegrationist pressures in the North European and Baltic Sea area - a region with over 50 million inhabitants and great economic and trading potentials. It brings together a wide range of scientific methods and perspectives in addition to a comprehensive statistical section with information found nowhere else. The result is a unique source of up-to-date knowledge of this increasingly important European region.

2000. XIV, 484 pp. 31 figs., 40 tabs. Hardcover * **DM 159**; £ 79.50; FF 590; Lit 175.600; öS 1161; sFr 144 ISBN 3-540-67909-X

L. Hoffmann, P. Bofinger, H. Flassbeck, A. Steinherr

Kazakstan 1993–2000

Independent Advisors and the IMF

By introducing a new national currency in November 1993, the Republic of Kazakstan took a decisive step towards establishing its political and economic independence. The cooperation between Kazak experts and independent international advisors such as the group of German economists under the leadership of Lutz Hoffmann played an important role for the Kazak government in choosing the most effective concepts and instruments for economic policy. Thus, the main topic of this book is the discussion of the macroeconomic problems during the first years of transition and the role of international financial institutions, in particular the International Monetary Fund.

2001. XI, 278 pp. 24 figs., 39 tabs. Softcover * **DM 89**; £ 45; FF 336; Lit 98.290; öS 650; sFr 81 ISBN 3-7908-1355-9

R.R. Stough, B. Johansson, C. Karlsson (Eds.)

Theories of Endogenous Regional Growth

Lessons for Regional Policies

The contributions in the book develop the advances into a theoretical framework for endogenous regional economic growth and explain the implications for regional economic policies in the perspective of the new century. Endogenous growth models can reflect increasing returns and hence refer more adequately to empirical observations than earlier models, and the models become policy relevant, because in endogenous growth models policy matters. Such policies comprise efforts to stimulate the growth of knowledge intensity of the labour supply and knowledge production in the form of R&D.

2001. X, 428 pp. (Advances in Spatial Science) Hardcover * **DM 169**; £ 58.50; FF 637; Lit 186.640; öS 1234; sFr 153 ISBN 3-540-67988-X

Please order from
Springer · Customer Service
Haberstr. 7 · 69126 Heidelberg, Germany
Tel: +49 (0) 6221 - 345 - 217/8
Fax: +49 (0) 6221 - 345 - 229
e-mail: orders@springer.de
or through your bookseller

* Recommended retail prices. Prices and other details are subject to change without notice. In EU countries the local VAT is effective. d&p · BA 41780-x